WITHDRAWN

Nematode Interactions

Nematode Interactions

Edited by

M. WAJID KHAN

Plant Pathology and Plant
Nematology Laboratories
Department of Botany
Aligarh Muslim University
Aligarh
India

CHAPMAN & HALL

London · Glasgow · New York · Tokyo · Melbourne · Madras

Published by Chapman & Hall, 2–6 Boundary Row, London SE1 8HN

Chapman & Hall, 2-6 Boundary Row, London SE1 8HN, UK

Blackie Academic & Professional, Wester Cleddens Road, Bishopbriggs, Glasgow G64 2NZ, UK

Chapman & Hall, 29 West 35th Street, New York NY10001, USA

Chapman & Hall Japan, Thomson Publishing Japan, Hirakawacho Nemoto Building, 6F, 1-7-11 Hirakawa-cho, Chiyoda-ku, Tokyo 102, Japan

Chapman & Hall Australia, Thomas Nelson Australia, 102 Dodds Street, South Melbourne, Victoria 3205, Australia

Chapman & Hall India, R. Seshadri, 32 Second Main Road, CIT East, Madras 600 035, India

First edition 1993

© 1993 Chapman & Hall

Typeset in 10/12 pt Palatino by Colset Pte Ltd, Singapore
Printed in Great Britain by St Edmundsbury Press Ltd,
Bury St Edmunds, Suffolk

ISBN 0 412 46130 7

A catalogue record for this book is available from the British Library

Library of Congress Cataloging-in-Publication data available

Printed on permanent acid-free text paper, manufactured in accordance with the proposed ANSI/NISO Z 39.48-199X and ANSI Z 39.48-1984

Contents

Contributors

M.K. Dasgupta
  Department of Plant Protection, Palli Shiksha Bhavana, Institute of Agriculture, Visva-Bharati, Sriniketan 731 236, India.

Jonathan D. Eisenback
  Department of Plant Pathology, Physiology and Weed Science, Virginia Polytechnic Institute and State University, Blacksburg, Virginia 24061–0331, USA.

K. Evans
  AFRC Institute of Arable Crops Research, Rothamsted Experimental Station, Harpenden, Herts AL5 2JQ, UK.

L.J. Francl
  Department of Plant Pathology, North Dakota State University, Fargo, ND 58105, USA.

Robin M. Giblin-Davis
  University of Florida, Fort Lauderdale Research and Education Center, 3205 College Avenue, Fort Lauderdale, FL 33314, USA.

A. Hasan
  Department of Nematology, N.D. University of Agriculture and Technology, Kumarganj, Faizabad 224 229, India.

P.P.J. Haydock
  Crop Environment Research Centre, Harper Adams Agricultural College, Newport, Shropshire TF10 8NB, UK.

M. Wajid Khan
  Plant Pathology and Plant Nematology Laboratories, Department of Botany, Aligarh Muslim University, Aligarh 202 002, India.

Rashid M. Khan
  Nematology Laboratory, Central Horticultural Experiment Station, Indian Institute of Horticultural Research, PO Hinoo, Ranchi 834 002, India.

Haddish Melakeberhan
   Department of Entomology, Michigan State University, East Lansing, MI
   48824, USA.

K.N. Pathak
   Department of Nematology, Rajendra Agricultural University, Pusa 848
   125, Samastipur, India.

Jean-Claude Prot
   International Rice Research Institute, Plant Pathology Division, PO Box
   933, 1099 Manila, Philippines.

P. Parvatha Reddy
   Division of Entomology and Nematology, Indian Institute of Horticultural Research, PO Hessarghatta Lake, Bangalore 560 089, India.

K. Sitaramaiah
   Department of Plant Pathology, Andhra Pradesh Agricultural University,
   Agricultural Research Station, Chinthalapudi, Ponnur 522 124, Guntur,
   India.

Abd-El-Samie H.Y. Taha
   Department of Plant Protection, Faculty of Agriculture, Ain Shams
   University, Shourba El Kheima, Cairo 13769, Egypt.

John M. Webster
   Department of Biological Sciences, Simon Fraser University, Burnaby,
   Vancouver, BC, Canada V5A 1S6.

Bernhard Weischer[*]
   Institute for Nematology and Vertebrate Research, Federal Biological
   Research Centre for Agriculture and Forestry, Munster, Germany.

T.A. Wheeler
   Department of Plant Pathology, Ohio Agricultural Research and
   Development Center, Ohio State University, Wooster, OH 44691, USA.

Giuseppe Zacheo
   Consiglio Nazionale delle Ricerche, Istituto di Nematologia Agraria
   Applicata ai Vegetali, Via Amendola 165/A, 70126 Bari, Italy.

---

[*]Current address: Westf. Museum für Naturkunde, Sentruperstrasse 285, D-4400 Munster, Germany.

Preface

Nematode interactions are important biological phenomena and of great significance in agriculture. It is a fascinating subject which is multidisciplinary by nature, and concerns any scientist involved with plant health. There have been marked advances in our knowledge of various aspects of the subject in the last two decades. This study area has been the subject of several reviews, but there was no exclusive text on the subject. This has stressed the need to document the information, developing a unifying theme which treated nematode interactions in a holistic manner. This book is about the interaction of plant-parasitic nematodes with other plant pathogens or root symbionts, the nature of their associations, their impact on the host and consequential interactive effects on the involved organisms. Since nematodes are at the centre of the theme, the responsibility of understanding of other plant pathogens dealt with in this book is largely delegated to the reader. I have limited the book content to interactions with biotic pathogens and root symbionts only, for various reasons.

The book embodies 16 chapters, and attempts to present balanced information on various aspects of nematode interactions with other plant pathogens and root symbionts. Some chapters describe general aspects of the subject. Interactions of nematodes with specific groups of organisms are addressed in the remaining chapters. The first chapter, while dealing with the historical perspective of the science of plant nematology, and general aspects of nematode parasitism on plants, introduces the subject of nematode interactions with other plant pathogens and root symbionts receive comprehensive treatment in subsequent chapters. Chapters 2, 3 and 4 describe the various physiological impacts of nematode parasitism on plants, concepts of nematode interactions and mechanisms of nematode interactions respectively. Exclusive chapters elucidate interactions of nematodes with viruses, plant pathogenic bacteria, wilt-inducing and root-rot fungi, insects, root-nodule bacteria and mycorrhizal fungi. Plant-parasitic nematodes as vectors of bacterial and fungal plant pathogens, the role of fungi in fungus–nematode interactions and interactions between nematodes in cohabitance are accounted in separate chapters. An independent chapter has been devoted to biochemical and genetic basis of fungus–nematode interactions.

The options available for the management of disease complexes and their utility, limitations and prospects are analysed in the last chapter. For obvious reasons the fungus–nematode interactions have occupied more chapters than any other group of plant pathogens, and as a consequence there may appear to be a certain amount of duplication. This was unavoidable, since similar materials were being used to elucidate different points.

Though the term 'interaction' should be used in the statistical sense, its use in the book chapters is synoptic. Nematode interaction research is at a crossroads. The theorists, experimenters and statistical analysts will have to broaden and deepen their imaginations, sharpen their tools and techniques and enliven the debate. The contributors in this book have enjoyed the freedom of retaining and developing their concepts of interaction in a responsibly democratic manner, rather than *laissez faire*.

This book is not an encyclopaedic review. However, an international emphasis has been placed on trends and probable future developments. The chapters incorporate both theoretical and practical aspects, and may serve as base-line information for future researches through which significant developments are expected. It is intended that the book will be useful to students, teachers and researchers, both in universities and research institutes, especially in relation to biology and agricultural sciences.

With great pleasure, I extend my sincere thanks to all the contributors for their timely response, their excellent and up to date contributions and consistent support and cooperation. I express my deep sense of gratitude to Professor Dr B. Weischer of Germany for his encouragement and valuable suggestions from the planning stage of this project to the finalization of the manuscript. Encouragement and advice from Professor Abrar M. Khan, Father of Plant Nematology in India, who initiated me to this discipline and has been a great source of inspiration, are also gratefully acknowledged. I acknowledge with thanks valuable assistance from my friends, colleagues and students. Special thanks are extended to Professor A.K.M. Ghouse, Chairman of the Department of Botany, Aligarh Muslim University, and to Dr Ziauddin Ahmad Siddiqui, my friend and colleague, for their encouragement and courtesies, to Dr Zaheeruddin, Department of Statistics, Aligarh Muslim University, for valuable comments and to Mr M. Halim Siddiqui, my Ph.D. student, for his untiring help during preparation of the manuscript. I gratefully acknowledge the financial support in part by Professor M.N. Faruqui, Vice-Chancellor of the Aligarh Muslim University, for preparation of the manuscript.

I am extremely thankful to Chapman & Hall in London, for receiving the book proposal and completing the review process expeditiously to grant acceptance for publication. Subsequent cooperation and understanding of the staff, especially Helena Watson, her predecessors Philippa MacBain and Craig Baxter, who were responsible for this project, and Lorraine Schembri, are thankfully acknowledged.

I express my sincere thanks to my family members, my wife Shabnam and sons especially the youngest, Rashid, for all the support they provided and the neglect and loss they suffered during the preparation of this book.

Finally, I am thankful to the Almighty God who provided and guided all the channels to work in cohesion and coordination since conception of the idea, to develop a book on nematode interactions until successful completion of the job.

M. WAJID KHAN

Introduction

Giuseppe Zacheo

Nematology is a very modern branch of science and was probably the last major discipline to establish its independence from the parent science of zoology in the middle of the last century. Nematodes are the most numerous multicellular animals on earth (more than 15 000 species described) and they represent two-thirds of the total fauna. They are categorized as being free living in a marine, freshwater or soil environment, and as parasites of animals and plants. Nematodes form a group of animals which, while known for more than 3000 years, has been greatly neglected by zoologists until recently.

The oldest record of nematodes among the ancient Mediterranean civilization occurs in Eber's papyrus dated 1553–1550 BC, in which there are indications about *Ascaris* and *Dracunculus medinensis* as human pathogens. The oldest reference to parasitic nematodes is found in *The Yellow Emperor's Classic of Internal Medicine* from China in 2700 BC. The description of symptomatology and treatment of the giant intestinal worm (*Ascaris*) is surprisingly accurate. Ch'en Yeu (*c*. 1174 BC) in his work *The Three Causes and One Effect of Disease* suggests that people are parasitized by worms through eating fruits, vegetables and animals' viscera, which contain their progenies. This represents a concept that was not accepted in the Western world until Louis Pasteur's experiments in 1864, when the theory of spontaneous generation had been discarded.

The major contribution to the advancement of nematology was the microscope. *Turbatrix aceti* was the first nematode observed under a microscope, in 1656 by Borellus, but the first account of plant-parasitic nematodes was given in 1743 by Needham, who discovered that fibres dissected from wheat galls took life. He did not realize the importance of his discovery. It represents the end of the concept of spontaneous generation and the start of plant nematology as a science.

Observations on nematodes which parasitize roots were made in 1855 by Berkeley, who described the symptoms of root galls induced by root-knot nematodes on cucumber in an English glasshouse. In 1859, soon after the discovery of root gall nematodes, Schacht found cysts of a nematode on declining sugar-beet crops. In the late nineteenth and early twentieth centuries, nematology experienced rapid growth in terms of species descriptions, biological investigations and control. In 1881, Kuhn was the first to use hydrogen sulphide in the treatment of soil for the control of the nematodes, and to suggest crop rotation as a method in nematode control. In the USA, Cobb almost alone publicized the existence and importance of nematodes as plant parasites. His book *Contributions to a Science of Nematology* is considered to be a milestone in nematology. His scientific contributions are notable, but even more important to the science is that he obtained independent recognition for the science in the United States Department of Agriculture. Derivations from the Greek word *nema*, such as nematology, nematologist and nematocide, were proposed by Cobb and widely accepted by researchers in that field. The names of Goodey in England and of Steiner and Christie in the USA and Filipjev in the USSR were associated with much of the basic work.

Later, various other researchers contributed to the knowledge of an increasing number of species and genera of plant-parasitic nematodes. The *Introduction to Nematology* by Chitwood and Chitwood (1950) is the most comprehensive treatise on the relationships, classification and characteristics of all nematodes. The worldwide distribution of economically important plant-parasitic nematodes indicates that they have been an important factor in the economy of ancient civilizations. They were possibly responsible for severe losses of crops, which determined famine and related migrations of entire populations.

Diseases induced by nematodes in crops have certainly been known for a long time, but the interrelationship between plant and parasites was not realized, also because of the absence of evident symptomatology. Crop failures, caused by specific nematode invasion, until recently have been attributed to other factors like soil exhaustion, climatic conditions or other phenomena known to be agents responsible for the decline of plants. Unexplained losses to crops, following old declining orchards, have sometimes been attributed to toxins excreted by tree roots.

During the long life of orchards, certain plant-parasitic nematodes can build up their populations. In some of the abandoned tobacco plantations of Virginia, overgrown with bush and forest, it was possible to recover nematodes belonging to the root-knot, root-lesion and other well-known groups. Apparently they were associated with losses of the crops and perhaps were as important as erosion and soil depletion in forcing the owners to abandon the land many decades before (Thorne, 1961). This agrees with Chitwood's statement: 'The most severe nematology problems occur where good host crops are grown too frequently for too long on the same land.' The annual losses incurred by the depredation of nematodes are almost incalculable. The rapid development of plant nematology in the last fifty years is not surprising when we consider that nematodes have invaded all cultivated areas in the world.

The study of nematodes as parasites greatly extended interest in their taxonomy, pathology, epidemiology and control. Investigations on their functional anatomy, physiology and biochemistry have been neglected until recently. Plant-parasitic nematodes are likely to become even more important to world agriculture in the future because of their spread by man. Contaminated seeds provide an ideal means of distribution from country to country and species may be disseminated in this manner, until they become cosmopolitan. Today plant nematology as a plant science is flourishing. The most modern techniques utilized in many disciplines of biology are now also applied to plant nematology and progress has been made in understanding the mechanisms of disease, the secondary effect of infection and the frequent appearance of new strains of pathogen, capable of infecting resistant plants. In the last twenty years, different branches of investigations have been developed on the functional anatomy of nematodes and on every aspect of interaction between these animals and plants, helped by the accumulated knowledge of biochemistry, physiology, breeding, genetics, immunology and biological control.

1.1 NEMATODES AS PLANT PARASITES

Plant-parasitic nematodes are found in all agricultural regions of the world and any crop is likely to suffer damage from these parasites. Nematodes cause diverse damage in plants, depending on feeding habit. Most of them attack roots and different nematodes may attack the same plants. They may also be additive to other stress factors and can induce predisposition of their hosts to attack by other pathogens, such as fungi and bacteria. Some nematodes are virus vectors and, in this case, the virus represents the major pathogen. There is considerable evidence that the combination of some of the factors is sufficient to induce crop losses and it is difficult to estimate the loss due solely to nematodes.

The effect of nematode infection is a general reduction in plant growth. A patch of poorly growing crop in an otherwise healthy crop is often the first evidence of nematode problems. Other effects of nematode feeding are excessive root branching, cessation of root elongation and retardation of root growth. Damage by nematodes feeding on above-ground plant parts is greatest in greenhouses and other situations of high humidity. Discoloration of foliage, twisted leaves and distorted shoots are also indicative of nematode infection. Poor growth may be caused by reduced translocation, inadequate nutrient absorption, abnormal production of toxic metabolites, etc.

Damage to sugar-beet factories in Germany in 1802, which had produced the first commercial sugar, represents the first evidence of the tremendous economic potential and the widespread destructiveness of plant-parasitic nematodes. Fifty years of intensive production, without adequate crop rotation, constituted the best conditions for the development of massive populations of parasites. Schacht, in 1859, detected for the first time the presence of a cyst nematode on sugar-beet, and Schmidt, in 1871, named it as *Heterodera schachtii* which inflicted enormous losses in the sugar-beet fields.

No accurate estimates are available of the total impact of nematode damage to the world's food crops, but in the USA alone the total annual loss from all crop disease exceeds $2 billion, part of which can be attributed both directly and indirectly to nematode action. In cotton alone, losses resulting from nematodes are estimated at more than $53 million annually.

Measurements of crop losses are difficult to make because of the various factors involved in damage affecting plants. In 1971, the Society of Nematologists estimated that average losses caused by plant-parasitic nematodes were about 10%. Losses in individual situations may run to 50%. Nematodes are serious pests for perennial tree and vine crops and for annual crops, such as vegetables and fruits. Although nematode problems occur in all the areas of the world, the most evident damage occurs in warm areas. The higher temperatures and longer growing season induce more nematode generations and consequently an increased nematode population.

Meloidogyne spp. represent a serious parasite of many crops, especially vegetable crops. The four species, *M. incognita*, *M. javanica*, *M. arenaria* and *M. hapla*, are widely distributed throughout the agricultural regions of the world and appear to be very adaptable to a wide range of agro-ecosystems. It is impossible to be precise concerning the losses caused by *Meloidogyne* spp. However, some estimates of individual crop losses have been made by various authors. In tomato plants, when *M. incognita* occurs 50% of fruit yield is suppressed (Lamberti, 1979). In tropical areas root-knot nematodes cause losses in potato crops estimated to be 24% (Sasser, 1979). Diomande (1981) also reported losses of 75% in upland rice. In cooler areas severe crop damage is caused by *Globodera rostochiensis* and *G. pallida* to potato, *Heterodera schachtii* to sugar-beet, *H. avenae* to oats, *Anguina tritici* to wheat and *Ditylenchus dipsaci* to a number of crops, including alfalfa and onion. The burrowing nematode *Radopholus similis* causes severe losses in tropical and subtropical crops such as citrus and bananas. Yield reduction has also been reported on numerous crops infested by *Pratylenchus* spp. The soya bean cyst nematode *H. glycines* is a nematode which spreads rapidly throughout the soya bean-growing areas and induces severe crop loss.

Other climatic or edaphic factors may be involved in the crop losses. *Meloidogyne incognita* causes severe damage on sugar-beet (70–80% reduction) (Lamberti, 1979), and to tomato in microplots in the coastal plain, whereas it effects only slight losses in yield in the mountains, where the heavier soil, higher rainfall and lower temperature favour plant growth (Barker *et al.*, 1976). The soil texture may favour one species of nematode more than another. *Pratylenchus coffeae* predominates in fine-textured soils and *Radopholus similis* has the advantage in coarse-textured soils (O'Bannon *et al.*, 1976). In a particle-size gradient, *Ditylenchus dipsaci* aggregates where the particles are the finest (Wallace, 1961). There are many reports and observations that associate the distribution and disease severity of *Meloidogyne* with sandy soils. The percentage of *M. incognita* juveniles able to migrate and penetrate roots decreases as the clay and silt fractions in the soil increase. Water potentials, carbon dioxide and oxygen concentration may alter the behaviour of nematodes. It was found that there is a linear relationship between the rate of oxygen diffusion and nematode activity. The effect of aeration on invasion, movement and hatch of *M. hapla* in organic soil was adversely affected by carbon dioxide concentration above 10%. In dry environments the gelatinous matrix of the egg sac of *M. incognita* appears to maintain a high moisture level and provides a barrier to water loss from eggs. The juveniles are able to survive in moist soil for a month without much physiological ageing, as measured by stored food reserves. *Meloidogyne* juveniles are contained in moist soil aggregates at or near anaerobic conditions, which demonstrates that nematodes can enter the state of anhydrobiosis and prolong their survival in dry soil (Van Gundy, 1985). In summary, the relationship of soil texture, water stress, aeration, rate of nematode reproduction and growth of the host plant is a complex ecological situation. Resistance has been an important factor for economical control of root-knot nematode.

1.1.1 Nematode pathogenesis and disease development

Plant-parasitic nematodes can be separated into different groups according to feeding habits. The development of mode of parasitism within nematodes was probably in a sequence from ectoparasites to endoparasites. Ectoparasites represent the primitive form of parasitism and include nematodes that are apparently capable of penetrating root tissues only by means of their stylet. They remain outside the roots and are epidermal or subepidermal feeders. Gregarious feeding by ectoparasites, e.g. Longidoridae, causes a decline in root growth. Attacked root tips usually stop growing and become transformed into terminal galls. Migratory endoparasites enter and migrate within roots, feeding on various tissues and inducing dramatic and usually lethal cell response of root tissue. Sedentary endoparasites have evolved intimate relationships within their hosts by the development of specialized feeding sites within plant tissues. Feeding sites are permanent structures and consist of modified cells with intense metabolic activity. The characteristic structure of syncytia and giant cells induced by various endoparasites develop as responses to specific substances introduced into the cells through their stylet. The structural and metabolic alterations occurring in plants, attacked by nematodes, have been recently reviewed (Wyss, 1981; Jones, 1981; Dropkin, 1979; Webster, 1985).

Disease in the plant host may be due to cellular response to nematode secretion, rather than to mechanical damage of the tissues. Galling of root tips induced by ectoparasites and modified and syncytia and giant cells associated with endoparasites should be considered as the direct consequences of secretions injected into host cells and the subsequent feeding activity. Nematode parasitism of a plant is a complex and dynamic interaction and may involve hatching stimuli, penetration of host tissues, recognition of tissue suitable for feeding-site formation, modification of host tissue, and an active response from the host (Hussey, 1989a, 1989b). 'As the parasite invades, the host responds' (Wallace, 1973). According to Dropkin (1969), cellular reaction to *Ditylenchus dipsaci*, *Meloidogyne*, *Heterodera* and *Nacobbus* was not a passive response to enzymes injected by the nematodes, but an active host participation in response to some controlling force from the parasite. The response of plants to parasites must be considered at the biochemical and physiological levels.

Plant-parasitic nematodes have large oesophageal glands with a complex secretory function, recently revealed by ultrastructural observations (Endo, 1984, 1987; Endo and Wergin, 1988). The tylenchid nematodes have three oesophageal glands – one dorsal and two subventral. The duct of the dorsal gland enters the oesophageal lumen near the base of the stylet. Secretions of the dorsal gland can pass into the host cells through the stylet. The sequence of feeding of parasitic nematodes has been reported by several authors (Doncaster and Seymour, 1973; Wyss and Zunke, 1986).

Subventral gland secretions are released into the oesophageal lumen and seem to function in the digestion of food. The morphology of glands of sedentary endoparasitic nematode changes during the processes of parasitism. Dorsal and subventral glands contain spherical membrane-bound secretory granules of different sizes. Granules in subventral glands of preparasitic juveniles are larger and have distinct membranes, whereas in juveniles that have entered the roots (two to three days) they are smaller, with no distinct membranes (Bird and Sauer, 1967). The dorsal gland, poor in granules in preparasitic juveniles, is very active in terms of the production of secretory granules, three days after root penetration (Bird, 1967; Endo, 1987). The granules become small and electron dense in contrast to their variable morphology in preparasitic juveniles. In adult females the subventral glands atrophy and the dorsal gland predominates (Bird, 1983).

Detailed studies on *Heterodera* and *Meloidogyne* spp. show that the dorsal gland is involved in pathogenesis and the subventral glands probably have no role in the infection process (Wyss and Zunke, 1986; Endo, 1987). The secretory granules of the dorsal gland appear to be involved in the formation of feeding tubes and are integral parts of host–parasite interactions (Rumpenhorst, 1984). Their variation in density and size may represent stages of degradation, related to the release of their content. Histochemical and ultrastructural cytochemical analyses demonstrated that secretions thought to be produced by the dorsal gland reacted positively for proteins. Proteins were found to be basic and three of these, when electrophoretically separated, were identified as glycoproteins (Bird, 1968; Veech et al., 1987). A new approach has been used to study the nature of oesophageal gland granules, considered the principal source of secretions involved in pathogenesis. Atkinson *et al.* (1988) developed a procedure to produce monoclonal antibodies to oesophageal gland secretory granules. They reported that a large number of antibodies generated using homogenates of preparasitic second-stage juveniles of *H. glycines* were specific for the two subventral glands and only one for the dorsal gland. Antibodies obtained from adult females were found to be specific for dorsal gland but not for subventral glands. These findings are consistent with previous reports on the involvement of dorsal gland secretions in the nematode–host interrelationship. Immunogold localization on ultrathin sections of oesophageal glands of *M. incognita* juveniles revealed that the antigen was segregated around an electron-transparent core in the matrix of secretory granules in the subventral glands. The antigens were also detected in the cytoplasm adjacent to secretory granules of the dorsal gland and in the stylet secretions of adult females. Purification of stylet secretory component by immunoaffinity revealed that it was glycoprotein (Hussey and Mims, 1990). The fine structure of the crystal-like secretions, observed in an open dorsal gland valve, showed the same features as the walls of feeding tubes formed in the host cells utilized as a feeding site by *M. incognita* (Hussey and Mims, 1990). The function of the secretions of oesophageal glands is probably related to the induction and maintenance of the feeding site in the host. The maintenance of a specific feeding site is very important for sedentary plant-parasitic nematodes. They induce unspecialized cells to form highly specialized cells, with specific functions, in response to various repressors and inducers injected into the host tissue. The morphology and development of the specialized feeding sites have been extensively reviewed (Bird, 1974). Jones (1981) compared the structural features of modified cells induced by different sedentary endoparasites. In every case, nuclei and nucleoli are enlarged, cytoplasm increases, and cell organelles are multiplied. Usually modified cells expand and their expansion may be accompanied by cell wall breakdown (syncytia) or nuclear mitosis (giant cells). The significance of syncytia and giant cells, in functional terms, is similar in sequestering nutrients for developing nematodes. Both structures resemble enlarged, multinucleate transfer cells (Jones and Northcote, 1972) and are induced, maintained and completely dependent on a continuous stimulus from the nematode (Bird, 1974). Removal of solutes is needed to maintain the structures. The result of removal of products is to stimulate the pathways to replace them and this leads to an increased metabolic rate. Nematode secretions elicit a normal plant cell process and result in plant disease development. The complex cellular modifications induced by species of *Meloidogyne* and *Heterodera* need to be further elucidated with the help of recent advances in immunology and molecular biology.

1.1.2 Biochemical defence mechanism of plants to nematode parasitism

A plant responds to the disturbance caused by nematodes in ways which appear to give an increased chance of survival. To do this the plant must maintain the capacity to coordinate its utilization of food and energy resources. The nature of the response depends on the plant and nematode species. Preformed barriers in the plant do not appear to be effective against plant-parasitic nematodes. All plant-parasitic nematodes possess a protrusible stylet that functions to rupture their egg shell during hatching, to penetrate plant cell walls during infection and to remove cell contents during the feeding process. The action of stylet penetration, combined with enzyme release by the nematodes (cellulases, pectinases and other cell wall-degrading enzymes) is capable of overcoming all plant barriers such as cell walls and cuticles (Kaplan and Keen, 1980). But not all nematodes, recognized as plant parasites, are pathogens for all plant species. A restricted number of plant-parasitic nematodes are capable of attacking only a given plant species. This is due to the presence in the plant of different types of defence mechanism. Because of their historical connotation with nematode development, the plants are classified as resistant or susceptible hosts. However, the use of the terms resistant and susceptible to describe plant response in relation to nematode reproduction should be avoided. For a complete evaluation of plant response to a nematode, more parameters should be measured, e.g. nematode reproduction, damage caused by the nematode, and host efficiency. Host susceptibility and resistance each results from a series of sequential events that are easily divided into pre- and post-infectional stages. The sequential events may or may not lead to the establishment of disease.

Resistance is the ability of plants to inhibit, restrict or retard disease development, upon challenge by a nematode. Thus, anything that prevents, restricts or retards disease development contributes to host resistance (Veech, 1981). Susceptibility is a compatible combination between nematode and host, resulting in the development of the parasite. If a nematode causes little injury, the plant is tolerant.

Experiments on preinfectional host–nematode interactions demonstrated that root exudates play an important role in both attracting and repelling nematodes (Kaplan and Keen, 1980; Huang, 1985; Veech, 1981). Griffin and Waite (1971) reported that when an egg mass of *Meloidogyne hapla* was placed on a water–agar plate, equidistant from germinating resistant and susceptible alfalfa seeds, approximately three-quarters of the hatched juveniles were attracted to the susceptible seedlings and only one-quarter were attracted to the resistant. A more direct linkage to a repulsion mechanism comes from cucumber plants carrying a single dominant bitter gene (*Bi*). These plants produce a triterpenoid compound (cucurbitacin) which is repellent to nematodes and toxic for a variety of insects. Different groups of naturally occurring compounds may exert stimulatory, repellent, nematicidal or nematostatic properties. *Tagetes, Asparagus, Crotolaria, Daphne, Carthamus, Physostigma, Eragrostis,* etc. are examples of groups of plants possessing preinfectional antinematode chemicals (Gommers, 1981). The compounds and the mechanisms responsible for repulsion and attraction, with few exceptions, have not been identified or characterized.

When nematodes are attracted to the roots of a plant, each species establishes its own distinctive relationship, which influences the physiology and cytology of the plant tissues. The migratory nematodes generally feed for a relatively short time and move from one feeding site to another. The sedentary nematodes establish specific feeding sites. Nematodes influence the cells at the feeding sites by injecting gland material and by removing the host cytoplasm or part of it. The susceptible response may therefore consist of a rapid accommodation by the plant to the removal of materials by the parasite. In a resistant plant, after nematode attack, several degrees of incompatibility have been reported (Canto-Saenz, 1985; Kaplan and Keen, 1980). In a highly resistant plant, after three or four days the number of nematodes in the roots sharply declines, presumably as a consequence of failure to establish a feeding relationship.

The most common response of resistant plants is hypersensitivity of cells damaged by migrating juveniles. However, the degree of necrosis and time of its appearance may differ from plant to plant. The metabolic changes and the physiological alterations of plants associated with nematode infection are not as well known as the visible changes in cell or tissue morphology. Most of the publications of biochemical changes compare infected tissue with non-infected. Changes in respiratory rate and in related enzymes, such as ascorbic acid oxidase, cytochrome c oxidase, catalase and peroxidase, have been reviewed by different authors (Dropkin, 1979; Kaplan and Keen, 1980; Canto-Saenz, 1985). There are also reports on increased activity of phenylalanine ammonia lyase (PAL), oxidoreductive, hydrolytic and oxidative enzymes (Veech and Endo, 1970). However, data on altered biochemistry of infected plants are too fragmentary to permit any understanding of the mechanism by which nematodes influence the plants. Apparently plants have a biochemical mechanism which operates differently in resistant or susceptible plants. In a resistant host, an endoparasitic nematode blocks the synthesis of growth hormones and promotes other factors inducing hypersensitive reactions instead of giant cells. Thus, around the area of the nematode head, necrotic cells are frequently found in resistant tissues. After inducing necrosis, however, the nematode may migrate to non-necrotic cells and recommence feeding; these cells subsequently become necrotic. Absence of nutrients may induce nematode death. The death of the nematode after appearance of necrosis suggests that the plant responds to the feeding of the nematode with a mechanism which induces the activation or synthesis of toxic substances. An alternative approach would be to consider the nematode to be capable of suppressing the defence mechanism of the plant in compatible combinations. The disease resistance mechanism induces in the plant a variety of responses such as lignification and toxin (phytoalexin) synthesis which are considered to result from the processes of secondary metabolism. With regard to primary metabolism there is little information. Evidence for a general operating resistance mechanism was found in the tomato/*Meloidogyne incognita* interaction. In a resistant tomato cultivar an increase of cyanide-resistant respiration was observed in root tissues (Zacheo and Bleve-Zacheo, 1988) and in isolated mitochondria (Molinari *et al.*, 1990). Proteins from cyanide-resistant mitochondria contain more hydroxyproline than those from cyanide-sensitive mitochondria (Zacheo *et al.*, 1977). The biosynthesis of hydroxyproline-containing proteins requires ascorbic acid (Arrigoni *et al.*, 1977). When susceptible tomato plants were artificially increased in ascorbic acid concentration, the population of *M. incognita* juveniles decreased significantly. A decrease in ascorbic acid, obtained by application of lycorine – an inhibitor of ascorbic acid synthesis – induced a reduction in tomato resistance to root-knot nematode. Ascorbic acid, through the increased synthesis of hydroxyproline-containing proteins, induces the activation of cyanide-resistant respiration (Arrigoni *et al.*, 1979). It seems likely that the cyanide-resistant respiration transfers electron flow to oxygen, in a univalent manner. This results in an increased rate of intracellular production of oxygen superoxides (O_2^-) or some product uniquely derived from superoxide. Zacheo and Bleve-Zacheo (1988) demonstrated that in resistant tomato plants showing activation of cyanide-resistant respiration, during nematode attack, stimulation of oxygen free radical production was detectable. This phenomenon was not observed in infested susceptible plants, because the superoxides were scavenged by the increased superoxide dismutase activity. The function of these enzymes is to remove superoxide anions by catalysing the disproportionation of superoxides to hydrogen peroxide, which was eliminated through the action of peroxidase and catalase. An interpretation of these results is that root cells in resistant plants react to the presence of a nematode by developing an NADPH-oxidase activity, which produces large amounts of free radicals. This process has been considered one of the earliest biochemical reactions in the hypersensitive response which may occur on the host plasma membrane. It has been suggested that superoxides produced by plants, after infection, may be directly or indirectly responsible for hypersensitive cell death and consequently for resistance to infecting agents. How superoxides, which may act as a stimulator of membrane functions, cause activation of various metabolic processes associated with hypersensitive reaction is still fragmented and remains to be investigated in detail. The role of superoxides in disease resistance was recently identified by Doke *et al.* (1987). The free radicals of oxygen elicit the accumulation of phytoalexins. Superoxides produce some endogenous activators of the phytoalexin synthetic pathway in the cells surrounding necrotized tissue. Different authors reported accumulation of phytoalexins in resistant plants infected by nematodes (Kaplan *et al.*, 1980a, 1980b· Veech, 1981). Recent studies (Preisig and Kuc, 1987) reported that salicylhydroxamic acid, an inhibitor of the cyanide-resistant respiration, failed to accumulate phytoalexins, by suppressing the hypersensitive response. These results suggest that cyanide-resistant respiration and superoxides are involved in biosynthesis of phytoalexins in resistant plants during pathogen infections.

1.1.3 Nematode management

Resistance has been an important means for the economical control of root-knot nematode. Considerable progress has been made in developing root-knot resistance in vegetable crops. Plant breeders have developed over 50 tomato cultivars, commercially available, containing the gene *Mi* for resistance to *Meloidogyne incognita*, *M. javanica* and *M. arenaria*. Breeding lines of legumes, with a high level of resistance to root-knot nematodes, have also been developed recently. Cyst nematode, if uncontrolled, induces losses of 100% in potato yields. When suitable resistant cultivars are grown, a decrease of 80% each year in the nematode population is detectable (Evans and Brodie, 1980). Resistant cultivars not only prevent nematode damage but also lower population densities and extend protection to the following susceptible crop.

The best combination of chemical, cultural, biological and genetic methods is the most effective control of pests (diseases, insects, nematodes and weeds) that yields an economic return. The most widely used method during the last 15 years, for controlling nematodes after proper identification, is chemical soil treatment before planting an annual crop. Nematicides are effective in destroying nematodes in the soil, but certain factors must be considered during their application, such as temperature, presence or absence of host plants and soil type. Fumigant, non-fumigant and systemic nematicides are available. Because they are expensive, harmful and can affect beneficial organisms in the soil, large yield increases are required to justify their usage.

A relatively new procedure for controlling nematodes is that of biological control. Nematodes are attacked, under natural conditions in the soil, by a wide variety of organisms, such as predators and parasites. Predators include fungi, nematodes, insects and mites. Parasites include viruses, protozoa, bacteria and fungi. Several reports attest to the dramatic increase of research effort towards biological control of plant-parasitic nematodes (Sayre and Starr, 1985; Jatala, 1986). Such associations result in a biological balance, manifested by attachment and penetration by one or more pathogenic microorganisms in the eggs, juveniles and adult nematodes, causing their death (Jansson and Nordbring-Hertz, 1980; Walsh *et al.*, 1983). The characteristics of natural enemies are mobility, synchronization with the host, and the ability to survive host-free periods. Many of the known predators and fungal enemies of nematodes are inadequate with regard to some of these characteristics. This may explain why they have rarely been used successfully against plant-parasitic nematodes (Mankau, 1980). Stirling and Wachtel (1980) greatly improved the mass-production method of the bacterium *Pasteuria penetrans*. Its ability to attack important nematode pests of several crops, to persist in the soil for a long time and to resist high temperatures, and its compatibility to several nematicides, represents a good tool as a successful biocontrol agent.

The suppression of a nematode population with biotic organisms is manifested slowly in time and the results are not as spectacular as those obtained by using nematicides. Successful results have recently been obtained with biocontrol agents in nematode management practice (Cayrol and Frankowski, 1979; Jatala 1986; Kerry *et al.*, 1982). However, biological control of nematodes by natural enemies per se is not yet very practical under field conditions because of the paucity of information about how to manage the microorganisms involved.

1.2 NEMATODE INTERACTIONS WITH OTHER PLANT PATHOGENS

Under natural conditions a plant is a potential host to various microorganisms and they can influence each other by occupying the same habitat. Infection by one pathogen can alter host response to subsequent infection by another pathogen. Different parasites on the same plant interact, which results in disease complexes, and these interactions may lead to susceptibility by predisposition or resistance through preinduction of resistance against a particular parasite (Sidhu and Webster, 1981).

In nature plants are rarely exposed to the influence of a single pathogen. Fawcett (1931) recognized that 'nature does not work with pure cultures' and that many plant diseases are influenced by associated organisms. Roots grow in soil containing a great number of microorganisms, whose action is often combined to induce damage. Plant-parasitic nematodes often play a major role in disease interactions. Interaction involving nematodes is important because they contribute substantially to variability in crop growth (Zadoks and Schein, 1979). It is possible that some dramatic crop losses involving nematodes are due to the interaction between several determinants that exacerbate the effects (Wallace, 1983). Nematodes interact with different groups of plant pathogens and root symbionts. The association between nematodes and other pathogens or symbionts in plant disease encompasses a wide array of species and results in a biological balance that may even lead to an understanding of a few interactions. It seems reasonable to expect that infection by one pathogen may alter the host response to subsequent infection by another (Taylor, 1990). The mechanism of interaction and the precise role of each component in a disease complex has yet to be widely exploited. The fungus–nematode and the bacterium–nematode interactions are numerous, and weakly parasitic fungal and bacterial parasites can cause considerable damage once they gain entry into plant roots in the presence of feeding nematodes.

1.2.1 Interactions with fungi

Much experimental evidence indicates a biological interaction between nematodes and certain soil-borne fungi. In most interactions the nematodes are not essential for the establishment and development of fungal pathogens. However, the nematodes usually assist and enhance the pathogenicity mechanism of the fungus towards modifications in the host plants. For many years it was thought that nematode wounds facilitate the entry of fungi. Perry (1963) and Starr *et al.* (1989) demonstrated that cotton roots infected by *Meloidogyne incognita* were not attractive for *Fusarium oxysporum*, but Mouza and Webster (1982) observed a synergistic disease interaction between *F. oxysporum* and *Pratylenchus penetrans* on alfalfa, with simultaneous inoculation. It is also reported that mechanical wounding increased *Fusarium* root invasion more than nematode attack. Evidence indicates that interactions between *Fusarium* and root-knot and cyst nematodes are biological and physiological rather than physical in nature (Roy *et al.*, 1989). Giant cells and syncytia induced by some endoparasites act as transfer cells and their activity is directly related to the physiological stage of the nematode. Most synergistic interactions between these endoparasitic nematodes and *Fusarium* occur after three to four weeks of nematode infection. Three to four weeks represents the time of the maximum level of concentration of total protein and other metabolites in transfer cells induced by the nematodes. Many fungi benefit from this enriched medium and *Fusarium* hyphae in root-knot galls are more extensive than those in non-infected roots (Melendez and Powell, 1967). Moreover, root exudates reflect the biochemical and physiological changes induced by nematode infection, and hence the development of fungi in roots is influenced by the presence of additional metabolic products in the rhizosphere and on the roots. Root exudates known to attract the motile stage of fungal pathogens (Zentmyer, 1961) represent a source of nutrients for the soil microflora and may be a stimulus for the germination of dormant spores. Thus changes induced by a nematode in the root exudates may be the first stage in the synergistic interaction between nematode and fungi (Taylor, 1990). Nordmeyer and Sikora (1983) reported that roots of *Trifolium subterraneum* exposed to culture filtrates of *F. avenacearum* were more easily penetrated by juveniles of *Heterodera daverti* than were control roots. Apparently short exposure to filtrates rendered the roots more attractive to nematode juveniles. This phenomenon did not occur after longer exposure, indicating that there may be an optimal level of root exudation to which the nematode is attracted (Freckman and Caswell, 1985). However, Griffin (1990) reported that *Ditylenchus dipsaci* and *F. oxysporum* f. sp. *medicaginis* synergistically affected the mortality and plant growth of Ranger alfalfa, a cultivar susceptible to stem nematode and *Fusarium* wilt. Disease severity did not differ with simultaneous or sequential inoculation of the two pathogens.

In plant pathology attempts are made to ascribe disease resistance to one or a few factors, to explain the physiology of diseased plants on the basis of a single toxin or a few enzymes, and to overlook the possibilities of multiple-pathogen aetiology. In many cases it is true, but it must be considered that nature rarely takes things 'one at a time' (Powell, 1979). An individual plant can be influenced by one or more potential pathogens. When a plant is jointly infected by more than one pathogen, it seems reasonable to expect that the activities of one pathogen influence the activity of other pathogen(s). An infected plant is altered in some way in its physiological functions and also its susceptibility to invasion by other pathogens is changed. It may result in a synergistic interaction in terms of combined effect of the pathogens, or in a negative interaction in terms of competitive exclusion.

Plant-parasitic nematodes are known to predispose some plants to fungal pathogens (Atkinson, 1892; Bergeson, 1972; Sidhu and Webster, 1977; Mai and Abawi, 1987) and this type of interaction was genetically evaluated in the *Meloidogyne incognita–Fusarium oxysporum lycopersici* disease complex on tomato host. Endoparasitic nematodes such as root-knot and cyst nematodes have long been known as primary pathogens for their ability to predispose plants to infection by secondary pathogens such as several species of *Fusarium*, *Phytophthora* and *Rhizoctonia*. However, the presence of fungal infection can sometimes alter nematode growth and reproduction (Jorgenson, 1970). Whitney (1971) attributed synergism between *Heterodera* and *Pythium* to an increase in the growth of the fungus when the nematode was present, leading as a final result to a more devastating fungal attack. Sidhu and Webster (1977) demonstrated that the resistance of tomato cultivars to *Fusarium* wilt was diminished following invasion by root-knot nematodes. Migratory nematodes also represent a predisposing factor to infection by certain fungi and bacteria. Mechanical wounding of the root favours the entry of other pathogens. For example, *Pratylenchus* sp. is reported to interact synergistically with *Verticillium* and *Belonolaimus* sp. and *Trichodorus* sp. with *Fusarium* (Faulkner *et al.*, 1970; Yang *et al.*, 1976; Vrain, 1987). The mechanism of synergism with migratory nematodes is different from that with sedentary endoparasites, because only necrosis and other processes of destruction develop rather than root physiological changes.

1.2.2 Interactions with bacteria

Nematodes may predispose plants to bacterial diseases. Nematodes and bacteria in association induce different symptoms in plants from those found when either pathogen is present alone. The association between gall-inducing nematodes and a species of bacteria results in toxin production in the host plant. Bacteria can only enter plants through wounds or through natural openings or by vectors. The role of nematodes in relation to bacterial pathogens has usually been regarded as providing wounds in the roots through which bacteria may enter the plant. This appeared to be the case in the association between *Pseudomonas solanacearum* and the root-knot nematodes *Meloidogyne incognita* and *Agrobacterium tumefaciens* associated with *M. hapla*. Nematodes can act as vectors by carrying bacteria on their body surface. In this case an obligate aetiological relationship is established between nematode and bacterium in the manifestation of the resulting disease (Taylor, 1990). In the association between *Rhodococcus fascians* and *Aphelenchoides ritzemabosi* there is evidence that the bacterium attaches specifically to the cuticle of the nematode and is thus transported to the strawberry host plant (Hawn, 1971; Riley *et al.*, 1988). Both pathogens together induce various grades of leaf deformation. Strawberry plants attacked by the nematode alone do not show leaf deformations and bacterial infection alone induces no severe disease. The associations between *Anguina tritici* and *Clavibacter tritici* and *Anguina funesta* and *Clavibacter* sp. cause a disease known as spike blight in wheat and ryegrass respectively (Cheo, 1946; Gupta and Swarup, 1972; Bird, 1981). No grains are formed in the spikelets but only a mass of bacteria. In the absence of bacteria, the nematodes induce blackened seed galls. The presence of *A. funesta* and a bacterium on ryegrass results in the production of a toxin fatal to sheep and cattle. Bacterial wilt caused by *Pseudomonas solanacearum* is more severe in resistant cultivars of tomato and eggplant in the presence of *M. incognita*. The combination of the two pathogens suppresses the survival rate of wilt-resistant tomato plants 33–36%. In other associations ecto- and endoparasitic nematodes can facilitate bacterial invasion by providing wounds in the host root (Poinar and Hansen, 1986).

1.2.3 Interactions with viruses

Some species of nematodes belonging to the families Longidoridae and Trichodoridae have a unique and important role as pathogens of plants, because they also transmit certain viruses to their host plants. The historical aspects and diseases caused by nematode-transmitted viruses have been extensively reviewed (Taylor, 1980; Martelli and Taylor, 1989). In the Trichodoridae 13 species of *Trichodorus* and *Paratrichodorus* are known to transmit rod-shaped, short and long (45–115 nm and 180–210 nm) particles classified as tobraviruses. Tobacco rattle, known from 1943, and pea early browning viruses, transmitted by Trichodoridae, infect a wide range of wild and cultivated plants. Tobacco rattle induces disease, economically important, for bulbous flower crops in Europe and potato crops in Europe and the USA. Browning disease is particularly diffuse in Europe and induces severe stunting and premature plant death in large areas of pea crops (Taylor, 1980). Hewitt *et al.* (1958) provided first experimental evidence that *Xiphinema index* acted as a vector of grapevine fanleaf virus. In the Longidoridae two genera – *Xiphinema* and *Longidorus* – transmit viruses, the nepoviruses, which are polyhedral, isodiametric particles about 28 nm in diameter. Nepoviruses are reported to infect a wide host range, from wild plants to annual and perennial crops. Nepoviruses are pathogens of economic importance, as they affect and damage major crops. *X. index* is distributed worldwide and is closely related to its natural host, grapevine. All vectors of grapevine virus are longidorids and 12 viruses of grapevine are known or suspected to be nematode transmitted. Grapevine degeneration is characterized by deformation and discoloration of the leaves, reduced vigour and poor fruit setting. The crop may be drastically affected, with average losses of above 60% (Rudel, 1985). In Great Britain and Central Europe raspberry and strawberry crops are stunted and often die, when mixed infection of nematode vectors and related nepoviruses are present (Murant *et al.* 1990).

1.2.4 Interactions with insects

Nematodes parasitize many groups of insects and play an important role in natural control of these pests (Nickle and Welch, 1984). In some cases, nematodes invade insects without causing them damage and also are transported for long distances. *Rhadinaphelenchus cocophilus*, which causes red-ring disease in coconut and other palms, is vectored by an insect *Rhynchophorus palmarum*. *Bursaphelenchus xylophilus*, the causal nematode of pine wilt, is vectored by an insect of the genus *Monochamus*. Up to 75% of the beetles have been found to be vectors and 15 000 to a maximum of 230 000 dauer larvae of nematodes per insect were recovered. Nematodes are located in the tracheae of adult insects, which appear not to be adversely influenced by their presence. During insect feeding, the nematodes enter the pine shoot, immediately invade resin canals, and induce reduction and cessation of oleoresin exudation and transpiration reduction. Quick death of affected trees is characteristic of the pine wilt disease (Mamiya, 1988). As cell death occurs in a very short time after nematode inoculation (24 hours), Oku *et al.* (1980) hypothesized that some metabolites of the pine wood nematode or bacterium associated with the nematode have a toxic effect on pine trees. Environmental factors modify disease incidence in pine. A strong correlation between high temperature and water stress and development of pine wilt disease was found in Japan, the most severely infected area in the world (Kishi, 1980). Dropkin and Foudin (1979) and Dropkin *et al.* (1981) observed, for the first time, pine wilt disease in the USA, and Nickle (1981) reported losses of Japanese black pine along the Atlantic coast. It appears that pines native to Japan are more susceptible to pine wood nematode than pine species distributed in other parts of the world.

1.2.5 Interactions with other nematodes

Interaction between different species of nematodes is usually antagonistic rather than beneficial. Two species can cohabit in the same host, but competition is severe when both species have similar feeding behaviour. Interaction between ectoparasites can be stimulatory, but competitive advantage increases as the host–parasite relationship becomes more complex. Combined infections of ecto- and endoparasitic nematodes are, in most cases, suppressive for ectoparasites as sedentary endoparasites are more advanced parasites. In any case two or more species of nematodes may interact and the outcome of the interaction is density and time dependent. The combined effect usually leads to increase of disease, which becomes more severe when two or more species of nematode interact with another pathogen (Eisenback, 1985).

Nematodes influence the plant response to other pathogens and are influenced by associated microorganisms. Nematode attack sometimes increases the effect of disease caused by other organisms. Thus plant breeders should be aware of the possibility of nematode attack and the pathological additive effect of two pathogens, when they develop cultivars with resistance to fungi and bacteria.

1.3 NEMATODE INTERACTIONS WITH ROOT SYMBIONTS

Plant-parasitic nematodes affect growth indirectly, by interfering with the plant's symbiotic relationship. Cyst nematodes affect root nodulation induced by *Rhizobium japonicum* on soya bean (Huang and Barker, 1983; Ko *et al.*, 1984). Nodules from nematode-infected plants had lower fresh weight, specific nitrogenase and leghaemoglobin content than nodules from uninfected plants. The nematode effect depends on the nematode population density and can be reversed by removing the pathogen from the root system (Ko *et al.*, 1984). Soya bean root nodules are also reported to be parasitized by *Rotylenchulus reniformis* (Meredith *et al.*, 1983). A strong negative correlation has been observed between nematode populations and nitrogen fixation in soya bean (Germani *et al.*, 1981) and peanut (Germani *et al.*, 1982). No deleterious effect on nodule growth was detected on cowpea when *Rhizobium* and *Rotylechulus reniformis* and *Meloidogyne javanica* were simultaneously inoculated, unless *Rotylechulus reniformis* was inoculated prior to *Rhizobium* (Taha and Kassab, 1980).

Another system in which nematodes seem to be important is the mycorrhizal association between certain fungi and plant roots. Vesicular–arbuscular mycorrhizal fungi (VAM) are obligate symbionts and benefit plants by increasing the flow of essential elements into roots, by increasing the root surface for uptake of soil constituents, through the hyphal network at the exterior of the root, or through vesicles and branching structures within cortical cells. Generally VAM have a stimulating effect on plant growth increasing shoot, root and seed weights as well as pod and seed numbers (Carling and Brown, 1980). Nematode infection can damage the roots sufficiently to block the symbiotic plant–fungal associations. Field observations have indicated that root-knot nematodes cause a reduction in mycorrhizal level in the plant system (Jatala, 1986). Mycorrhizal fungi usually protect plants from pathogenic fungi but not from nematodes. Sikora (1979) suggested that the presence of the mycorrhizal fungus in the plant root system prior to nematode attack may exert its influence on nematodes by altering root attractiveness, reduction of juvenile penetration, impeding juvenile development and retarding giant cell formation. These effects probably result from complex physiological changes associated with the mycorrhizal infection, and are not caused by direct competition between the two organisms. Attraction could be adversely influenced by changes in the chemicals of root exudates, by hardening of the root tissue due to increased lignin level in the endo- and exodermis of mycorrhizal plants. Increased phenolics and decreased auxin levels in the roots may impede juvenile growth and giant cell formation (Dehne, 1982). In general, mycorrhiza–plant-parasitic nematodes result in a less deleterious effect of nematodes on plants. Roots of tobacco, carrot, soya bean and tomato colonized by VAM resulted in a decreased population of *Globodera solanacearum*, *Meloidogyne hapla* and *M. incognita* respectively (Sikora, 1979; Carling *et al.*, 1989). Suppression of soya bean cyst nematode reproduction and development by VAM fungi was detected in soya bean grown in greenhouse and in field microplots. Nematode suppression might be the result of competition for plant nutrients between VAM fungi and nematodes. Vesicular–arbuscular mycorrhizal fungi are very dependent on host photosynthetates during the early stages of root colonization when fungal structures are forming. This period of VAM establishment coincides with the entry and selection of feeding site by the cyst juveniles. Insufficient nutrients for nematodes could result in decreased reproduction of the early generation (Tylka *et al.*, 1991). Conversely *M. arenaria* on grape plants increased in the presence of VAM. Population densities were greater in mycorrhizal than in non-mycorrhizal plants (Atilano *et al.*, 1981).

1.4 FUTURE DIRECTIONS

There have been significant advances over the last few years in understanding nematode pathogenesis. Nevertheless knowledge in most areas of plant nematology is fragmentary and control measures for plant diseases caused by nematodes are limited. Basic research is needed in important areas such as physiology, biology and biochemistry of the nematode host–parasite relationship, including resistance and susceptibility, virus transmission and biological control.

Despite the intensive efforts to analyse how nematode secretions are involved in pathogenesis, the precise mode of action of specific compounds in plant tissues is still lacking. Further analysis of the initial events in the plant–parasite association, involving nematode secretory components, will permit statements about the influence of these pathogens on the physiology of the plants. Moreover, the characterization of the secretions may provide useful information on their role in plant pathogenesis and biological activity. The combined approaches of enzymology and physiology, at the cellular level, may offer more precise insights on what is happening in the living cells. New or improved methods of measuring the concentration of various types of defence-related compounds in situ (RNA, enzymes, proteins) and final products of gene activation (phytoalexins, wall-bound phenolics, other cell-wall components, etc.) or metabolites resulting from other activation mechanisms should be developed.

During the past twenty years, a considerable amount of work centred around screening plants for resistance to nematodes. Up to the present time, breeding for resistance to plant pathogens has been based on identification of certain traits of resistance in a given cultivar or line and their subsequent introduction, by genetic crossing, into a crop species. This technique has been successful in some cases. However, with certain crops it is quite difficult to combine nematode resistance with desirable agronomic characteristics. Recent results have shown that, because of the discovery of appropriate plant vectors (e.g. *Agrobacterium* plasmids), rapid introduction of new traits into the plant nuclear genome can be obtained. The ability to clone and to sequence genes and to identify specific gene products opens a new era in plant pathology. In view of their practical importance in providing inexpensive and generally stable control of many diseases in important food and fibre plants, cloning techniques seem to be a good way for the production of new plants resistant to nematodes. The importance of these approaches is that some degree of resistance can also be obtained by incorporating viral genes into the host genome that may improve the degree of resistance in plants attacked by virus-vector nematodes.

Biological control strategies must be developed with the intention of discouraging the use of nematicides, because they are expensive and accumulate in the soil and in ground water. Moreover, the use of nematicides has led to the increase of pathogen resistance through natural selection. A new procedure for the future for alleviating the nematode problems is that of biological control. The possibility of using the biological complexity of the soil mass, in the proper way, may significantly aid control of parasites. Much of the work on biological control is yet only descriptive and more research must be done. Biological control acts slowly and could be used as an adjunct to other control methods such as chemicals, solarization, etc. The solarization offers economic feasibility such as the possibility of a long-term effect, reduced usage of herbicides, safety and reduced hazards to the environment. Solarization suffers some restriction in terms of climatic areas and destruction of all microfauna. Researchers must further investigate the effect of solarization on the biological component of the soil. A combination of the strategies of molecular biology and genetic engineering, to induce plant resistance and biological control associated with solarization, can effectively manage plant diseases of complex aetiology involving nematodes.

REFERENCES

Arrigoni, O., Arrigoni-Liso, R. and Calabrese, G. (1977) Ascorbic acid requirements for biosynthesis of hydroxy proline-containing proteins in plants. *FEBS Letters*, **81**, 135–8.

Arrigoni, O., Zacheo, G., Arrigoni-Liso, R., Bleve-Zacheo, T. and Lamberti, F. (1979) Relationship between ascorbic acid and resistance in tomato plants to *Meloidogyne incognita*. *Phytopathology*, **69**, 579–81.

Atilano, R.A., Menge, J.A. and Van Gundy, S.D. (1981) Interaction between *Meloidogyne arenaria* and *Glomus fasiculatus* in grape. *Journal of Nematology*, **13**, 52–7.

Atkinson, G.F. (1892) Some diseases of cotton. *Bulletin of Alabama Agricultural Experiment Station*, **41**, 61–5.

Atkinson, H.J., Harris, P.D., Halk, E.U. and Novitski, C. (1988) Monoclonal antibodies to the soybean cyst nematode *Heterodera glycines*. *Annals of Applied Biology*, **112**, 459–69.

Barker, K.R., Shoemaker, P.B. and Nelson, L.A. (1976) Relationship of initial population densities of *Meloidogyne incognita* and *M. hapla* to yield of tomato. *Journal of Nematology*, **8**, 232–9.

Bergeson, G.B. (1972) Concepts of nematode–fungus association in plant disease complexes: A review. *Experimental Parasitology*, **32**, 301–14.

Bird, A.F. (1967) Changes associated with parasitism in nematodes. I. Morphology and physiology of preparasitic and parasitic larvae of *Meloidogyne javanica*. *Journal of Parasitology*, **53**, 768–76.

Bird, A.F. (1968) Changes associated with parasitism in nematodes. IV. Cytochemical studies on the ampulla of the dorsal espophageal gland of *Meloidogyne javanica* and on exudations from the buccal stylet. *Journal of Parasitology*, **54**, 879–90.

Bird, A.F. (1974) Plant response to root-knot nematode. *Annual Review of Phytopathology*, **12**, 69–85.

Bird, A.F. (1981) The *Anguina–Corynebacterium* association, in *Plant Parasitic Nematodes*, Vol. III (eds B.M. Zuckerman and R.A. Rohde), Academic Press, New York, pp. 303–23.

Bird, A.F. (1983) Changes in the dimensions of the oesophageal glands in root-knot nematodes during the onset of parasitism. *International Journal for Parasitology*, **13**, 343–8.

Bird, A.F. and Sauer, W. (1967) Changes associated with parasitism in nematodes. II. Histochemical and microspectrophotometric analysis of preparasitic and parasitic larvae of *Meloidogyne javanica*. *Journal of Parasitology*, **53**, 1262–9.

Canto-Saenz, M. (1985) The nature of resistance to *Meloidogyne incognita* (Kofoid and White, 1919) Chitwood, 1949, in *Advanced Treatise on Meloidogyne, Vol. I: Biology and Control* (eds J.N. Sasser and C.C. Carter), North Carolina State University Graphics, Raleigh, pp. 225–31.

Carling, D.E. and Brown, M.F. (1980) Relative effect of vesicular–arbuscular mycorrhizal fungi on growth and yield of soybeans. *Soil Science Society of America Journal*, **44**, 528–32.

Carling, D.E., Roncadori, R.W. and Hussey, R.S. (1989) Interactions of vesicular–arbuscular mycorrhizal fungi, root-knot nematode, and phosphorus fertilization on soybean. *Plant Disease*, **73**, 730–3.

Cayrol, J.C. and Frankowski, J.P. (1979) Une metode de lutte biologique contre les nematodes a galles des racines appartenant au genre *Meloidogyne*. *Revue de Horticulture*, **193**, 15–23.

Cheo, C.C. (1946) A note on the relation of nematodes (*Tylenchus tritici*) to the development of the bacterial disease of wheat caused by *Bacterium tritici*. *Annals of Applied Biology*, **33**, 446–9.

Chitwood, B.G. and Chitwood, M.B. (1950) *An Introduction to Nematology*, Monumental Printing, Baltimore.

Dehne, H.W. (1982) Interaction between vesicular–arbuscular mycorrhizal fungi and plant pathogens. *Phytopathology*, **72**, 1115–19.

Diomande, M. (1981) Root-knot nematodes on upland rice (*Oryza sativa* and *O. glaberrima*) and cassava (*Manihot esculenta*) in Ivory Coast, in *Proceedings of Third IMP Research Planning Conference on Root-knot Nematodes, Meloidogyne spp.*, Regions IV and V, Ibadan, Nigeria, pp. 37–45.

Doke, N., Chai, H.B. and Kawaguchi, A. (1987) Biochemical basis of triggering and suppression of hypersensitive cell response, in *Molecular Determinants of Plant Diseases* (eds S. Nishimura, C.P. Vance and N. Doke), Springer-Verlag, Berlin, pp. 235–51.

Doncaster, C.C. and Seymour, M.K. (1973) Exploration and selection of penetration site by Tylenchida. *Nematologica*, **19**, 137–45.

Dropkin, V.H. (1969) Cellular responses of plants to nematode infections. *Annual Review of Phytopathology*, **7**, 101–22.

Dropkin, V.H. (1979) How nematodes induce disease, in *Plant Disease: An Advanced Treatise*, Vol. IV (eds J.G. Horsfall and E.B. Cowling), Academic Press, New York, pp. 219–38.

Dropkin, V.H. and Foudin, A.S. (1979) Report of the occurrence of *Bursaphelenchus lignicolus*-induced pine wilt disease in Missouri. *Plant Disease Reporter*, **63**, 904–5.

Dropkin, V.H., Foudin, A., Kondo, E., Linit, M.J., Smith, M. and Robbins, K. (1981) Pinewood nematode: A threat to U.S. forests. *Plant Disease*, **65**, 1022–7.

Eisenback, J.D. (1985) Interaction among concomitant populations of nematodes, in *An Advanced Treatise on Meloidogyne, Vol. I: Biology and Control* (eds J.N. Sasser and C.C. Carter), North Carolina State University Graphics, Raleigh, pp. 193–213.

Endo, B.Y. (1984) Ultrastructure of the esophagus of larvae of the soybean cyst nematode, *Heterodera glycines*. *Proceedings of the Helminthological Society of Washington*, **51**, 1–24.

Endo, B.Y. (1987) Ultrastructure of esophageal gland secretory granules in juveniles of *Heterodera glycines*. *Journal of Nematology*, **19**, 469–83.

Endo, B.Y. and Wergin, W.P. (1988) Ultrastructure of the second-stage juvenile of the root-knot nematode, *Meloidogyne incognita*. *Proceedings of the Helminthological Society of Washington*, **55**, 286–316.

Evans, K. and Brodie, B.B. (1980) The origin and distribution of the golden nematode and its potential in the U.S.A. *American Potato Journal*, **57**, 79–89.

Faulkner, L.R., Bolander, W.J. and Skotland, C.B. (1970) Interaction of *Verticillium dahliae* and *Pratylenchus minyus* in Verticillium wilt of the nematode as determined by a double root technique. *Phytopathology*, **60**, 100–3.

Fawcett, H.S. (1931) The importance of investigations on the effect of known mixtures of organisms. *Phythopathology*, **21**, 545–50.

Freckman, D.W. and Caswell, E.P. (1985) The ecology of nematodes in agroecosystems. *Annual Review of Phytopathology*, **23**, 275–96.

Germani, G., Ollivier, B. and Diem, H.G. (1981) Interaction of *Scutellonema cavenessi* and *Glomus mosseae* on growth and N-fixation of soybean. *Revue de Nematologie*, **4**, 277–80.

Germani, G., Cuany, A. and Merny, G. (1982) L'analyse factorielle des correspondences appliquée à l'influence deux nematode sur la croissance de l'arachide et sa fixation symbiotique de l'azote. *Revue de Nematologie*, **5**, 161–8.

Gommers, F.J. (1981) Biochemical interactions between nematodes and plants and their relevance to control. *Helminthological Abstract*, **50**, 9–24.

Griffin, G.D. (1990) Pathological relationship of *Ditylenchus dipsaci* and *Fusarium oxysporum* f. sp. *medicaginis* on alfalfa. *Journal of Nematology*, **22**, 333–6.

Griffin, G.D. and Waite, W.W. (1971) Attraction of *Ditylenchus dipsaci* and *Meloidogyne hapla* by resistant and susceptible alfalfa seedlings. *Journal of Nematology*, **3**, 215–19.

Gupta, P. and Swarup, G. (1972) Ear-cockle and yellow ear-rot diseases of wheat, 2. Nematode bacterial association. *Nematologica*, **18**, 320–4.

Hawn, E.J. (1971) Mode of transmission of *Corynebacterium insidiosum* by *Ditylenchus dipsaci*. *Journal of Nematology*, **3**, 420–1.

Hewitt, W.B., Raski, D.J. and Goheen, A.C. (1958) Nematode vector of soil-borne fanleaf virus of grapevine. *Phytopathology*, **48**, 586–95.

Huang, J.S. (1985) Mechanism of resistance to root-knot nematodes, in *An Advanced Treatise on Meloidogyne, Vol. I: Biology and Control* (eds J.N. Sasser and C.C. Carter), North Carolina State University Graphics, Raleigh, pp. 165–74.

Huang, J. and Barker, K.R. (1983) Influence of *Heterodera glycines* of leghemoglobins of soybean nodules. *Phytopathology*, **73**, 1002–4.

Hussey, R.S. (1989a) Disease-inducing secretions of plant-parasitic nematodes. *Annual Review of Phytopathology*, **27**, 123–41.

Hussey, R.S. (1989b) Monoclonal antibodies to secretory granules in esophageal glands of *Meloidogyne* species. *Journal of Nematology*, **21**, 392–8.

Hussey, R.S. and Mims, C.W. (1990) Ultrastructure of esophageal glands and their secretory granules in the root-knot nematode *Meloidogyne incognita*. *Protoplasma*, **156**, 9–18.

Jansson, H.B. and Nordbring-Hertz, B. (1980) Interactions between nematophagous fungi and plant-parasitic nematodes: Attraction, induction of trap formation and capture. *Nematologica*, **26**, 383–9.

Jatala, P. (1986) Biological control of plant-parasitic nematodes. *Annual Review of Phytopathology*, **24**, 453–89.

Jones, M.G.K. (1981) The development and function of plant cells modified by endoparasitic nematodes, in *Plant Parasitic Nematodes*, Vol. III (eds B.M. Zuckerman and R.A. Rohde), Academic Press, New York, pp. 255–79.

Jones, M.G.K. and Northcote, D.N. (1972) Multinucleate transfer cells induced in *Coleus* roots by the root-knot nematode, *Meloidogyne arenaria*. *Protoplasma*, **75**, 381–95.

Jorgenson, E.C. (1970) Antagonistic interaction of *Heterodera schachtii* Schmit and *Fusarium oxysporum* (Woll.) on sugarbeets. *Journal of Nematology*, **2**, 393–8.

Kaplan, D.T. and Keen, N.T. (1980) Mechanism conferring plant incompatibility to nematodes. *Revue de Nematologie*, **3**, 123–34.

Kaplan, D.T., Keen, N.T. and Thomason, I.J. (1980a) Association of glyceollin with the incompatible response of roots to *Meloidogyne incognita*. *Physiological Plant Pathology*, **16**, 309–18.

Kaplan, D.T., Keen, N.T. and Thomason, I.J. (1980b) Studies on the mode of action of glyceollin in soybean incompatibility to the root knot nematode, *Meloidogyne incognita*. *Physiological Plant Pathology*, **16**, 319–25.

Kerry, B.R., Crump, D.H. and Mullen, L.A. (1982) Studies of the cereal cyst-nematode *Heterodera avenae* under continuous cereals, 1975–1978. II. Fungal parasitism of nematode females and eggs. *Annals of Applied Biology*, **100**, 489–94.

Kishi, Y. (1980) Mortality of pine trees by *Bursaphelenchus lignicolus* M. and K. (Nematoda: Aphelenchoididae) in Ibaraki Prefecture and its control (Japanese with English summary). *Bulletin Ibaraki Prefecture Foreign Experimental Station*, **11**, 1–83.

Ko, M.P., Barker, K.R. and Huang, J.S. (1984) Nodulation of soybeans as affected by half-root infection with *Heterodera glycines*. *Journal of Nematology*, **16**, 97–105.

Lamberti, F. (1979) Economic importance of *Meloidogyne* spp. in subtropical and mediterranean climates, in *Root-Knot Nematodes (Meloidogyne Species), Systematics, Biology and Control* (eds F. Lamberti and C.E. Taylor), Academic Press, New York, pp. 341–57.

Mai, W.F. and Abawi, G.S. (1987) Interactions among root-knot nematodes and Fusarium wilt fungi on host plants. *Annual Review of Phytopathology*, **25**, 317–38.

Mamiya, Y. (1988) History of pine wilt disease in Japan. *Journal of Nematology*, **20**, 219–26.

Mankau, R. (1980) Biological control of nematode pests by natural enemies. *Annual Review of Phytopathology*, **18**, 415–40.

Martelli, G.P. and Taylor, C.E. (1989) Distribution of viruses and their nematode vectors, in *Advances in Disease Vectors Research* (ed. K.F. Harris), Springer-Verlag, New York, pp. 151–89.

Melendez, P.L. and Powell, N.T. (1967) Histological aspects of the Fusarium wilt–root-knot complex in flue-cured tobacco. *Phytopathology*, **57**, 286–92.

Meredith, J.A., Inserra, R.N. and Monzon de Fernandez, D. (1983) Parasitism of *Rotylenchulus reniformis* on soybean root rhizobium nodules in Venezuela. *Journal of Nematology*, **15**, 211–14.

Molinari, S., Zacheo, G. and Bleve-Zacheo, T. (1990) Effects of nematode infestation on mitochondria isolated from susceptible and resistant tomato roots. *Physiological and Molecular Plant Pathology*, **7**, 27–37.

Mouza, B.E. and Webster, J.M. (1982) Suppression of alfalfa growth by concomitant populations of *Pratylenchus penetrans* and two *Fusarium* species. *Journal of Nematology*, **14**, 364–7.

Murant, A.F., Jones, A.T., Martelli, G.P. and Stace-Smith, R. (1990) Nepoviruses: Diseases and virus identification, in *The Plant Viruses*, Vol. 5 (eds B.D. Harrison and A.F. Murant), Plenum Press, New York.

Nickle, W.R. (1981). Research on the pine wood nematode in the United States. *XVII IUFRO World Congress*, Japan, Division 2, pp. 269–71.

Nickle, W.R. and Welch, H.E. (1984) History, development and importance of insect nematology, in *Plant and Insect Nematodes* (ed. W.R. Nickle), Marcel Dekker, New York, pp. 627–53.

Nordmeyer, D. and Sikora, R.A. (1983) Effect of a culture filtrate from *Fusarium avenaceum* on the penetration of *Heterodera daverti* into the roots of *Trifolium subterraneum*. *Nematologica*, **29**, 88–94.

O'Bannon, J.H., Radewald, J.D., Tomerlin, A.T. and Inserra, R.N. (1976) Comparative influence of *Radopholus similis* and *Pratylenchus coffeae* on citrus. *Journal of Nematology*, **8**, 58–63.

Oku, H., Shiraishi, T., Ouchi, S., Kurozumi, S. and Ohta, H. (1980) Pine wilt toxin, the metabolite of a bacterium associated with a nematode. *Naturwissenschaften*, **67**, 198–9.

Perry, D.A. (1963) Interaction of root-knot and Fusarium wilt of cotton. *Empire Cotton Growers Review*, **40**, 41–7.

Poinar, G.O., Jr and Hansen, E.C. (1986) Associations between nematodes and bacteria. *Helminthological Abstract*, **55**, 62–81.

Powell, N.T. (1979) Internal synergisms among organisms inducing disease, in *Plant Disease: An Advanced Treatise*, Vol. IV (eds J.G. Horsfall and E.B. Cowling), Academic Press, New York, pp. 113–33.

Preisig, C.L. and Kuc, J.A. (1987) Phytoalexins, elicitors, enhancers, suppressors, and other considerations in the regulation of R-gene resistance to *Phytophthora infestans* in potato, in *Molecular Determinants of Plant Diseases* (eds S.

Nishimura, C.P. Vance and N. Doke), Japan Scientific Society Press/Springer-Verlag, Berlin, pp. 203–21.

Riley, I.T., Readon, T.B. and McKay, A.C. (1988) Genetic analysis of plant pathogenic bacteria in the genus *Clavibacter* using alozyme electrophoresis. *Journal of General Microbiology*, **134**, 3025–30.

Roy, K.W., Lawrence, G.W., Hodges, H.H., McLean, K.S. and Killebrew, J.F. (1989) Sudden death syndrome of soybean: *Fusarium solani* as incitant and relation of *Heterodera glycines* to disease severity. *Phytopathology*, **79**, 191–7.

Rudel, M. (1985) Grapevine damage induced by particular virus vector combinations. *Phytopathologia Mediterranea*, **24**, 183–5.

Rumpenhorst, H.J. (1984) Intracellular feeding tubes associated with sedentary plant-parasitic nematodes. *Nematologica*, **30**, 77–85.

Sasser, J.N. (1979) Economic importance of *Meloidogyne* in tropical countries, in *Root-Knot Nematodes (Meloidogyne Species), Systematics, Biology and Control* (eds F. Lamberti and C.E. Taylor), Academic Press, New York, pp. 359–74.

Sayre, R.M. and Starr, M.P. (1985) *Pasteuria penetrans* (ex-Thorne, 1940) nom. rev. comb. n. sp. n., a mycelial and endospore-forming bacterium parasitic in plant-parasitic nematodes. *Proceedings of the Helminthological Society of Washington*, **52**, 49–65.

Sidhu, G. and Webster, J.M. (1977) Predisposition of tomato to the wilt fungus (*Fusarium oxysporum lycopersici*) by the root-knot nematode (*Meloidogyne incognita*). *Nematologica*, **23**, 436–42.

Sidhu, G.S. and Webster, J.M. (1981) Genetics of plant–nematode interaction, in *Plant Parasitic Nematodes*, Vol. III (eds B.M. Zuckerman and R.A. Rohde), Academic Press, New York, pp. 61–87.

Sikora, R.A. (1979) Predisposition to *Meloidogyne* infection by the endotrophic mycorrhizal fungus *Glomus mosseae*, in *Root-knot Nematodes (Meloidogyne Species), Systematics, Biology, and Control* (eds F. Lamberti and C.E. Taylor), Academic Press, New York, pp. 399–404.

Society of Nematologists (1971) Estimated crop losses due to plant-parasitic nematodes in the United States. Special publication No. 1 (supplement to the *Journal of Nematology*).

Starr, J.L., Jeger, M.J., Martyn, R.D. and Schilling, K. (1989) Effects of *Meloidogyne incognita* and *Fusarium oxysporum* f. sp. *vasinfectum* on plant mortality and yield of cotton. *Phytopathology*, **79**, 640–6.

Stirling, G.R. and Wachtel, M.F. (1980) Mass production of *Bacillus penetrans* for the biological control of root-knot nematodes. *Nematologica*, **26**, 308–12.

Taha, A.H.Y. and Kassab, A.S. (1980) Interrelations between *Meloidogyne javanica*, *Rotylenchulus reniformis*, and *Rhizobium* sp. on *Vigna sinensis*. *Journal of Nematology*, **12**, 57–62.

Taylor, C.E. (1980) Nematodes, in *Vectors of Plant Pathogens* (eds K.F. Harris and K. Maramorosch), Academic Press, New York, pp. 375–416.

Taylor, C.E. (1990) Nematode interactions with other pathogens. *Annals of Applied Biology*, **116**, 405–16.

Thorne, G. (1961) *Principles of Nematology*, McGraw-Hill, New York.

Tylka, G.L., Hussey, R.S. and Roncadori, R.W. (1991) Interactions of vesicular-arbuscular mycorrhizal fungi, phosphorus and *Heterodera glycines* on soybean. *Journal of Nematology*, **23**, 122–33.

Van Gundy, S.D. (1985) Ecology of *Meloidogyne* spp.: Emphasis on environmental factors affecting survival and pathogenicity, in *Advanced Treatise on Meloidogyne, Vol. I: Biology and Control* (eds J.N. Sasser and C.C. Carter), North Carolina State University Graphics, Raleigh, pp. 177–82.

Veech, J.A. (1981) Plant resistance to nematodes, in *Plant Parasitic Nematodes*,

Vol. III (eds B.M. Zuckerman and R.A. Rohde), Academic Press, New York, pp. 377–403.

Veech, J.A. and Endo, B.Y. (1970) Comparative morphology and enzyme histochemistry in root knot resistant and susceptible soybeans. *Phytopathology*, 60, 896–902.

Veech, J.A., Starr, J.L. and Nordgren, R.M. (1987) Production and partial characterization of stylet exudate from adult females of *Meloidogyne incognita*. *Journal of Nematology*, 19, 463–8.

Vrain, T.C. (1987) Effect of *Ditylenchus dipsaci* and *Pratylenchus penetrans* on Verticillium wilt on alfalfa. *Journal of Nematology*, 19, 379–83.

Wallace, H.R. (1961) The orientation of *Ditylenchus dipsaci* to physical stimuli. *Nematologica*, 6, 222–36.

Wallace, H.R. (1973) *Nematode Ecology and Plant Disease*, Edward Arnold, London.

Wallace, H.R. (1983) Interaction between nematodes and other factors on plants. *Journal of Nematology*, 15, 221–7.

Walsh, J.A., Shepherd, A.M. and Lee, D.L. (1983) The distribution and effect of intracellular rickettsia-like microorganisms infecting second-stage juveniles of the potato cyst nematode *Globodera rostochiensis*. *Journal of Zoology*, 199, 395–419.

Webster, J.M. (1985) Interaction of *Meloidogyne* with fungi on crop plants, in *Advanced Treatise on Meloidogyne, Vol. I: Biology and Control* (eds J.N. Sasser and C.C. Carter), North Carolina State University Graphics, Raleigh, pp. 183–92.

Whitney, E.D. (1971) Synergistic effect between *Heterodera schachtii* and *Pythium ultimum* on damping-off of sugarbeet vs. additive effect of *H. schachtii* and *P. aphanidermatum*. *Phytopathology*, 61, 917.

Wyss, U. (1981) Ectoparastic root-nematodes: Feeding behaviour and plant cell responses, in *Plant Parasitic Nematodes*, Vol. III (eds B.M. Zuckerman and R.A. Rohde), Academic Press, New York, pp. 325–51.

Wyss, U. and Zunke, U. (1986) Observations on the behavior of second-stage juveniles of *Heterodera schachtii* inside host roots. *Revue de Nematologie*, 9, 153–65.

Yang, H., Powell, N.T. and Barker, K.R. (1976) Interaction of concomitant species of nematodes and *Fusarium oxysporum* f.sp. *vasinfectum* on cotton. *Journal of Nematology*, 8, 81–6.

Zacheo, G. and Bleve-Zacheo, T. (1988) Involvement of superoxide dismutase and superoxide radicals in the susceptibility and resistance of tomato plants to *Meloidogyne incognita* attack. *Physiological and Molecular Plant Pathology*, 32, 313–22.

Zacheo, G., Lamberti, F., Arrigoni-Liso, R. and Arrigoni, O. (1977) Mitochondrial protein-hydroxyproline content of susceptible and resistant tomatoes infected by *Meloidogyne incognita*. *Nematologica*, 23, 471–6.

Zadoks, J.C. and Schein, R.D. (1979) *Epidemiology and Plant Disease Management*, Oxford University Press, Oxford.

Zentmyer, G.A. (1961) Chemotaxis of zoospores for root exudates. *Science*, 133, 1595–6.

2

The phenology of plant–nematode interaction and yield loss

Haddish Melakeberhan and John M. Webster

Plant-parasitic nematodes, often referred to as 'hidden enemies', are among the most widespread and important pathogens causing crop loss. They are major pathogens in their own right and through their interaction with other disease-causing agents (Sidhu and Webster, 1977, 1981b). However, the accurate assessment of crop loss data and the nature of its cause are difficult to determine and are best described by the following quote:

> Our present estimates and determinations of crop loss from any given factor leave much to be desired. Experiments to measure crop loss are difficult to design because of the many overlapping interactions involved. Nonetheless, it is necessary to do the best job we can in making such estimates and determinations. Dowler and Van Gundy (1984)

This is particularly true in developing countries where there are few nematologists and where the prevalence of nematode pathogens is high. The diminutive size of these nematodes, their edaphic habitat and the non-specific nature of the symptoms they induce make detection and control of nematode problems difficult. Such symptoms as patchiness, stunting, chlorosis and/or root deformation, which may vary with species of host and nematode parasite, and nematode population size (Barker and Olthof, 1976) and time, may be followed by a decrease in biomass, quality of fruit and fibre and overall crop yield. Crop yield, therefore, is a result of the interaction of biotic and abiotic factors over time on the physiological processes in the plant. In this chapter we will consider the influence of nematodes during growth and development of the crop plant and some of the physiological mechanisms of nematode pathogenesis that result in crop loss.

2.1 THE INFLUENCE ON PLANT GROWTH AND ASSOCIATED YIELD

In understanding the influence of nematodes and associated organisms on host growth and yield, it is important to consider the phenology of the host–

parasite interaction. As plant parasites, nematodes affect plant growth by breaking down cell structure, removing cell contents, disrupting physiological processes and by modifying the genetic expression of the host (Sidhu and Webster, 1981a). Symptom expression at the cellular, tissue and whole plant level has been reviewed extensively (Dropkin, 1969; Paulson and Webster, 1970, 1972; Bird, 1974; Webster, 1975; Jones, 1981; Hussey, 1989). The nature of these changes may differ depending on the location of the parasite in the plant host and the influence of other organisms and physical factors. The types of the structural change nematodes induce at the feeding sites in the roots can conveniently be grouped into destructive, adaptive or neoplastic changes (Dropkin, 1980). Destructive changes, where host cells are killed, are a common effect of migratory endoparasites and ecto-parasites. The adaptive changes to host cells, caused by nematodes such as *Tylenchulus semipenetrans*, *Heterodera schachtii* and *Rotylenchulus reniformis*, is where host cells are structurally and physiologically modified to satisfy the nutritional requirements of the sedentary nematodes. *Meloido-gyne* and *Nacobbus* induce new growth in host tissues and represent the neoplastic changes. The nature of the impact of plant-parasitic nematodes on plant growth and crop yield takes various forms, ranging from modification of nutrient uptake and diminished energy availability to holistic or discrete physiological effects.

2.2 PHYSIOLOGICAL BASIS OF YIELD LOSS

When nematodes infect one part of the plant host, e.g. roots, they rapidly disrupt host physiological processes throughout the whole plant, either directly or indirectly. The physiological basis for decreased host biomass and crop yield loss caused by nematodes is widespread in the plant host and involves most of the major processes in the plant, including host respiration, photosynthesis, nutrient translocation and availability, water relations and phytohormone balance, as well as the nematodes' energy demand. Often these physiological processes are interrelated and it is difficult to separate the precise sequence of events that result in crop loss.

2.2.1 Respiration

Both forms of respiration – the one that takes place in the dark (dark respiration) and the other in the light (photorespiration) – are likely to be influenced by nematode infections. The rate of dark respiration in most plant–pathogen interactions almost invariably increases (Daly, 1976), whereas there is not a consistent trend in plant–nematode interactions (Bird and Millerd, 1962; Melakeberhan *et al.*, 1986). However, total respiration (growth and maintenance) decreases as a function of plant biomass accumulation (Melakeberhan and Ferris, 1989; Melakeberhan *et al.*, 1990).

Photorespiration – the uptake of oxygen and the evolution of carbon dioxide in the light – results from glycolate synthesis in the chloroplasts and subsequent glycolate and glycine metabolism in peroxisomes and mitochondria (Nobel, 1983). The photorespiration rate in C_3 plants, which synthesize their carbohydrates via the Calvin cycle and to which most cultivated crops belong, can be as high as 30% of the rate of photosynthesis (Nobel, 1983). Photorespiration is known to be increased in beans by rust (*Uromyces appendiculatus* var. *typica*) infection (Raggi, 1978), an indication of increased physiological inefficiency of an infected plant. An increase in the amount of energy unaccounted for and in the physiological inefficiency of *Meloidogyne incognita*-infected grapevines has been reported (Melakeberhan and Ferris, 1989; Melakeberhan *et al.*, 1990). Although not measured, it is likely that nematodes, like other pathogens (Daly, 1976), increase the wasteful processes of the pentose pathway or photorespiration.

2.2.2 Photosynthesis

Before considering the effect of nematodes on photosynthesis – one of the determinant physiological processes affecting plant growth and crop yield – it is necessary briefly to mention the photosynthetic process. Photosynthesis involves the light reaction in the thylakoid and dark reaction in the stroma side of a chloroplast (Salisbury and Ross, 1978). The light reaction starts with the splitting of water at the photosystem (PS) II reaction centre, where chlorophyll gives rise to the electron flow system that passes through PS I. The electrons donated by NADPH and energy provided by ATP drive the dark reaction (Calvin cycle) and the combination of carbon dioxide and water in the presence of ribulose biphosphate gives rise to 3-phosphoglyceric acid, which starts the carbohydrate synthesis. Thus, factors that affect chlorophyll and nitrogen content or nutrient and water uptake directly influence photosynthesis.

Information about the influence of nematode parasitism on photosynthesis, however, is limited. Nematodes affect carbon dioxide assimilation and the partitioning and translocation of photoassimilates. The total amount of carbon dioxide entering the plant and being assimilated is decreased in tomatoes (*Lycopersicon esculentum* Mill.) by *Meloidogyne javanica* (Loveys and Bird, 1973; Wallace, 1974; Meon *et al.*, 1978), in potatoes (*Solanum tuberosum* L.) by *Globodera rostochiensis* (Franco, 1980), in French beans (*Phaseolus vulgaris* L.) (Melakeberhan *et al.*, 1984, 1985a) and in grapes (*Vitis vinifera* L.) by *M. incognita* (Melakeberhan and Ferris, 1989; Melakeberhan *et al.*, 1990). The diminished photosynthesis, in turn, leads to decreased biomass and crop yields in nematode–infected plants. The debilitating effect of *M. incognita* on photosynthesis of *P. vulgaris* is more important in younger than older plants, and in infections prior to seed production. The effect increases with duration of infection and increasing nematode inoculum (Melakeberhan *et al.*, 1986).

With the exception of a few (e.g. *Anguina* spp., some *Aphelenchus* spp.,

Aphelencoides spp., some *Bursaphelenchus* spp. and some *Ditylenchus* spp.), most plant-parasitic nematodes are root parasites. Thus, when and how nematodes affect shoot photosynthesis, and whether the effect on photosynthesis is direct (primary) or indirect (secondary), are of interest. A decrease in photosynthetic rate of *Meloidogyne incognita*-infected beans has been observed as early as three days after inoculation (Melakeberhan *et al.*, 1984, 1986). As the photosynthetic apparatus in the leaves is not physically contacted by most nematodes, it is reasonable to assume that photosynthesis is affected indirectly. Nematodes have a primary effect on the uptake of water (Evans *et al.*, 1977; Melakeberhan *et al.*, 1991) and nutrient elements (Melakeberhan *et al.*, 1985a, 1987, 1988).

Besides the effect on photosynthesis through an effect on nutrient and water relations, some sedentary nematodes have a specialized effect on photosynthate partitioning. Bird and Loveys (1975) and McClure (1977) demonstrated an accumulation of a significant amount of $^{14}CO_2$-labelled material in *Meloidogyne incognita*-induced root galls. Studies of the concentration of sugars in *M. incognita*-induced galls of moderately resistant and susceptible *Vitis vinifera* cultivars showed an increase in the concentration of non-reducing sugars with duration of infection and increasing size of nematode inoculum in a susceptible compared with a resistant cultivar (Melakeberhan *et al.*, 1990). It may be that in nematode-infected plants there is a more rapid photosynthate translocation to the nematode feeding site in susceptible than in resistant cultivars, and this may be one of the reasons why susceptible cultivars suffer more physiological damage than the resistant cultivars (Melakeberhan *et al.*, 1990). It appears that *M. incognita* acts as a metabolic sink (Bergeson, 1966; McClure, 1977), by diverting photoassimilates for its own growth from the plant. In turn, this may lead to an unbalanced root : shoot ratio (Dropkin, 1980) as the parasitic *M. incognita* competes with seeds and vegetative growth of the host for photoassimilates. In effect, the nematode induces a hidden cost to the host in terms of diminished physiological efficiency and material transportation to the roots.

2.2.3 Water relations

The effect of nematodes on water relations and stomatal conductance, transpiration and root conductance were reviewed by Wilcox-Lee and Loria (1987). Although the intensity of the effect on water relations may vary with the type of host–parasite interaction, nematodes decrease water uptake by destroying plant roots. Examples of this are seen in the effect of *Globodera rostochiensis* on potato (Evans *et al.*, 1977; Fatemy and Evans, 1986a, 1986b) and *Meloidogyne javanica* on tomatoes (Meon *et al.*, 1978).

The physiological effects of decreased water availability include decreased nutrient uptake and translocation of solutes, decreased chloroplast activity, decomposition of proteins and nucleic acids, and increased hydrolytic enzymes (Schoeneweiss, 1978). Some nematode infection processes have the same effect as severing the root or shoot in that it results in loss of turgor and wilting above the point of severance, and may lead to death of the whole or part of the plant. The sequence in which these physiological processes occur or how they are linked is not fully understood. *Bursaphelenchus xylophilus* kills mature *Pinus* species by girdling the trunk (Tamura *et al.*, 1987) and cavitation (Kuroda, 1989; Kuroda *et al.*, 1988), which blocks water translocation (Ikeda and Susaki, 1984; Ikeda *et al.*, 1990) distal to the location of nematode inoculations (Melakeberhan *et al.*, 1991). Consequently, the plant fails to photosynthesize normally (Melakeberhan and Webster, 1990), loses turgor, wilts and dies. Wilt symptoms in pine due to *B. xylophilus* appear as early as three days after inoculation (Mamiya and Tamura, 1977; Melakeberhan *et al.*, 1991). Using a chlorophyll *a* fluorescence (*Fvar*) technique that assays the activity of the water-splitting complex of PS II (Govindjee *et al.*, 1981; Peeler and Naylor, 1988; Toivonen and Vidaver, 1988), Melakeberhan *et al.* (1991) showed a decline in PS II complex activity within 24 hours of inoculating this nematode. Water relations therefore appear to be a primary process that nematodes affect to influence the carbon dioxide uptake process directly.

2.2.4 Host nutrition

Normal plant growth requires an optimum balance of nutrient elements together with their normal uptake and distribution within the plant. The general notion has been that nematode infections that cause root deformation directly diminish absorption and nutrient elements, resulting in an inadequate supply to the plant (Elkins *et al.*, 1979; Price *et al.*, 1982; Price and Sanderson, 1984; Rawthorne and Hague, 1986). The apparent effect of nematodes on host nutrition varies depending on the type of plant tissue measured (Ibrahim *et al.*, 1982; Nasr *et al.*, 1980; Trudgill *et al.*, 1975), nematode and host species, stage of infection as well as experimental conditions (Evans *et al.*, 1977; Melakeberhan *et al.*, 1987; Oteifa, 1952). However, the degree to which nematodes modify the availability, uptake and translocation of nutrients to plants, and the physiological mechanisms that influence crop yield, are not understood. The threshold concentration of an element below which there is an adverse effect on host nutrition varies with the element and type of host–parasite interaction. However, a statistically significant change in the concentration of a given element may not be necessary in order to have a significant impact on host physiology (Melakeberhan *et al.*, 1987) as other stress factors may modify host response. Typical chlorosis symptoms associated with decreased leaf chlorophyll and leaf nitrogen (Melakeberhan *et al.*, 1985a) and potassium concentration (Melakeberhan *et al.*, 1987) with parallel decreases in photosynthetic rates (Melakeberhan, 1986) are indications of some of the complex interactions involving nutrients and host physiology. A decrease in soil nitrogen directly influences plant chlorophyll content, and together with decreased potassium influences photosynthesis. Nematodes may induce changes in concentration of these elements in the host by impeding absorption or their metabolism. In agricultural fields inadequate soil management practices may lead to the unavailability of key elements. A decrease in leaf potassium concentration can result in poor stomatal regulation (Outlaw, 1983), consequently diminishing photosynthesis and crop yield (Melakeberhan *et al.*, 1985a, 1985b, 1986, 1987). There appears to be some inverse relationship in the distribution among certain elements within

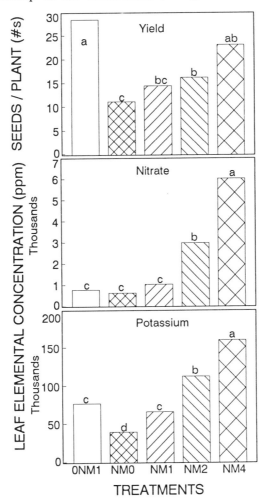

Figure 2.1 The effect of normal-strength Hoagland solution without nematodes (0NMI), or with 4000 second-stage *Meloidogyne incognita* juvenile nematodes (NM1) and without (NM0), or twice (NM2) or four times (NM4) the concentration of KNO_3 in Hoagland solution, on yield, leaf NO_3 and K concentration of *Phaseolus vulgaris* measured four weeks after inoculation. Plants were one week old at the time of inoculation. After Melakeberhan (1986).

a nematode-infected plant. For example, a decreased ratio of K : Ca in *Globodera rostochiensis*-infected potatoes (Fatemy and Evans, 1986a), and higher root concentration of K than Ca and vice versa in the shoot of *Meloidogyne incognita*-infected beans (Melakeberhan *et al.*, 1987), suggest that the nematodes may be influencing their metabolism differently in different parts of the plant or may be impeding K translocation to the shoot. Thus, the increase in shoot Ca concentration may indicate either its ability to compete with K for absorption (Johansen *et al.* 1968; Torimitsu *et al.*, 1985), or its ease of translocation to the shoot but diminished mobility once in the shoot (Ayres, 1984), or its ability to link and modify cell wall structures (Clarkson, 1984; Ferguson, 1984; Loughman, 1981; McEuen *et al.*, 1981).

From a crop management standpoint, it is desirable to incorporate the physiological database on host–parasite interactions either to control nematodes or to reduce their impact on crop production. The use of fertilizers to compensate for nematode damage (Kirkpatrick *et al.*, 1964; Spiegel *et al.*, 1982; Trudgill, 1980) is a common crop husbandry practice. Knowing that nitrogen and potassium are key elements affecting host physiology, it was possible to improve the yield and the leaf concentrations of K and NO_3 of nematode-infected plants by altering the concentration of KNO_3 in Hoagland and Arnon (1939) solution (HS) (Melakeberhan *et al.*, 1988). It took four times the normal strength of KNO_3 in HS for nematode-infected plants to yield similar to the uninfected controls receiving the normal-strength solution (Fig. 2.1). The improved yield was associated with a delay in the onset of chlorosis and an increased leaf K concentration which probably improved stomatal regulation and photosynthesis (Geiger and Conti, 1983). Associated with these increases in photosynthesis were changes in the growth medium pH, decreased root–galling, and increased shoot Ca concentration of infected bean plants. The increase of Ca and K in the shoot and the decreased galling by increasing the KNO_3 concentration in the growth medium along with changes in pH and ionic balance may indicate an effect on the nematode's ability to invade, thereby increasing leaf K concentration and host physiological efficiency.

2.2.5 Phytohormones

The concentration of indole acetic acid (Viglierchio and Yu, 1968), cytokinins (Bird *et al.*, 1980), abscisic acid (Volmar, 1991) and ethylene in plants (Glazer *et al.*, 1983, 1984) are affected by nematode infections. Root-knot nematodes induce an increase in phytohormone concentration which modifies plant tissue, especially around the nematode feeding sites, and plant host physiology (Sawhney and Webster, 1975, 1979). Although it has been suggested that nematodes may induce phytohormone production by injecting proteolytic enzymes into a host (Webster, 1967) or by coding the plant hosts to produce phytohormones (Setty and Wheeler, 1968), the precise mechanism of these changes in phytohormone balance is still unknown.

Gibberellins, required for cell elongation and cytokinins for cell division (Kefeli, 1978), translocation of photoassimilates (Thorpe and Lang, 1983) and nutrients (Mauk and Nooden, 1983; Neumann and Stein, 1983) and for chlorophyll synthesis (Maunders *et al.*, 1983), are particularly important in the plant–nematode interaction. These phytohormones are produced in root tissues (Scott and Horgan, 1984) and the amount produced is proportional to root size (Carmi and Heuer, 1981). Thus, irrespective of the type of nematode feeding behaviour, a nematode infection that damages root tips or results in smaller root biomass may decrease the concentration of cytokinins and gibberellins. Hence, the common symptoms of stunting and leaf senescence might be indications that the balance of these phytohormones has been affected by nematodes. Moreover, the root concentration of cytokinins varies with the life cycle of the nematode (Van Staden and Dimalla, 1977; Bird *et al.*, 1980) and is high in the galls and low in the shoot (Dimalla and Van Staden, 1977; Van Staden and Dimalla, 1977). This suggests that nematode infection impedes the translocation of cytokinins to the shoot and perhaps explains why there is an uneven distribution and decreased concentration of nutrient elements in the shoots of *Meloidogyne incognita*-infected beans (Melakeberhan *et al.*, 1987).

2.2.6 Nematode energy demand

Measuring nematode energy (food) consumption is essential for quantifying the amount the nematode removes from the host (Ferris, 1981; Atkinson, 1985) and the amount remaining for host growth and development (Melakeberhan and Ferris, 1989). The accurate measurement of the amount of energy obligate plant-parasitic nematodes consume from a host during their life cycle is limited by the lack of appropriate techniques to measure nematode growth. Some of the techniques used to estimate nematode growth outside a host root are volumetric (Robinson, 1984), gravimetric (Klekowski *et al.*, 1972; Robinson, 1984), or a combination of volumetric and gravimetric (Andrassy, 1956; Melakeberhan and Ferris, 1988), area (Bird, 1959) and oxygen consumption (Klekowski *et al.*, 1972; Reversat, 1987).

Nematode energy demand is a function of individual nematode size and its reproductive potential. Melakeberhan and Ferris (1988) measured the growth and oviposition of *Meloidogyne incognita* on French Colombard (susceptible) and Thompson Seedless (moderately resistant) *Vitis vinifera* cultivars for up to 1000 degree days (DD-base 10°C) or at 17.5 DD/day. Adult female *M. incognita* weighed 29–32 μg at *c.* 550 DD in both cultivars, whereas including oviposition at 1000 DD the weight was 127 and 211 μg in Thompson Seedless and French Colombard respectively. The difference was attributed to more egg production in the latter (309.76 $\pm$ 125.7) compared with the former (195 $\pm$ 92.4) cultivar (Melakeberhan and Ferris, 1988). According to these authors, the cost to the host of supporting a single nematode from second-stage juvenile to reach the adult stage and the end of oviposition, respectively, was 0.535 and 1.176 calories on French Colombard and 0.486 and 0.834 calories in Thompson Seedless.

Some studies indicate that energy consumption is proportional to nematode body size (Ingham and Detling, 1986; Melakeberhan and Webster, 1992), which suggests that smaller nematodes may have to be present in large numbers in order to have a significant energy demand. Although inadequate information is available on all nematodes, it is probable that cyst and root-knot nematodes have greater energy demands than nematodes of smaller size and less specialized feeding habits. Although parasite size and feeding habits may influence energy demand, nematodes such as *Bursaphelenchus xylophilus* seem to have more influence on host physiology than on energy demand (Melakeberhan and Webster, 1992).

It is important to note that total yield loss depends on the type and age of host plant as well as the nematode species. For example, if a nematode consumes 1 calorie in grapes, by conversion this is equivalent to *c.* 0.213 mg plant tissue dry weight (Lieth, 1968) and contains a carbon fraction of 0.4 mg carbon dioxide (Penning de Vries, 1974). However, the impact of such a calorific loss may vary with type and age of a given host, and the nematode energy demand may be different on an annual than on a perennial host. For example, the impact of *Meloidogyne incognita* on French bean physiology was most significant during the leaf expansion and flowering stages when it coincided with the egg production stages of the nematode (Melakeberhan, 1986). Thus, it appears that nematode energy demands are greatest during the egg production phase and are the greatest burden to the host during the equivalent period of growth, namely fruit and seed production.

2.3 CONCLUSIONS AND FUTURE TRENDS

As we look into the future, the latest statistics warn us of the increasing world population, which is not matched by increasing food production (Webster, 1987). The parameters that affect the balance of population increase and food production and food distribution go beyond the discipline of nematology. Nematodes, causing an estimated annual crop loss of 100 billion US dollars worldwide (Sasser and Freckman, 1987), however, will remain an important component of the factors diminishing crop yield. This, along with the ever-increasing ecological safety awareness, presents us with a tremendous challenge of improving existing traditional methods and looking for better alternative methods of nematode management that are biotechnologically based.

Environmental and public safety concerns appear to be limiting the use of nematicides, and force us to improve farming systems and the traditional methods such as crop rotation as alternative nematode management methods. Thus, understanding the physiological mechanisms of the plant–nematode interaction – the key to formulating holistic hypotheses (Wallace, 1987) – may be useful for developing better and more durable nematode

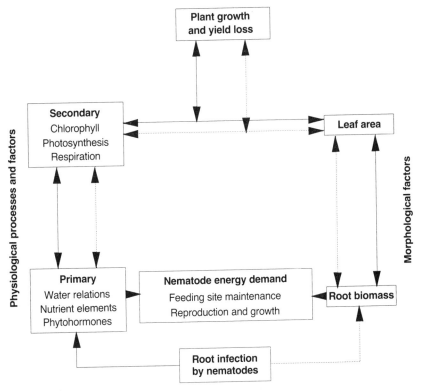

Figure 2.2 Diagram showing the possible relationship between nematode energy demand, physiological processes and factors and morphological factors in host root and shoot that affect plant growth and yield. Solid line represents above-threshold nematode infection and broken line represents below-threshold nematode infection.

management alternatives. The physiological and morphological parameters that collectively affect crop yield are complex and interrelated (Fig. 2.2). Actively feeding nematodes in a plant root probably affect first the primary physiological processes of water relations, nutrient uptake and translocation and the phytohormone balance, all of which, in turn, rapidly influence a range of secondary physiological processes and factors such as chlorophyll synthesis, photosynthesis and, to a lesser extent, respiration in the shoot. However, the intensity of host response is influenced by many plant and nematode related factors that combine to give a threshold level. When nematode infection is below the threshold, represented by the dotted line in Fig. 2.2, nematode energy demand for growth and reproduction and maintenance of feeding sites, the impact on physiological processes and shoot and root biomass, host symptoms and crop yield loss is minimal. When nematode infection is above the threshold, represented by the solid line in Fig. 2.2, nematode energy demand and the impact on the physiological and morpho-metric parameters is high. Consequently, host productivity is diminished. Under such circumstances, an understanding of the host–parasite interaction as shown by nutritionally based compensatory programmes (Melakeberhan *et al.*, 1988) could be useful in future management strategies.

Although nematode resistance might be a preferred control method and there is some advance such as the development of the *Mi* gene, it is limited by the lack of available cultivars. However, the use of biotechnology as it relates to nematode identification and the genes that code for pathogenesis offers some hope in the future (Hussey, 1989). Irrespective of the nematode management method we adopt, however, we need to appreciate that nature does not recognize purity and nematodes are only one of many abiotic fac-tors, pathogens and parasites in the ecosystem (Wallace, 1978). The inter-action of all these factors influences the phenology of any one of them.

REFERENCES

Andrassy, I. (1956) Die Rauminhalts- und Gewichtsbestmmung der Fadenwurmer (Nematoden). *Acta Zoologica Academiae Hungaricae*, **2**, 1–15.

Atkinson, H.J. (1985) The energetics of plant parasitic nematodes: A review. *Nematologica*, **31**, 62–71.

Ayres, P.G. (1984) Effects of infection on root growth and function: Consequences for plant nutrient and water relations, in *Plant Disease: Infection, Damage and Loss* (ed. R.K.S. Wood), Blackwell Scientific Publications, Oxford, pp. 105–17.

Barker, K.R. and Olthof, T.H.A. (1976) Relationships between nematode population densities and crop response. *Annual Review of Phytopathology*, **14**, 327–53.

Bergeson, G.B. (1966) Mobilization of minerals to the infection site of root-knot nematodes. *Phytopathology*, **56**, 1287–9.

Bird, A.F. (1959) Development of the root-knot nematodes *Meloidogyne javanica* (Treub) and *Meloidogyne hapla* Chitwood in tomato. *Nematologica*, **4**, 31–42.

Bird, A.F. (1974) Plant response to root-knot nematode. *Annual Review of Phytopathology*, **12**, 69–85.

Bird, A.F. and Loveys, B.R. (1975) The incorporation of photosynthates by *Meloidogyne incognita*. *Journal of Nematology*, **7**, 111–13.

Bird, A.F. and Millerd, A. (1962) Respiration studies on *Meloidogyne*-induced galls in tomato roots. *Nematologica*, **8**, 261–6.

Bird, A.F., Sauer, R.N., Chapman, R.N. and Loveys, B.R. (1980) The influence of light quality on the growth and fecundity of *Meloidogyne* infecting tomato. *Nematologia Mediterranea*, **8**, 59–66.

Carmi, A. and Heuer, B. (1981) The role of roots in control of bean shoot growth. *Annals of Botany*, **48**, 519–27.

Clarkson, D.T. (1984) Calcium transportation between tissues and its distribution in the plant. *Plant, Cell and Environment*, **7**, 449–56.

Daly, J.M. (1976) The carbon balance of diseased plants: Changes in respiration, photosynthesis and translocation. *Encyclopedia of Plant Physiology*, **4**, 450–79.

Dimalla, G.G. and Van Staden, J. (1977) Cytokinins in the root-knot nematode, *Meloidogyne incognita*. *Plant Science Letters*, **10**, 25–9.

Dowler, W.M. and Van Gundy, S.D. (1984) Importance of agricultural plant nematology, in *Plant and Insect Nematodes* (ed. W.R. Nickle), Marcel Dekker, New York, pp. 1–12.

Dropkin, V.H. (1969) Cellular changes of plants to nematode infections. *Annual Review of Phytopathology*, 7, 101–22.

Dropkin, V.H. (1980) *Introduction to Plant Nematology*, John Wiley, New York.

Elkins, C.B., Haaland, R.L., Rodriguez-Kabana, R. and Haveland, C.S. (1979) Plant-parasitic nematode effects on water uptake of a small- and large-rooted tall fescue genotype. *Agronomy Journal*, 71, 497–500.

Evans, K., Trudgill, D.L. and Brown, N.J. (1977) Effects of potato cyst-nematodes on potato plants. V: Root system development in light and heavy-infected susceptible and resistant varieties, and its importance in nutrient and water uptake. *Nematologica*, 23, 153–64.

Fatemy, F. and Evans, K. (1986a) Growth, water uptake and calcium content of potato cultivars in relation to tolerance of cyst nematodes. *Revue de Nematologie*, 9, 171–9.

Fatemy, F. and Evans, K. (1986b) Effects of *Globodera rostochiensis* and water stress on shoot and root growth and nutrient uptake of potatoes. *Revue de Nematologie*, 19, 181–4.

Ferguson, I.B. (1984) Calcium in plant senescence and fruit ripening. *Plant, Cell and Environment*, 7, 477–89.

Ferris, H. (1981) Mathematical approaches to the assessment of crop damage, in *Plant Parasitic Nematodes*, Vol. III (eds B.M. Zuckerman and R.A. Rohde), Academic Press, New York, pp. 405–20.

Franco, J. (1980) Effect of potato cyst-nematode, *Globodera rostochiensis*, on photosynthesis of potato plants. *Fitopatologia*, 15, 1–6.

Geiger, D.R. and Conti, T.R. (1983) Relation of increased potassium nutrition to photosynthesis and translocation of carbon. *Plant Physiology*, 71, 141–4.

Glazer, I., Orion, D. and Apelbaum, A. (1983) Interrelationships between ethylene production, gall formation, and root-knot nematode development in tomato plants infected with *Meloidogyne javanica*. *Journal of Nematology*, 15, 539–44.

Glazer, I., Apelbaum, A. and Orion, D. (1984) Reversal of nematode-induced growth retardation in tomato plants, *Lycopersicon esculentum*, by inhibition of ethylene action. *Journal of American Society of Horticultural Science*, 109, 886–9.

Govindjee, Downton, W.J.S., Fork, D.C. and Armond, P.A. (1981) Chlorophyll *a* fluorescence transient as an indicator of water potential of leaves. *Plant Science Letters*, 20, 191–4.

Hoagland, D.R. and Arnon, D.I. (1939) The water culture method for growing plants without soil. University of California, Berkeley, California. *Circular* no. 347.

Hussey, R.S. (1989) Disease-inducing secretions of plant-parasitic nematodes. *Annual Review of Phytopathology*, 27, 123–41.

Ibrahim, I.K.A., Razek, M.A. and Khalil, H.A.A. (1982) Effects of *Meloidogyne incognita* and *Fusarium oxysporum* f. *vasinfectum* on plant growth and mineral content of cotton, *Gossypium barbadense* L. *Nematologica*, 28, 298–302.

Ikeda, T. and Susaki, T. (1984) Influence of pine wood nematodes on hydraulic conductance and water relations in *Pinus thunbergii*. *Japanese Journal of Forestry Society*, 66, 412–20.

Ikeda, T., Kiyohara, T. and Kusunoki, M. (1990) Changes in water status of *Pinus thunbergii* Parl. inoculated with species of *Bursaphelenchus*. *Journal of Nematology*, 22, 132–5.

Ingham, R.E. and Detling, J.K. (1986) Effect of defoliation and nematode consumption on growth and leaf gas exchange in *Bouteloua curtipendula*. *Oikos*, 46, 23–8.

Johansen, C., Edwards, D.G. and Loneragan, J.F. (1968) Interactions between potassium and calcium in their absorption by intact barley plants: I. Effects of potassium on calcium absorption. *Plant Physiology*, 43, 171–2.

Jones, M.G.K. (1981) Host cell response to endoparasitic nematode attack: Structure and function of giant cells and syncytia. *Annals of Applied Biology*, **97**, 353–72.

Kefeli, V.I. (1978) *Nature of Plant Growth Inhibitors and Phytohormones*, Dr W. Junk Publications, The Hague.

Kirkpatrick, J.D., Mai, W.F., Parke, K.G. and Fisher, E.G. (1964) Effect of phosphorus and potassium nutrition of sour cherry on the soil population levels of five plant parasitic nematodes. *Phytopathology*, **54**, 706–12.

Klekowski, R.Z., Wasilawiska, L. and Papliniska, E. (1972) Oxygen consumption by soil inhabiting nematodes. *Nematologica*, **18**, 391–403.

Kuroda, K. (1989) Terpenoids causing tracheid-cavitation in *Pinus thunbergii* infected by the pine wood nematode (*Bursaphelenchus xylophilus*). *Annals of the Phytopathological Society of Japan*, **55**, 170–8.

Kuroda, K., Yamada, T., Minio, K. and Tamura, H. (1988) Effects of cavitation on the development of pine wilt disease caused by *Bursaphelenchus xylophilus*. *Annals of the Phytopathological Society of Japan*, **54**, 606–15.

Lieth, H. (1968) The measurement of calorific values of biological material and the determination of ecological efficiency, in *Functioning of Terrestrial Ecosystem at the Primary Production Level* (ed. F.E. Eckardt), *Proceedings of the Copenhagen Symposium*, UNESCO, pp. 233–42.

Loughman, B. (1981) Metabolic aspects of transport of ions by cells and tissues of roots. *Plant and Soil*, **63**, 47–55.

Loveys, B.R. and Bird, A.F. (1973) The influence of nematodes on photosynthesis in tomato plants. *Physiological Plant Pathology*, **11**, 11–21.

Mamiya, Y. and Tamura, H. (1977) Transpiration reduction of pine seedlings inoculated with the pine wood nematode, *Bursaphelenchus lignicolus*. *Journal of Japanese Forest Society*, **59**, 59–63.

Mauk, C.S. and Nooden, L.D. (1983) Cytokinin control of mineral nutrient distribution between the foliage and seeds in soybean explants. *Plant Physiology*, **72**, 167.

Maunders, M.J., Brown, S.B. and Woolhouse, H.W. (1983) The appearance of chlorophyll derivatives in senescing tissue. *Phytochemistry*, **22**, 2443–6.

McClure, M.A. (1977) *Meloidogyne incognita*: A metabolic sink. *Journal of Nematology*, **9**, 88–90.

McEuen, A.R., Hart, J.W. and Sabnis, D.D. (1981) Calcium binding protein in sieve tube exudate. *Planta*, **151**, 531–4.

Melakeberhan, H. (1986) The influence of *Meloidogyne incognita* (Nematoda) on the physiology and yield of *Phaseolus vulgaris*. PhD thesis, Simon Fraser University, Vancouver, Canada.

Melakeberhan, H. and Ferris, H. (1988) Growth and energy demand of *Meloidogyne incognita* in susceptible and resistant *Vitis vinifera* cultivars. *Journal of Nematology*, **20**, 545–54.

Melakeberhan, H. and Ferris, H. (1989) Impact of *Meloidogyne incognita* on physiological efficiency of *Vitis vinifera*. *Journal of Nematology*, **21**, 74–80.

Melakeberhan, H. and Webster, J.M. (1990) Effect of *Bursaphelenchus xylophilus* on the assimilation and translocation of ^{14}C in *Pinus sylvestris*. *Journal of Nematology*, **22**, 506–12.

Melakeberhan, H. and Webster, J.M. (1992) The insignificance of the energy requirements of *Bursaphelenchus xylophilus* as a causal factor in *Pinus sylvestris* seedling mortality. *Fundamentals and Applied Nematology*, **15**, 179–82.

Melakeberhan, H., Webster, J.M. and Brooke, R.C. (1984) Improved techniques for measuring the CO_2 exchange rate of *Meloidogyne*-infected beans. *Nematologica*, **30**, 213–21.

Melakeberhan, H., Webster, J.M. and Brooke, R.C. (1985a) Response of *Phaseolus vulgaris* to a single generation of *Meloidogyne incognita*. *Nematologica*, **31**, 190–202.

Melakeberhan, H., Brooke, R.C., Webster, J.M. and D'Auria, J.M. (1985b) The influence of *Meloidogyne incognita* on growth, nutrient content and physiology of *Phaseolus vulgaris*. *Physiological Plant Pathology*, 26, 259–68.

Melakeberhan, H., Webster, J.M. and Brooke, R.C. (1986) Relationship between physiological response and yield loss of different age French bean to *Meloidogyne incognita*. *Plant Pathology*, 35, 203–13.

Melakeberhan, H., Webster, J.M., Brooke, R.C., D'Auria, J.M. and Cackette, M. (1987) Effect of *Meloidogyne incognita* on plant nutrient concentration and its influence on the physiology of beans. *Journal of Nematology*, 19, 324–30.

Melakeberhan, H., Webster, J.M., Brooke, R.C. and D'Auria, J.M. (1988) Effect of KNO_3 on CO_2 exchange rate, nutrient concentration and yield of *Meloidogyne incognita*-infected beans. *Revue de Nematologie*, 11, 391–7.

Melakeberhan, H., Ferris, H. and Dias, J. (1990) Physiological response of resistant and susceptible *Vitis vinifera* cultivars to *Meloidogyne incognita*. *Journal of Nematology*, 22, 224–30.

Melakeberhan, H., Toivonen, P.M.A., Vidaver, W.E., Webster, J.M. and Dube, S. (1991) Effect of *Bursaphelenchus xylophilus* on water potential and the water-splitting complex of photosystem II of *Pinus sylvestris* seedlings. *Physiological and Molecular Plant Pathology*, 38, 83–91.

Meon, S., Fisher, J.M. and Wallace, H.R. (1978) Changes in free proline following infection of plants with either *Meloidogyne incognita* or *Agrobacterium tumefaciens*. *Physiological Plant Pathology*, 12, 251–6.

Nasr, T.A., Ibrahim, I.K.A., El-Azeb, E.M. and Hassan, M.W.A. (1980) Effect of root-knot nematodes on the mineral, amino acid and carbohydrate concentrations of almond and peach rootstocks. *Nematologica*, 26, 133–8.

Neumann, P.M. and Stein, Z. (1983) Xylem transport and the regulation of leaf metabolism. *What's New in Plant Physiology*, 14, 33–6.

Nobel, P.S. (1983) *Biophysical Plant Physiology and Ecology*, W.H. Freeman, New York.

Oteifa, B.A. (1952) Potassium nutrition of the host in relation to infection by a root-knot nematode, *Meloidogyne incognita*. *Proceedings of the Helminthological Society of Washington*, 19, 99–104.

Outlaw, W.H., Jr (1983) Current concepts on the role of potassium in stomatal movements. *Physiologia Plantarum*, 59, 301–2.

Paulson, R.E. and Webster, J.M. (1970) Giant cell formation in tomato roots caused by *Meloidogyne incognita* and *Meloidogyne hapla* (Nematoda) infection: A light and electron microscope study. *Canadian Journal of Botany*, 48, 271–6.

Paulson, R.E. and Webster, J.M. (1972) Ultrastructure of the hypersensitive reaction in roots of tomato, *Lycopersicon esculentum* L., to infection by the root-knot nematode, *Meloidogyne incognita*. *Physiological and Molecular Plant Pathology*, 2, 227–34.

Peeler, T.C. and Naylor, A.W. (1988) The influence of dark adaptation temperature on the reappearance of variable fluorescence following illumination. *Plant Physiology*, 86, 152–4.

Penning de Vries, F.W.T. (1974) Use of assimilates in higher plants, in *Photosynthesis and Productivity in Different Environments* (ed. J.P. Cooper), Cambridge University Press, London, pp. 459–79.

Price, N.S. and Sanderson, J. (1984) The translocation of calcium from oat roots infected by the cereal cyst nematode *Heterodera avenae* (Woll.). *Revue de Nematologie*, 7, 239–43.

Price, N.S., Clarkson, D.T. and Hague, N.G.M. (1982) The uptake of potassium and phosphorus in oat infected with the cereal cyst nematode, *Heterodera avenae* Woll. *Revue de Nematologie*, 5, 321–5.

Raggi, V. (1978) The CO_2 compensation point, photosynthesis and respiration in rust infected bean leaves. *Physiological Plant Pathology*, **13**, 135–9.

Rawthorne, D. and Hague, N.G.M. (1986) The effect of *Heterodera avenae* on the root system of susceptible and resistant oat seedlings. *Annals of Applied Biology*, **108**, 89–98.

Reversat, G. (1987) Increase in chemical oxygen demand during the growth in *Heterodera sacchari*. *Revue de Nematologie*, **10**, 115–17.

Robinson, A.F. (1984) Comparison of five methods for measuring nematode volumes. *Journal of Nematology*, **16**, 343–7.

Salisbury, F.B. and Ross, C.W. (1978) *Plant Physiology*, Wadsworth, Belmont, CA.

Sasser, J.N. and Freckman, D.W. (1987) A worldwide perspective on nematology: The role of the society, in *Vistas on Nematology* (eds J.A. Veech and D.W. Dickson), Society of Nematologists, Hyattsville, pp. 7–14.

Sawhney, R. and Webster, J.M. (1975) The role of plant growth hormones in determining the resistance of tomato plants to the root-knot nematode, *Meloidogyne incognita*. *Nematologica*, **21**, 95–103.

Sawhney, R. and Webster, J.M. (1979) The influence of some metabolic inhibitors on the response of susceptible/resistant cultivars of tomato to *Meloidogyne incognita*. *Nematologica*, **25**, 86–93.

Schoeneweiss, D.F. (1978) Water stress as a pre-disposing factor in plant disease, in *Water Deficits and Plant Growth*, Vol. 5 (ed. T.T. Kozlowiski), Academic Press, London, pp. 61–99.

Scott, I.M. and Horgan, R. (1984) Root cytokinins of *Phaseolus vulgaris* L. *Plant Science Letters*, **34**, 81–7.

Setty, K.G.H. and Wheeler, A.W. (1968) Growth substances in roots of tomato infected with root-knot nematodes (*Meloidogyne* species). *Annals of Applied Biology*, **61**, 495–501.

Sidhu, G.S. and Webster, J.M. (1977) The role of amino acid fungal auxotrophs to study the predisposition phenomena in the root-knot–wilt fungus disease complex of tomato. *Physiological Plant Pathology*, **11**, 117–27.

Sidhu, G.S. and Webster, J.M. (1981a) The genetics of plant-nematode parasitic systems. *Botanical Review*, **47**, 387–419.

Sidhu, G.S. and Webster, J.M. (1981b) Influence of population levels of root-knot nematode on Fusarium wilt severity of tomato. *Phytoprotection*, **62**, 61–6.

Spiegel, Y., Cohen, E., Kafkafi, U. and Sulami, M. (1982) Influence of potassium and nitrogen fertilization on parasitism by the root-knot nematode *Meloidogyne javanica*. *Journal of Nematology*, **14**, 530–5.

Tamura, H., Mineo, K. and Yamada, T. (1987) Blockage of water conduction in *Pinus thunbergii* inoculated with *Bursaphelenchus xylophilus*. *Japanese Journal of Nematology*, **17**, 23–30.

Thorpe, M.R. and Lang, A. (1983) Control of import/export of photosynthate in leaves. *Journal of Experimental Botany*, **34**, 231–9.

Toivonen, P.M.A. and Vidaver, W. (1988) Variable chlorophyll *a* fluorescence and CO_2 uptake in water stressed white spruce seedlings. *Plant Physiology*, **86**, 744–8.

Torimitsu, K., Hayashi, M., Ohta, E. and Sakata, M. (1985) Effect of K and H stress and role of Ca in the regulation of intercellular K concentration in mung bean roots. *Physiologia Plantarum*, **63**, 247–52.

Trudgill, D.L. (1980) Effects of *Globodera rostochiensis* and fertilizers on the mineral nutrient content and yield of potato plants. *Nematologica*, **26**, 243–54.

Trudgill, D.L., Evans, K. and Parrott, D.M. (1975) Effects of potato cyst-nematodes on potato plants: Effects of haulm size, concentration of nutrients in haulm tissue and tuber yield of nematode resistant and a nematode susceptible potato variety. *Nematologica*, **21**, 183–91.

Van Staden, J. and Dimalla, G.G. (1977) A comparison of the endogenous cytokinins in the roots and xylem exudates of nematode resistant and susceptible tomato cultivars. *Journal of Experimental Botany*, **28**, 1351-6.

Viglierchio, D.R. and Yu, P.K. (1968) Plant growth substances and plant parasitic nematodes. II: Host influence on auxin content. *Experimental Parasitology*, **23**, 88-95.

Volmar, K.M. (1991) Abscisic acid and ethylene increase in *Heterodera avenae*-infected tolerant or intolerant oat cultivars. *Journal of Nematology*, **23**, 425-31.

Wallace, H.R. (1974) The influence of root-knot nematode, *Meloidogyne javanica*, on photosynthesis and on nutrient demand by roots of tomato plants. *Nematologica*, **2**, 27-33.

Wallace, H.R. (1978) The diagnosis of plant diseases of complex etiology. *Annual Review of Phytopathology*, **16**, 379-402.

Wallace, H.R. (1987) Effects of nematode parasites on photosynthesis, in *Vistas on Nematology* (eds J.A. Veech and D.W. Dickson), Society of Nematologists, Hyattsville, pp. 253-9.

Webster, J.M. (1967) The influence of plant-growth substances and their inhibition on the host-parasite relations of *Aphelenchoides ritzemabosi* in culture. *Nematologica*, **13**, 256-62.

Webster, J.M. (1975) Aspects of the host-parasite relationships of plant parasitic nematodes. *Advances in Parasitology*, **13**, 225-50.

Webster, J.M. (1987) Introduction, in *Principles and Practice of Nematode Control in Crops* (eds R.H. Brown and B.R. Kerry), Academic Press, New York, pp. 1-11.

Wilcox-Lee, D. and Loria, R. (1987) Effects of nematode parasitism on plant-water relations, in *Vistas on Nematology* (eds J.A. Veech and D.W. Dickson), Society of Nematologists, Hyattsville, pp. 260-6.

3

The concept of interaction

M. Wajid Khan and M.K. Dasgupta

The interactions of plant-parasitic nematodes present a complicated subject encompassing diverse host–parasite systems. The word 'concept' literally means abstract idea or notion. The germ theory was the most significant concept ever developed to explain the cause of a disease. It laid the foundation of pathology. The theory considered the germ (parasite, pathogen), along with the favourable environment responsible for disease. By applying this concept, plant diseases were comprehended. Koch's postulates provided the means by which specific aetiological agents of diseases were identified and their pathogenicity established. With time, plant pathologists, however, became obsessed with the germ theory and Koch's postulates in plant disease comprehension. One of the obsessions, that plant diseases are induced by single plant pathogens working alone, i.e. specific aetiology, assumed the status of an empirical rule. The presidential address of Fawcett (1931) to the American Phytopathological Society at Cleveland, Ohio, USA, however, was a landmark in the history of plant pathology. His assertion that nature does not work in pure cultures alone, but most frequently with associations, impelled plant pathologists to study the effect on the development of disease by inoculation of plants with known mixtures of microorganisms. The tremendous input of information since then has brought several new facts to light and has led many to question the dogma of specific aetiology of Koch's postulates (Wallace, 1978). From the vast research and literature on the subject, certain generalizations have emerged. The validity of some of these generalizations is being questioned too (Sikora and Carter, 1987). This chapter mainly examines certain concepts of interactions of plant-parasitic nematodes with other pathogens in plants that have been developed over the years.

3.1 CAUSE-AND-EFFECT RELATIONSHIP: UNI- AND MULTICAUSAL DISEASES

A plant is rarely infested, affected or infected by a single organism, although only a few may reach economic injury level. The nature of economic injury, including the nature of biological association of the host with each pest as well as their mutual influences, is largely unknown. Aggravation of overall economic injury of the host is a common observation. Such observations make ecological pathologists look for interaction as the commonest phenomenon of association. Ecological associations between nematodes and other organisms have, however, been recognized and a recent account of the biology of associations of nematodes is available (Dasgupta, 1992). These associations can be grouped into the following:

1. aetiological associations often causing disease complexes;
2. competitive associations;
3. nematodes as vectors of plant viruses and other plant pathogens;
4. nematodes as parasites and predators;
5. nematodes as prey and host to other predators and parasites.

Among the plethora of organisms responsible for economic injury to plants, only a few can become pests in the context of a given space and time, and among them still fewer are pathogens in their own right, whether or not they are involved in any 'disease complex' – be it severally or jointly caused.

The tradition of scientific enquiry that we have inherited from Francis Bacon's *Novum Organum* and René Descartes's *Discourse on the Method of Rightly Conducting the Reason, and Seeking Truth in the Sciences* (see Commins and Linscott, 1954) is the belief in the one cause–one effect relationship. Disease complexes are known not only between nematodes and other pathogens but also among other pathogens without any nematodes being involved. Thus, in place of the long-held belief of one pathogen–one disease, the new concept of multiple pathogens acting in concert causing economic injury to an organism has emerged. An organism lives in an ecological network in which the biotic and abiotic factors interact.

3.1.1 Complex disease aetiology

Aetiology is the science of cause and includes the study of all factors involved directly or indirectly in the initiation and development of disease (Roberts and Boothroyd, 1975). Specific aetiology is a two-component system of host and pathogen. A plant infected by a pathogen becomes greatly altered and effects of the pathogen and host response are governed by various factors, including the components of environment. Therefore, the plant diseases of specific aetiology represent the disease triangle relating to host plant, pathogen and environment.

Diseases of complex aetiology present a different picture. When a host plant is infected by more than one pathogen, the activities and effects of one are likely to be influenced by the activities and effects of the other. The overall relations of the pathogens may become interactive. The alterations of the host plant have been shown to be of great importance in diseases of of complex aetiology. Powell (1979) described diseases of complex aetiology involving nematodes as a three-component system consisting of host, nematode(s) and other pathogen(s). The nematode(s) and the other pathogen(s) become engaged with the host either sequentially or concomitantly.

In sequential aetiology of disease complexes, one pathogen of the disease complex infects and alters the host in advance. Infection by the other pathogen follows. Powell (1979) termed the two pathogens engaging in such a sequence in the complex as primary and secondary pathogens respectively. According to his characterization, a primary pathogen invades the host prior to invasion by a secondary pathogen and governs the development of the disease complex, by altering the host in such a way that it becomes more suitable for the secondary pathogen. The time interval in this sequential engagement may vary from a day to several weeks. Primary pathogens have the inherent, specific and independent capability of causing disease. In other words, they can induce disease without involvement of any other pathogen. Secondary pathogens may or may not have the inherent or specific and independent capability of inducing disease. They may be independent or dependent on primary pathogens for disease development in the diseases of complex aetiology. Nematodes and viruses are commonly recognized as primary pathogens and fungi and bacteria as secondary pathogens (Powell, 1979).

Powell (1979) cautioned prudence in distinguishing the secondary pathogen from secondary invaders. Secondary pathogens are active participants in pathogenesis and determine its course. Secondary invaders, on the other hand, simply colonize the dead cells in the lesions caused by primary pathogens. While primary pathogens are generally obligate parasites, secondary pathogens are mostly facultative parasites (Powell, 1979).

Sequencing of the two pathogens, i.e. infections in a chain, does not appear always to be natural. Similarly, the terms primary and secondary are not adequate for all the reported situations in nematode interaction studies. If the secondary pathogen possesses inherent pathogenic capability, it must not wait for nematode infection and host alteration to occur because the nematode is not essential for its infection. The possibility of such a fungal pathogen to engage with the host prior to nematode infection and becoming a primary pathogen in the sequencing sense exists in nature. The fungal pathogen may derive benefit from its presence within the host, if the infection of the nematode brings about favourable changes in the host substrate; as a result disease severity may increase and a synergistic interaction may develop. Several reports indicate this kind of synergistic interaction. It is true that in most artificial inoculation studies, after prior inoculation of the nematode (two to four weeks), particularly with root-knot nematodes, synergistic interaction occurs. But this does not present the entire picture and is not fully natural. Hasan (1985) showed that infection of chilli roots by some fungal pathogens transformed the resistant host cultivars susceptible to root-knot nematodes. This terminology of primary and secondary pathogens in nematode interaction studies deserves further consideration. Sikora and Carter (1987) and Wallace (1989) have also questioned it. Wallace

(1989) stated that the plant is the centre of concern and biotic (including pathogens) and abiotic factors are part of the plant's environment. The environment is a network of factors. Some distal environmental factors may have a more important influence than proximal factors. Therefore, attaching the term primary or secondary to proximal or distal factors is not appropriate.

Disease complexes resulting from simultaneous infection of a host by two or more pathogens can be described as diseases of concomitant aetiology (Powell, 1979). In diseases of concomitant aetiology, the involved pathogens obviously cannot be designated as primary or secondary.

3.1.2 Biopredisposition

Biopredisposition is a special term used to describe predisposition caused by biotic factors. The term predisposition has been used in different ways in the literature. Predisposition is a modification caused by various factors (biotic or abiotic) in the expression of the host response to a pathogen. According to Powell (1979) biopredisposition is another term that is applicable to diseases of sequential aetiology. However, it is difficult to apply it to diseases of concomitant aetiology. Wallace (1989) called predisposition a type of synergism where the ability of a determinant to produce disease may depend on the influence of another determinant. Disease complexes of sequential aetiology, where nematode infection modifies the host in some way to the benefit of the fungal or bacterial pathogens, are examples of biopredisposition.

The nature and extent of biopredisposition in diseases of sequential aetiology are determined by the primary pathogen – the pathogen that infects first. Similarly, whether a synergistic interaction will occur or not depends upon the inoculum potential of the secondary pathogen – the pathogen that invades later – and on the nature and extent of the biopredisposition. The mechanism of biopredisposition may be mechanical or physiological or a combination of both. In biopredisposition the host plant is altered and becomes more vulnerable to attack by another pathogen, and consequently synergistic interaction occurs. Plants wounded before invasion of secondary pathogens usually become more susceptible to them. Most bacterial pathogens require openings, natural or artificial, for entry. Wounding of roots by nematodes generally helps bacterial plant pathogens. Physiological changes in the host induced by nematodes render the host more susceptible or modify host resistance to fungal pathogens.

3.2 COMPONENTS OF AN INTERACTING SYSTEM

The interacting system of disease complexes involving nematodes has at least three basic components: host plant, nematode(s) and other pathogen(s). It is the host that provides the site for interaction between the other two components: nematode(s) and other pathogen(s). Pitcher (1978) presented a scheme that indicated the principles underlying three-component interactions. Both the pathogens play important roles for interaction to develop. The relative importance of their roles may be different at different stages of development of the diseases of complex aetiology. In most interactions involving nematodes, it is the nematode that plays the primary role and modifies the host (Pitcher, 1978; Powell, 1979). Nematode infection makes the host susceptible or more suitable for the other pathogen. A few recent studies, however, show that this is not always true. Some fungal pathogens have also been shown to render the host susceptible to the nematode in such a system (Hasan, 1985).

Infection by the other pathogen depends upon its inherent pathogenic capability in the host. The other pathogen will not be able to infect the host in the absence of the nematode if it lacks pathogenicity on a given host. If it is pathogenic, the host substrate modified by the nematode becomes so suitable for the other pathogen that it inflicts greater pathogenic damage. In this system, the interaction between the two pathogens is thus host mediated and indirect. The inhibitory or stimulatory effects of one pathogen over the other, or mutually inhibitory or stimulatory effects, are also therefore indirect.

The three components of this system, all living, remain exposed to the environment. Artificial inoculation studies have adopted this three-component system mostly using four conventional treatments: (1) nematode alone; (2) other pathogen alone; (3) nematode and other pathogen (treated sequentially or concomitantly); and (4) control (no treatment), to determine the combined effects of both the pathogens on the host. This approach generally does not fully approximate or simulate all the existing natural situations. The role of environment, which may be crucial for interaction to occur or may be critical for either of the components, has not been considered. Some recent studies have, however, given attention to this inadequacy. To determine the impact of environmental factors on this three-component system, showing an interaction appears to be imperative.

3.3 ASSOCIATION OR INTERACTION: SOME RELEVANT QUESTIONS

Terminology is the science of correct application and usage of names given to concepts, structures and phenomena. Precise terminology is indispensable to any science (Roberts and Boothroyd, 1975). The term 'interaction' is used widely in plant nematology for various kinds of associations and its precise usage is rather rare. Both quantitative and qualitative interactions between two or more factors involved in plant diseases have been described as interaction in nematology literature. Nematodes, for that matter all organisms, come into association with other organisms in the course of their existence. Without any consideration whether or not 'interaction' (in the mathematical sense) operates, biologists are prone to see 'interaction' when merely terms like 'association', 'relationship' or just 'relation' suffice (Burrows, 1987). Pathological ecologists tend to describe various kinds of ecological associations in their own jargon, such as non-interfering coincidence, non-interfering co-infection, slight predominance and unilateral aggravation, clear predominance and ability to overcome handicap, unilateral aggravation, unilateral retardation, mutual aggravation, occupational privilege of the pioneer settler and co-pathogenicity of independently non-pathogenic partners (Dasgupta and Mandal, 1989).

Wallace (1983) suggested that use of the term 'interaction' should be restricted to quantitative plant disease interactions showing synergism or antagonism. This would be an 'interaction' in the statistical sense rather than in the descriptive sense. According to Burrows (1987), all non-statistical qualitative interactions can be simply termed 'relationship' or 'relation'.

Use of the terms 'synergism' and 'antagonism' has been variable, with different connotations. Dickinson (1979) has given an account of how the word synergism was introduced in plant pathology and was defined as 'an association of two or more organisms acting at one time and affecting a change that only one is not able to make'. Powell (1979) defined synergism as 'the concurrent or sequential pathogenesis of host plant by two (or more) pathogens in which the combined effects of the two pathogens are greater than the sum of the effects of each pathogen alone'. The first definition is qualitative whereas the definition of Powell (1979) is quantitative and more useful in nematology. In statistics, the term synergism is used to indicate a positive interaction, i.e. the sum of the treatment effects is not simply additive (Wallace, 1983, 1989). Wallace (1983) proposed the avoidance of the terms synergism and antagonism altogether and pleaded that the events could be adequately described by using the terms positive and negative interactions, respectively.

Sikora and Carter (1987) have argued that the literature on the relations between nematodes and other pathogens presents an inconsistent picture. Some of the relevant questions that arise are as follows.

1. Should all states of ill health be considered complex interactions (many causes–one effect) merely because the plants do not grow as pure cultures?
2. Have all aetiological interactions claimed been proven under field situations?
3. Have the inoculum levels and environmental factors been arbitrarily chosen for experiments to produce interactions?
4. Do the highly sensitive greenhouse tests in sterile sandy loam soils merit generalization?
5. Was the two to four-week nematode preinoculation design a fixation with the experimenters?
6. Was sequential inoculation also a prejudice over concomitant inoculation which could be a fact of nature?

7. Why should 'interactions' always lead to greater fungal damage?
8. Were the experimental designs and statistical techniques acceptable and the interpretations correct?

From a scanning of 39 articles published between 1976 and 1986, Sikora and Carter (1987) found that the relative frequencies of the more careful experiments were as follows: field-level research, 15%; natural partners, 31%; staggered population densities, 59%; concomitant inoculation, 59%; multifactorial analysis, 13%. Therefore, not more than three or four of the 39 experiments may have been precise enough.

On the other hand, well-designed greenhouse and field studies could not detect the often-reported synergism between *Meloidogyne incognita* and *Fusarium oxysporum* in tomatoes, nor was its resistance-breaking effect verified (Abawi and Barker, 1984). Other attempts to confirm the claimed 'disease complexes' such as between *Meloidogyne* and *Phytophthora cryptogea* (safflower) and *Rhizoctonia solani* (pea) also remained inconclusive (Sikora and Carter, 1987).

While interactions can no doubt occur where significant alterations are caused in the balance of biotic and abiotic factors under growing conditions of a plant, often the effects are merely additive. It is unnecessary and of no consequence to look for anything other than a simple explanation.

Use of fumigant nematicides in suppressing the nematode partner to claim interaction between nematode and fungus in root diseases ignores the biocidal effects of the fumigants on the fungus partner and other beneficial soil microflora. Instead, the use of non-fumigant and specific nematicides which can suppress disease complexes perhaps more clearly supports interaction, as in soya bean (*Fusarium* and *Meloidogyne*) and in potato (*Verticillium* and *Pratylenchus*) (Minton *et al.*, 1985; Morsink, 1967). The amount of yield response directly related to nematode control could not be determined because of the apparent interaction of the nematode with the expression of fusarial wilt of soya bean (Minton *et al.*, 1985).

Interactions between pathogenic organisms should be expected only when the units of inocula of both organisms are present at high enough population levels. Therefore, future experimenters will have to choose expected natural levels of population (Garber *et al.*, 1979).

Greenhouse and laboratory tests do not simulate nature and the conclusions are at best indicative. Sterile sandy loam soils distinctly favoured the nematodes and most often disfavoured the fungi. Steam sterilization could favour certain fungal pathogens because of release of ammonia. Similarly, misinterpretations could be possible from the two to four-week nematode preinoculation design, which is often not 'natural', or inoculation is sequential rather than concomitant. Timing should be determined by the observation of actual disease development in the field. Another area of negligence was the possible presence of seed-borne pathogens, including bacteria. Interaction claims should also elaborate the disease syndrome and epidemiology.

3.4 AETIOLOGICAL RELATIONS BETWEEN NEMATODE AND FUNGUS AS PARTNERS OF DISEASE COMPLEXES

Discovering the causes of ill health, i.e. the result of disease (Dasgupta, 1977, 1988), must involve looking synoptically into the relations between the various factors either by simulation models or by examination of hypotheses by the hypothetico-deductive approach. Such effects are likely to be continuous (Wallace, 1983). This approach is excellent in terms of identifying the factors – biotic or abiotic – no matter whether or not they are causes of disease (= pathogenesis *sensu* Dasgupta, 1977, 1988), or are responsible for ill health in a crop or cropping system, as well as for taking management decisions. However, Wallace's (1983) approach was not intended to establish the causal role of the factors in pathogenesis, although ill health or lack of homeostasis was considered 'disease' by Wallace (1983), which does widen the concept of pathogenesis. Wallace (1983) has treated the factors in a statistical sense. The same argument has also been controverted on the basis that the factors of ill health or yield loss may be additive (including substitutive) or interactive, and that a synoptic assessment is based on wrong statistical assumptions (Sikora and Carter, 1987). Thus, a disease or disease complex will be considered multicausal only if the causal nature of the factors are both biologically and statistically established.

The kinds of effects observed in the nematode–fungus aetiological relationship are:

1. fungus disease aggravated;
2. host growth suffered;
3. resistance to fungus reduced;
4. fungus suppressed by nematode;
5. nematode suppressed by fungus;
6. susceptibility to fungus increased (Norton, 1978).

Webster (1985) has reviewed the nature of changes in the host in terms of ultrastructure and translocatable factors in *Meloidogyne–Fusarium* disease complexes in contrast to independent infections. Interestingly, no specificity could be observed since *Meloidogyne* spp. could promote all kinds of root pathogens. No specific histological changes could be attributed to the so-called 'disease complexes' not occurring to any one of the partners. The metabolites released by the pathogens have largely remained uncharacterized but no characteristic metabolite has been identified. The reported breakdown of resistance to vascular wilt by the root-knot nematodes has also been questioned. On the contrary, there are reports that the monogenic resistance to fusarial wilt in tomato is not altered. There had been no field report until now to indicate any 'breakdown of resistance' owing to the nematode partner. In looking for nutritional and genetic bases of *Meloidogyne–Fusarium* relations, Webster has observed that the genetically epistatic and hypostatic parasitism between the partners of pathogenesis may be looked for. No such claim has so far appeared in the literature.

3.5 NEMATODE–NEMATODE INTERACTIONS: HOW TO DETERMINE?

The most important aspect has been the use of inappropriate statistics, as elaborated by Wallace (1983). According to Wallace (1983), multifactorial analysis in place of analysis of variance (ANOVA) should have been the choice, such as factorial designs requiring multiple densities of two or more organisms. Further, it needs to be emphasized that statistics merely indicate whether or not there is an interaction and will not explain the nature of interaction. A physiological experiment ought to follow. Wherever physiological experiments did follow, interaction could not be substantiated in most of the cases investigated (Webster, 1985). Analytical approaches such as synoptic multiple regression analysis are also limited but can be used to a great extent with practical advantage (Dasgupta and Chakraborti, 1990). The systems-analytical approach is the finest mathematical tool yet available for the modelling of multifactorial interactions (Dasgupta and Rama, 1987). However, it is necessary to point out that no amount or finest of statistical tools can replace the need for establishing the biological nature of interactions. The remainder of this section describes the necessary statistical protocol for the establishment of 'interaction' among pathogens in a disease complex.

The statistical definition of an interaction between two norms N_i and N_j is that $N_{ij} \neq N_{ii}$ and $N_{ji} \neq N_{jj}$ or both, whereas association exists when $N_{ij} \equiv N_{ii}$ and $N_{ji} \equiv N_{jj}$. The definition may be extended to events involving more than two partners. Such relations, however, exclude predator–prey, symbiosis, parasitism and other obligate associations. The nature of interaction may depend on density, abiotic and microclimatic factors, host genotype and physiological conditions, sequence of establishment, etc.

Burrows (1987) has discussed the possibilities of utilizing appropriate statistical techniques to determine the interactive associations between the parasitic nematodes. Instead of the norms explained in the previous paragraph, for the paired associations at specified population densities under given conditions, the possible observations could be: Y_i, response to type i alone; Y_j, response to type j alone; Y_{ij}, response to combined i and j types; Y_0, absence of both types. To be more practical, Burrows (1987) has identified two kinds of interactive associations (i.e. interactions) – cooperative and disruptive – and both through host effects alone. The design was based on the projection of response to mixtures of nematode populations from the observations of response to each component of the mixtures alone. Such a projection should be possible for non-interactive and simple forms of interactive associations between the components. Burrows (1987) has experimented with pathogen mixtures on yield and other host variables and has calculated dependences of response as percentages and determined the marginal responses over one factor alone.

For the projection of joint response from the marginal responses and for defining interactions on the response scale, the marginal responses need to be determined.

3.5.1 Marginal responses

For determining marginal responses as any addition to minimum response at zero inoculum level, some pattern functions, viz. exponential decay, mass action law, Weibull function, logistic in log (x), Burr's function or Richard's function in log (x) were chosen for fitting.

3.5.2 Threshold parameter

A threshold or tolerance dosage ξ below which there is no decline in response (not an economic threshold) can be introduced in all the pattern functions. When this is done for exponential decay, the well-known Seinhorst function is obtained. An alternative pattern was described by Ferris *et al.* (1981):

$$y(x) = \beta + (1 - \beta)(\delta)(x - \xi)$$

$$\text{if } x \geqslant \xi, \, 0 < \delta < 1$$

where $y(x)$ is proportionate response to dosage x, β is minimum response, δ is small change and ξ threshold dosage.

Burrows (1987) has questioned Ferris *et al.*'s (1981) assumption of $(1 - \delta)$ as a proportionate damage parameter per individual nematode for the scale of δ changes with changes in units of dosage.

3.5.3 Testing of hypothesis

According to the conditional error principle such as applied to likelihood ratio tests, the observed ratio of mean squares, MSH/MSE (mean hypothesis sum of squares/mean squared residuals), which is an estimate of the appropriate experimental error variance, is referred to as the central F distribution with degrees of freedom corresponding to MSH and MSE.

From a set of experimental data on interaction to be tested between *Meloidogyne incognita* (Mi) and *Hoplolaimus columbus* (Hc) on soya bean, the hyperbolic decay pattern is judged adequate for both Mi and Hc (Burrows, 1987). Future experimenters will do well to fit a specific nematode–nematode or a given suspected pathogenic interaction to these and other pattern functions before they can claim 'interactions'.

3.5.4 Associate effects and joint response

More commonly, joint responses to associations of two or more types of nematodes are analysed as factorial designs using ANOVA (analysis of variance) methods (Wallace, 1983). As ANOVA assumes the response being additive as a non-interactive phenomenon, the detection of 'statistical interaction' by ANOVA and subsequent inference of 'biological interaction' mean two different things. In such cases, ANOVA may lead to wrong conclusions merely because the assumptions of ANOVA do not fit into the event of a possible reality of interaction without additive effects. Essentially, in ANOVA, it is hypothesis of 'response additivity' as a non-interactive expectation while it is precisely that which deserves to be rejected in order to assume (mathematical-statistical) 'interaction'.

3.5.5 Product response models

It was R.A. Fisher who devised a linear additive product model for a factorial experiment along with ANOVA. It was the latter that received a good press and still enjoys a great following owing to its simplicity and analytical ability. However, it is the wrong ship to the port of interaction. Burrows (1987) explains some suitable product response models for two or more components in a community. Although simple and perhaps successful too, a product model is not free from conceptual objections as to its application in biological situations.

3.5.6 Combined dosage equivalence (CDE)

As opposed to joint response formulations derived from marginal responses, CDE formulation provides a simple projection of the surface of joint responses, given only the marginal calibrations and assuming non-interactive association, because all possible isopleths are straight lines containing known points on the dosage axes. From isopleth analysis, an interactive association (on the basis of marginal response) may turn out to be a cooperative association.

Burrows (1987) also suggests application of the alias criterion, such as testing if nematode type 1 is an alias of type 2, etc. The response additivity and response product models do not satisfy the alias criterion. Hence, Burrows considers that this is a strong argument for rejecting them as formulations of non-interactive associations while all functions $y(x_1, x_2)$ satisfying the alias criterion are necessarily CDE formulations. Further, CDE formulations are amenable to simple transformations of dosage.

3.6 MODEL FOR NEMATODE-MYCORRHIZA INTERACTIONS

For interaction of nematodes with mycorrhizal fungi, Hussey and Roncadori (1982) developed a descriptive model to characterize the effect of vesicular-arbuscular mycorrhiza (VAM) fungi on nematode–host relations. The model considered mycorrhizal effects on host efficiency and host sensitivity. Host efficiency was measured in terms of nematode or egg densities in roots or development of nematodes on a mycorrhizal plant in comparison to a non-mycorrhizal plant. Host sensitivity was determined in terms of growth suppression or yield reduction. This model was used as a framework in studies on interaction of nematodes with mycorrhizal fungi. Smith (1987) suggested employing factorial or regression analysis and desisting from single-factor descriptive associations commonly used to examine nematode–mycorrhiza interactions. He further advised testing the Seinhorst damage function, $Y = m + (1 - m) Z^{p - t}$; where Y is yield, m is minimum yield, Z is nematode virulence, p is nematode population density and t is tolerance limit, in order to determine the difference between VAM fungi and phosphorus on host response. These quantitative models may establish whether or not mycorrhizal plants or non-mycorrhizal plants grown in P-rich soils respond differently to varying initial nematode population densities.

3.7 CONCLUDING REMARKS

The precise use of terms in nematode interaction studies is very important and should not be evaded. The interaction needs to be illustrated statistically when we speak of interaction between nematodes and other plant pathogens in diseases of complex aetiology. This statistical proof of interaction should be supported by operative physiological or biochemical mechanisms.

Options of statistical solutions available to handle biological associations vis-à-vis interactions are not limited. There are no simple blanket solutions to all biological situations of interaction per se. Statistical assumptions often violate against biological reality. While statistical researchers will continue to develop finer and more situation-specific tools, the biologists will have to be careful against using a square tool for a round analysis. Pragmatism demands that the experiments should be well designed with clear-cut objectives to answer specific questions, and then analysed with a number of parallel methods such as Burrows's (1987) exercise with growth pattern functions if their assumptions are only close to reality, which should hopefully be the case.

REFERENCES

Abawi, G.S. and Barker, K.R. (1984) Effects of cultivar, soil, temperature and population levels of *Meloidogyne incognita* on root necrosis and *Fusarium* wilt of tomatoes. *Phytopathology*, **74**, 433–8.

Burrows, P.M. (1987) Interaction concepts for analysis of responses to mixtures of nematode populations, in *Vistas on Nematology* (eds J.A. Veech and D.W. Dickson), Society of Nematologists, Hyattsville, pp. 82–93.

Commins, S. and Linscott, R.N. (eds) (1954) *Man and the Universe: The Philosophies of Science*, Modern Pocket Library, New York.

Dasgupta, M.K. (1977) Concept of disease in plant pathology and its applicability elsewhere. *Phytopathologische Zeitschrift*, **88**, 136–9.

Dasgupta, M.K. (1988) *Principles of Plant Pathology*, Allied Publishers, New Delhi.

Dasgupta, M.K. (1992) *Phytonematology*, Naya Prakash, Calcutta (in press).

Dasgupta, M.K. and Chakraborti, S. (1990) Mathematical and statistical epidemiology, in *Basic Researches in Plant Disease Management* (ed. P. Vidhyasekaran), Daya Publishing, New Delhi, pp. 213–32.

Dasgupta, M.K. and Mandal, N.C. (1989) *Postharvest Pathology of Perishables*, Oxford and IBH Publishing Co., New Delhi.

Dasgupta, M.K. and Rama, K. (1987) Mathematical and statistical ecology of plant parasitic nematodes in the tropics. *International Journal of Tropical Plant Diseases*, **5**, 1-9.

Dickinson, C.H. (1979) External synergisms among organisms inducing disease, in *Plant Disease*, Vol. IV (eds J.G. Horsfall and E.B. Cowling), Academic Press, New York, pp. 97-111.

Fawcett, H.S. (1931) The importance of investigations on the effects of known mixtures of organisms. *Phytopathology*, **21**, 545-50.

Ferris, H.B., Turner, W.D. and Duncan, L.W. (1981) An algorithm for fitting Seinhorst curves to the relationship between plant growth and pre-plant nematode densities. *Journal of Nematology*, **13**, 300-4.

Garber, R.H., Jorgenson, E.C., Smith, S. and Hyer, A.H. (1979) Interaction of population levels of *Fusarium oxysporum* f. sp. *vasinfectum* and *Meloidogyne incognita* on cotton. *Journal of Nematology*, **11**, 133-7.

Hasan, A. (1985) Breaking resistance in chilli to root-knot nematode by fungal pathogens. *Nematologica*, **31**, 210-17.

Hussey, R.S. and Roncadori, R.W. (1982) Vesicular-arbuscular mycorrhizae may limit nematode activity and improve plant growth. *Plant Disease*, **66**, 9-14.

Minton, N.A., Parker, M.B. and Sumner, D.R. (1985) Nematode control related to Fusarium wilt in soybean root-rot and zinc deficiency in corn. *Journal of Nematology*, **17**, 314-21.

Morsink, F. (1967) *Pratylenchus penetrans*: Its interaction with *Verticillium albo-atrum* in the Verticillium wilt of potatoes and its attraction by various chemicals. PhD thesis, University of New Hampshire, Durham.

Norton, D.C. (1978) *Ecology of Plant Parasitic Nematodes*, John Wiley, New York.

Pitcher, R.S. (1978) Interactions of nematodes with other pathogens, in *Plant Nematology* (ed. J.F. Southey), Her Majesty's Stationery Office, London, pp. 63-77.

Powell, N.T. (1971) Interactions between nematodes and fungi in disease complexes. *Annual Review of Phytopathology*, **9**, 253-74.

Powell, N.T. (1979) Internal synergisms among organisms inducing disease, in *Plant Disease*, Vol. IV (eds J.G. Horsfall and E.B. Cowling), Academic Press, New York, pp. 113-33.

Roberts, D.A. and Boothroyd, C.W. (1975) *Fundamentals of Plant Pathology*, W.H. Freeman, San Francisco.

Sikora, R.A. and Carter, W.W. (1987) Nematode interactions with fungal and bacterial plant pathogens: fact or fantasy, in *Vistas on Nematology* (eds J.A. Veech and P.W. Dickson), Society of Nematologists, Hyattsville, pp. 307-12.

Smith, G.S. (1987) Interactions of nematodes with mycorrhizal fungi in *Vistas on Nematology* (eds J.A. Veech and D.W. Dickson), Society of Nematologists, Hyattsville, pp. 292-300.

Wallace, H.R. (1978) The diagnosis of plant diseases of complex etiology. *Annual Review of Phytopathology*, **16**, 379-402.

Wallace, H.R. (1983) Interactions between nematodes and other factors on plants. *Journal of Nematology*, **15**, 221-7.

Wallace, H.R. (1989) Environment and plant health: A nematological perception. *Annual Review of Phytopathology*, **27**, 59-75.

Webster, J.M. (1985) Interactions of *Meloidogyne* with fungi on crop plants, in *An Advanced Treatise on Meloidogyne, Vol. I: Biology and Control* (eds J.N. Sasser and C.C. Carter), North Carolina State University Graphics, Raleigh, pp. 183-92.

4

Mechanisms of interactions between nematodes and other plant pathogens

M. Wajid Khan

All interactions of plant-parasitic nematodes with other plant pathogens have three components: nematode, host and other pathogen. The plant pathogens known to interact with nematodes are mainly viruses, bacteria and fungi. Plant-parasitic nematodes also interact with each other in a mixed community. The mechanism of interaction mainly involves the operative role of the nematode and the other pathogen in relation to disease development and it depends upon the host–parasite relationships of the nematode and the other pathogen. Some plant-parasitic nematodes act as carriers of the pathogens and facilitate disease development while feeding on the host. In a large number of instances nematodes increase host susceptibility. The interaction between the nematode and fungal pathogen is often indirect and occurs owing to induced modifications in the host plant. Ectoparasitic nematodes cause minimal modifications but the modifications caused by endoparasitic nematodes, particularly sedentary endoparasites like root knot nematodes, are extensive and complex. Interactions of nematodes with symbiotic root associates disrupt the association to the disadvantage of the host. In interactions with other nematodes in cohabitance they compete for resources in short supply. In a given interaction, the nematode often plays more than one role (Pitcher, 1978). The roles of the other pathogens in the interactions vary depending upon their taxonomic nature, parasitic behaviour and host–parasite relationship. The mechanisms of interactions suggested by most of the authors in their studies are speculative, without experimental evidence. This chapter is focused on operative roles of plant-parasitic nematodes in their interactions with other plant pathogens. It presents an overall view of the mechanisms of interaction of nematodes with other plant pathogens, including nematodes, as well as their interactions with root symbionts. The roles that nematodes play in disease complexes have been the subject of several reviews, including those of Powell (1963, 1971a, 1971b), Pitcher (1965, 1978), Bergeson (1972) and Taylor (1979).

4.1 NEMATODES AS VECTORS OF PLANT PATHOGENS

4.1.1 Plant viruses

Species of *Xiphinema, Longidorus, Trichodorus* and *Paratrichodorus* transmit a number of plant viruses. *Xiphinema* and *Longidorus* transmit nepoviruses, and *Trichodorus* and *Paratrichodorus* vector tobraviruses. Nematode transmission of viruses is non-circulative. The transmission mechanism is of the ingestion–egestion type (Harris, 1981). The transmission process involves three steps: acquisition (ingestion) of virus particles while feeding on the roots of virus-infected plants; retention of the particles at the specific sites within the nematode vector; and egestion of the particles by dissociation from the sites of their retention. The transmission mechanism is very selective and specific association occurs between a virus and its nematode vector.

During nematode feeding virus particles present in the cell sap are ingested. Ingested virus particles are selectively and specifically adsorbed at the retention sites. Viruses not transmitted by the nematode, if present in the cell sap and ingested, are not retained at the sites. In general, one virus is transmitted by one nematode species. There are a few exceptions to this general rule. Some serologically related strains of a virus have different nematode vectors. At the same time, serologically unrelated viruses have a common vector. *Xiphinema diversicaudatum* transmits arabis mosaic virus but not grapevine fanleaf virus, which are serologically related. Arabis mosaic and strawberry latent ringspot viruses, not related serologically, are vectored by *X. diversicaudatum*. Serologically distinct strains of raspberry ringspot and tomato black ring viruses have different vectors. Scottish strains of the viruses are transmitted by *Longidorus elongatus*. The English strain of raspberry ringspot virus is vectored by *L. macrosoma* and tomato black ring virus by *L. attenuatus*. The specificity between tobraviruses and their trichodorid vectors is not as distinct as in longidorids. Some evidence from recent studies, however, shows such specificity (Brown *et al.*, 1989; Taylor, 1990).

Virus retention sites differ among the nematode genera. In *Xiphinema* species, virus particles are found associated with cuticular lining of the odontophore, the oesophagus and oesophageal pump (Taylor and Robertson, 1970). In *Longidorus* species, virus particles remain associated with the inner surface of the cuticular odontostyle. In *L. elongatus*, particles are also found between the odontostyle and its guiding sheath (Taylor and Robertson, 1969, 1975; Trudgill *et al.*, 1981). In *Trichodorus* and *Paratrichodorus* species tobacco rattle virus particles are found in association with the lining of the oesophagus from the most anterior region of the oesophostome to the oesophageal-intestinal valve and are not associated with solid onchiostyle (Taylor and Robertson, 1975). All these surfaces are shed during moulting. Therefore, virus particles do not pass from one stage of the nematode to another during development of adults. Ingested virus particles are retained within the nematodes for long periods. The period of retention, however, differs among the genera. Virus particles in *Xiphinema* and trichodorid vectors persist for longer periods than in *Longidorus*.

Egestion, the final step in the transmission, involves dissociation of the virus particles from retention sites and their transfer within the host plant tissue. Dissociation of virus particles is accomplished during nematode feeding. It is assumed that oesophageal gland secretions passing anteriorly during feeding help in dissociation of the particles. The pH within the oesophagus lumen is modified by the secretions. The modified lumen pH, by altering the surface charge of the virus particles, brings about their dissociation from the retention site (Taylor and Robertson, 1977; Taylor and Brown, 1981).

4.1.2 Plant bacteria

Bacteria can be transmitted by nematodes externally on their body surface or internally within the alimentary canal. Some nematodes have been implicated as disseminators of plant-pathogenic bacteria, carrying them internally within the alimentary tract by ingesting bacterial cells. Kalinenko (1936) demonstrated that *Helicotylenchus multicinctus* and *Pratylenchus pratensis* possessed plant-pathogenic bacteria like *Erwinia carotovora*, *Xanthomonas phaseoli*, *Pseudomonas fluorescence* and *Bacterium necrosis* internally. Isolates of the bacteria developed in both tubes from the surface sterilized nematodes. The exact importance of the internal transmission of plant-pathogenic bacteria by nematodes is not known.

External transmission of the plant-pathogenic bacteria by some nematodes is significant in the development of some disease complexes in plants. The nematodes act as vectors of the bacteria and carry bacterial cells over their body. Some species of *Anguina* and *Aphelenchoides* vector bacteria parasitic on aerial plant parts. Yellow ear rot or 'tundu' disease of wheat results from joint infection of *Clavibacter tritici* (= *Corynebacterium michiganense* pv. *tritici*) and *Anguina tritici* (Gupta and Swarup, 1968, 1972). *A. tritici* alone causes ear-cockle of wheat and *C. tritici* alone is incapable of causing yellow ear rot disease. Gupta and Swarup (1972) found that *C. tritici* did not cause the yellow ear rot when inoculated alone in soil or at the growing point of wheat seedlings. Surface-sterilized juveniles of *A. tritici* alone caused ear-cockle. No yellow ear rot symptoms developed when the bacterium was inoculated at the growing point of the seedlings infected with surface-sterilized juveniles at planting time. Unsterilized juveniles without further addition of bacteria caused yellow ear rot. The bacterium was, therefore, originally associated with the juveniles inside the galls. Gupta (1966) had shown the presence of the bacteria on the body surface of juveniles. Recent studies (Pathak and Swarup, 1984; Gokte and Swarup, 1988) also confirmed this observation. Seed galls were found to be contaminated with the bacterium and bacterial cells sticking to the body surface of juveniles. Ray (1987) has shown that infection of the wheat at seedling stage by *Anguina tritici* is essential for development of yellow ear rot through *Anguina tritici* juveniles contaminated by the bacteria.

Anguina funesta interacts with *Clavibacter* sp. on ryegrass (*Lolium rigidum*), causing a problem to grazing animals known as annual ryegrass toxicity (Riley and McKay, 1990). Spikelets of ryegrass become full of bacteria, with glumes covered with a bright yellow bacterial mass. In the absence of the bacterium, the nematode causes brown seed galls (Bird *et al.*, 1980). Seed galls colonized by the bacterium are the cause of toxicity syndrome in the animals (Bird, 1981). Seed galls with nematode alone show a dark brown colour and contain coiled anhydrous nematodes. The galls colonized by bacteria become bright yellow. Both nematodes and bacteria remain packed in the dry galls. Bacteria are found packed closely both along the walls of cells making up the gall walls and within the gall itself. The close contact between the infective juveniles and bacteria in anhydrous galls plays an important role in survival under extreme environmental conditions (Bird *et al.*, 1980).

Rhodococcus fascians (= *Corynebacterium fascians*) and *Aphelenchoides ritzemabosi* or *A. fragariae* together cause cauliflower disease of strawberry (Crosse and Pitcher, 1952). This complex disease requires the active contribution of both organisms for full disease syndrome. The nematode alone causes growth inhibition of apical meristem, producing short-lived alaminate leaves and feeding areas. The symptoms develop more rapidly in plants inoculated with bacteria. Therefore, the nematode is benefited initially. But then cauliflower symptoms develop and nematode population declines.

The bacterium is capable of stimulation of dormant meristems if transported at the site. The nematode acts as a vector and efficient inoculant. Since the bacterium eventually dominates over the nematode, it is likely that the nematode directly stimulates bacterial growth either by providing a useful metabolite or by modifying the host substrate (Pitcher, 1963).

Ditylenchus dipsaci (stem and bulb nematode) transmits *Clavibacter michiganense* subsp. *insidiosum* (= *Corynebacterium michiganense* pv. *insidiosum*), which causes wilt in alfalfa. The bacteria are carried on the body of the nematodes into the crown buds and are placed in conducive infection courts. Numerous wilt bacteria are found associated with the nematode cuticles and are transmitted from wilt-infected to healthy alfalfa plants (Hawn, 1963, 1971).

The role of the nematodes in these associations was considered to be restricted to localized transport of the pathogens (Pitcher, 1965, 1978). Undoubtedly, the transport is the essential step of the interactions. Adhesin-receptor theory has been proposed to explain the mechanism of adhesion of bacterial cells to nematode cuticle and the specificity between the bacterium and the nematode involved (Riley and McKay, 1990). But the mode of contamination of juveniles of nematodes with bacteria and their nature of association differs. Further, the mechanism of interaction between these pathogens seems to be much more complex than simple transport of bacterial cells. As the disease complex does not develop in the absence of the nematode, further interaction between the bacterium and nematode to produce the disease syndrome is logically expected. The nature of this interaction may be physiological, which might be helpful in establishment of the bacterium at their respective infection sites. Our existing bank of knowledge is not sufficient to explain the mechanism of this aspect of the interactions.

4.1.3 Fungal pathogens

Nematodes can vector fungal propagules internally, passing them through their guts in viable conditions or externally on the body surface. Internal vectoring of fungal propagules by plant-parasitic nematodes has not been reported. External transport of spores of fungi that infect plant roots on the nematode cuticle has also not been observed (Mai and Abawi, 1987). A foliar nematode, *Anguina tritici*, was recognized as a vector of spores of *Dilophospora alopecuri*, which attacks aerial parts of cereals. The nematodes, while moving between the leaf sheaths to reach the growing point, take the fungal conidia and deposit them on the growing point. Further, the nematode by feeding on the tender leaves helps in penetration and establishment of the fungus (Atanasoff, 1925). The role of the nematode is essential for successful parasitism of the fungus. Involvement of physiological alternation by nematode feeding that might be helping the fungus in its establishment may not be precluded. This has not been explored. Some saprozoic nematodes are reported to ingest fungal propagules of *Fusarium* and *Verticillium* and pass them through their guts in a viable state (Jensen and Siemer, 1971). The significance of this transport of fungal spores by saprozoic nematodes in the disease epidemiology is not known.

4.2 NEMATODES AS WOUND AGENTS

4.2.1 Fungal pathogens

Nematode parasitism of plants is associated with wounding of their hosts. The feeding process of all plant-parasitic nematodes produces a wound of some kind in the host plant, either by simple micropuncture or by rupturing or separating cells (Taylor, 1979). Ectoparasitic nematodes cause micropuncture on the plant root surface. Migratory endoparasites produce lesions on roots. Second-stage juveniles of sedentary endoparasites after root invasion migrate intercellularly through the cortex and establish contact with vascular tissue to induce giant cells or syncytia.

For a long time, after observation of interaction between *Meloidogyne* and *Fusarium* on cotton leading to increased wilting by Atkinson (1892), wounding was considered of paramount importance in facilitating the entry of fungal pathogens, wounds acting as passage or portal. Anatomical analysis of cotton seedling roots infected with *M. incognita* by Perry (1963), however, showed that juvenile penetration did not facilitate invasion of *F. oxysporum* and the wounded roots did not attract the fungus. The wound-facilitation concept in fungus–nematode interaction, especially between *Fusarium* and *Meloidogyne*, was disproved by other studies as well. Modification of host substrates was found to be primarily responsible for such interactions (Pitcher, 1965; Powell, 1971a, 1971b, 1979). Split-root (Bowman and Bloom, 1966), grafting (Sidhu and Webster, 1977) and stem-layering techniques further demonstrated that predisposition to infection by *Fusarium* wilt fungi occurs through systematically translocatable metabolites produced at the site of nematode infection.

Powell (1979), however, remarked that wounding by nematodes in increasing the susceptibility of plants to invasion by other pathogens still appears logical and applies in some cases, particularly in diseases involving migratory nematodes. Westerlund *et al.* (1974) observed that *Fusarium oxysporum* f. sp. *ciceri* may require wounding for efficient infection of chickpeas. It is most probable that wounding coupled with physiological changes in the host are important in such interactions.

4.2.2 Plant bacteria

Nematodes provide wounds to facilitate bacterial entry and modify the host tissue to enrich the substrate nutritionally to the advantage of plant bacteria. Plant bacteria enter their host plants mostly through wounds. Wounds caused by ecto- and endoparasitic nematodes on underground plant parts favour bacterial pathogens of plants. Lucas *et al.* (1955) found that tobacco plants exhibited increased wilt and earlier appearance of the symptoms when *Pseudomonas solanacearum* and *Meloidogyne incognita* were together in the soil. A comparable situation was noticed in plants that had been artificially wounded. The role of the nematode was therefore considered to be as a wounding agent. The study of Fukudome and Sakasegawa (1972), however, suggested a synergistic interaction between the two pathogens because the wilt symptoms were more severe when nematodes and bacteria were inoculated together than inoculation of bacteria alone on artificially wounded roots. Similar interactions where nematodes were claimed to serve as wounding agents have been observed in a number of studies (Stewart and Schindler, 1956; Schuster, 1959; Libman and Leach, 1962).

Modified host substrate is also claimed to favour bacterial pathogens. *Meloidogyne*-modified tissue acts as a more favourable substrate (Johnson and Powell, 1969). Moura *et al.* (1975) showed that on tomato cultivars, susceptible and resistant to *Corynebacterium michiganense*, the presence of *M. incognita* increased the bacterial canker only when the nematode was inoculated prior to the bacterium. Such an interaction did not occur when both pathogens were inoculated simultaneously or the bacterium was inoculated prior to the nematode. This suggested that the mechanism of interaction is a complex one where physiological changes in the host substrate caused by the nematode are important to develop synergistic relationships.

4.3 INCREASED HOST SUSCEPTIBILITY

4.3.1 Wilt fungi

A vast majority of nematode interactions involve fungal pathogens, especially wilt fungi. In general, the role of nematodes is not essential for fungal invasion, establishment and disease development caused by the fungus in fungus–nematode interactions, particularly in root diseases. Plant-parasitic nematodes play an assisting role by way of enhancing host susceptibility, leading to increased rate of development and severity of the disease caused by the fungal pathogens.

Root-knot nematodes

The bulk of studies on interactions between fungi and nematodes involve species of *Fusarium* and *Meloidogyne*. *Fusarium* spp. are highly destructive pathogens of a number of field, vegetable, fruit and other horticultural crops. All the wilt-causing formae speciales and races belong to the species *F. oxysporum* (Schlecht.) Synder & Hansen (Booth, 1971; Armstrong and Armstrong, 1981). Most *Fusarium* spp. penetrate the root tissue directly, often near the root tip (Nelson, 1981).

MacHardy and Beckman (1981) described the details of infection and pathogenesis of wilt-causing fusaria. Bell and Mace (1981), MacHardy and Beckman (1981) and Pegg (1981) have reviewed the chemical and structural changes in plants infected by wilt-causing fusaria. Mechanical plugging, toxins, enzymes and growth regulators singly or jointly have been implicated as causes of wilting (Mai and Abawi, 1987).

The overall mechanism of the interactions between *Fusarium* wilt fungi and root-knot nematodes is not fully understood (Mai and Abawi, 1987). Attack of root-knot nematodes of plants infected by *Fusarium* spp. increases the wilt symptom expression and death rate of the plants. The overall scenario of interactions between two pathogens can be visualized in different stages. Evidence suggests that the initial phase of the interaction occurs in rhizosphere, where the root exudates from root-knot infected plants stimulate the fungal pathogen. The exudates also suppress actinomycetes, the antagonists of the wilt fungus (Cook and Baker, 1983; Walter, 1965; Bergeson, 1972). More research is, however, needed to provide sufficient information on this aspect.

The next phase in these interactions involves the effect of root-knot infection on penetration of wilt fungi. Initially it was thought that micropunctures caused by nematodes on plant root facilitate entry of the fungal plant pathogens. Later, it was demonstrated that severity of fungal-induced wilt diseases increased greatly when root-knot nematodes were added three to four weeks prior to fungus inoculation of the host in comparison to simultaneous inoculations of both the pathogens. This led to the hypothesis that the nature of interactions between root-knot nematodes and *Fusarium* wilt fungi are physiological rather than physical.

The ultimate phase of the interaction between the two pathogens occurs during the pathogenesis of the wilt fungus. Modifications in the host plant by root-knot nematodes is the key factor to this phase of interaction leading to increased wilt severity.

The sedentary females of *Meloidogyne* spp. establish their feeding sites in the xylem parenchyma cells, bringing about significant changes in the morphology, anatomy and biochemistry of the host plant. Thus it is probably the major site of interaction between these pathogens (Mai and Abawi, 1987). Giant cells induced by *Meloidogyne* remain in a state of high metabolic activity through the continuous stimulation by the nematode (Webster, 1975). They contain maximum DNA, RNA and photosynthates about three to four weeks after infection (Bird, 1972). Concentration of sugars is also greater during the late stages of infection (Wang, 1973). The concentration of hemicellulose, organic acids, free amino acids, proteins and lipids are much higher in giant cells. This enriched medium benefits the fungal pathogens in their interactions with root-knot nematodes. The giant cells remain in a perpetual juvenile state which delays maturation and suberization of other vascular tissues, and thus *Fusarium* successfully penetrates and establishes in the xylem elements. Noguera (1982) observed that purified extracts of tomato roots infected with *M. incognita* did not show rishistin, an antifungal substance normally present in healthy plants. Loss of resistance of the tomato variety to wilt fungus was attributed to inhibition of rishistin production.

Inhibition of tylose formation by root-knot nematodes is also offered as a possible mechanism for increased wilt severity. Tyloses formed in the xylem vessels by expansion of xylem parenchyma through the pits do not develop from xylem parenchyma cells which are transformed into giant cells or physiologically altered adjacent cells (Webster, 1985). Nematode-free plants of tomato showed the formation of well-developed tyloses in the xylem vessels of stem, while in infected plants these were absent or much reduced.

Root lesion nematodes

Pratylenchus spp. are migratory endoparasites. They enter and migrate within roots, feeding on various tissues. Feeding and migration of the nematode cause considerable damage to root tissue and necrotic lesions appear on the root surface. *Pratylenchus* spp. have been implicated in synergistic interactions with other plant pathogens, particularly with *Verticillium* spp., on a number of crops. The mechanisms of these interactions are related to the development of necrotic infection courts and biochemical changes in the attacked plants. These alterations enhance the pathogenic capabilities of *Verticillium* spp. The necrotic lesions on roots serve as infection courts which facilitate the establishment of the fungus in the court and subsequent invasion. *V. dahliae* showed a distinct liking for the lesions on eggplant root caused by *P. penetrans*. According to Conroy *et al.* (1972) lesions in tomato roots were more important than any general physiological changes. Internal predisposing factors in *Pratylenchus*-infected plants which are also claimed to be translocatable from the site of infection to other parts of the plant are also involved (Faulkner *et al.*, 1970; Powell, 1979). The exact nature of translocatable factors or 'messenger' as called by Powell (1979) is not known. The incubation period of *Verticillium* is shortened in *Pratylenchus*-infected plants (Bergeson, 1963; Faulkner *et al.*, 1970). *P. minyus* was shown to shift susceptibility of peppermint to *Verticillium* wilt towards its own optimum temperature (Faulkner and Bolander, 1969). Therefore, physical and physiological changes in the plants infected with *Pratylenchus* are collated together to enhance the susceptibility of plants to *Verticillium* wilt.

Cyst nematodes

The host–parasite relationship of cyst nematodes is similar to root-knot nematodes. They form syncytia in the root tissue but do not cause extensive changes in root morphology as found in the case of root-knot nematodes. Reports of interaction of *Heterodera* and *Globodera* with *Fusarium* and *Verticillium* (Jorgenson, 1970; Hide and Corbett, 1973; Ross, 1965; Hide *et al.*, 1984; Evans, 1987; Storey and Evans, 1987) are relatively very few in comparison to root-knot nematodes. Details of the mechanism of interaction of cyst nematodes with wilt-causing fungi have not been analysed. The mechanism of their interaction may be similar to root-knot nematodes because of their great similarity in the host–parasite relationship (Powell, 1979).

4.3.2 Root-rot fungi

Some species of *Pythium* (*P. aphanidermatum*, *P. ultimum*), *Rhizoctonia* (*R. solani*, *R. bataticola*) and *Fusarium* (*F. solani*) and *Phytophthora parasitica*, *Scleroltium rolfsii* and *Colletotrichum coccodes* are prominent amongst the root-rot fungi that are known to interact with different plant-parasitic nematodes. Like wilt fungi, root-rot fungi are capable of causing root diseases on their own and have their own inherent mechanism of root penetration. The role of the nematode in root-rot diseases in general is related to assisting the fungal pathogen in its pathogenesis and increasing host susceptibility. Nematodes, by wounding, provide the fungal pathogen access to root tissue (Inagaki and Powell, 1969). The lesions caused by lesion or burrowing nematodes or invasion tracks formed by penetrating juveniles of root-knot or cyst nematodes provide a better substratum for establishment and colonization of the fungal pathogens (Booth and Stover, 1974; Polychrono-poulos *et al.*, 1969). Further, physiological alterations by the nematode improve the nutrient status of the host for the fungal pathogen. Golden and Van Gundy (1975) showed that galled regions of tomato roots infected with *Meloidogyne incognita* were heavily colonized by *R. solani*, indicating its preference for the region due to the presence of rich nutrient medium in the galls. A similar pattern of colonization by *R. solani* was observed by Khan and Müller (1982) on radish roots infected with *M. hapla*. Therefore, physiological alterations which ensure better nutrient availability to the penetrated fungal pathogen are the key factor of the synergistic damage caused to the host.

4.3.3 Saprophytic fungi

Fungi non-pathogenic to a host plant may become pathogenic in the presence of nematodes. Some investigators have shown that saprophytic fungi which are non-pathogenic on plants caused extensive decay of the roots infected with root-knot nematodes. *Rhizoctonia solani* and *Pythium ultimum*, minor pathogens on tobacco, caused rapid necrosis of roots invaded by *Meloidogyne incognita* (Melendez and Powell, 1970; Powell and Batten, 1967). Transformation of such minor pathogens into highly destructive ones by the root-knot nematodes led to further studies involving soil-borne saprophytic fungi, not regarded as pathogens on tobacco. Strikingly, any of the saprophytic fungi like *Curvularia trifolii*, *Botrytis cinerea*, *Aspergillus ochraceous*, *Penicillium martensii* and *Trichoderma harzianum* used in the study invaded roots infected with *M. incognita* and caused extensive decay (Powell *et al.*, 1971). This observation showed that root-knot possessed outstanding abilities to cause physiological changes in plants that can induce susceptibility in plants to attack by fungi present in the rhizosphere, whether pathogenic or non-pathogenic (Powell, 1979). The mechanism in these interactions was obviously physiological because the fungi were added to the roots that had been exposed to *M. incognita* for three to four weeks, the period necessary for attaining a state of high metabolic activity in giant cells.

4.3.4 Modifications in host substrate

In fungal–nematode interactions modifications in host substrates which enhance host susceptibility to fungal pathogens may be localized to the site of nematode infection, or systemic physiological changes occur in the host that render the sites away from the nematode infection favourable to fungal pathogens.

Localized modifications

Powell and Nusbaum (1960) were the first to provide evidence of localized modifications in the host tissue attacked by the nematode. Hyphae of *Phytophthora parasitica* var. *nicotianae* near the galls on tobacco roots were found to grow more luxuriantly in the hyperplastic parenchyma induced by *Meloidogyne incognita acrita*. Hyphae colonized giant cells and destroyed their contents. The hyphae then could enter the vascular tissue, both in susceptible and resistant cultivars. The localized substrate modification provided enriched medium for growth, establishment and further invasion. Similar evidence was provided by Melendez and Powell (1967) while studying interactions between *M. incognita* and *Fusarium oxysporum* f. sp. *nicotianae* on tobacco cultivars, susceptible and resistant to the fungus. The host substrate was found to be most favourable three to four weeks after nematode inoculation. Nutrient-rich galls caused by root-knot nematodes also favour luxuriant growth of other fungi as well as *Rhizoctonia solani*.

Rhizoctonia solani preferred *Meloidogyne hapla*-induced galls on radish, the mycelium accumulated over them showing vigorous growth and abundant sclerotial formation. Extensive necrosis of the galls occurred and roots became obliterated. Non-galled regions of roots did not show sclerotial formation (Khan and Müller, 1982). Galled roots of okra and tomato infected with *M. incognita* were highly susceptible to *R. solani*. Fungal sclerotia were formed only on gall tissues. Prepenetration studies showed that the fungus was specifically attracted to nematode gall tissue and the sclerotia were selectively formed on the galls. *R. solani* responded to stimuli which originated from the nematode-infected roots and passed through semi-permeable cellophane membranes, by producing sclerotia on the membranes just opposite to the galls. The leakage of nutrients from the roots attracted the fungus to the galls and initiated sclerotial development (Golden and Van Gundy, 1975).

Translocatable substances

Evidence for the involvement of translocatable substances or factors in the predisposition of plants to some fungal pathogens was provided by using split-root, double-root and layered plant techniques. Bowman and Bloom (1966) were the first to observe the presence of a translocatable substance or factor. Adopting split-root techniques, one part of the root system of a wilt-resistant tomato plant was inoculated with *Fusarium oxysporum* f. sp. *lycopersici* and the other with *Meloidogyne incognita*. The plant showed wilting even though the nematode and fungus were added to separate root systems. Faulkner *et al.* (1970) used peppermint stem rooted at two places and inoculated one root system with *Verticillium dahliae* and the other with *Pratylenchus minyus*. *Verticillium* wilt showed an increase even when the nematode infected the other root system. Sidhu and Webster (1977) using layered tomato plants demonstrated a mobility of factors that predisposed plants to wilt fungus *F. oxysporum* f. sp. *lycopersici*. Powell (1979) described it as a message of the nematode-infected host sent to the fungus through a messenger, perhaps a growth-regulating substance. The exact nature of the translocatable substances is not yet determined.

According to Webster (1985), giant cells being in a state of high metabolic activity contain most DNA, RNA and photosynthates about three to four weeks after infection, and a greater concentration of sugars during the late stage of infection. Food availability may be a significant factor for predisposition of fungus resistance. Introduction or initiation of additional metabolic products in the host by the nematode induces the fungus to develop in the plant. Substances introduced by the nematode or produced in the roots in response to the nematode are translocated throughout the plant. Recently, Hillocks (1986) showed that on a divided root system of cotton, wilt caused by *Fusarium oxysporum* f. sp. *vasinfectum* was increased by *Meloidogyne incognita* when both organisms were together on the same parts of the root system. However, in a second experiment in which plants growing in *Meloidogyne*-infested soil were stem inoculated with the wilt fungus the nematode increased wilt severity, despite the physical separation of the two organisms.

4.3.5 Breaking of disease resistance

Breaking of disease resistance by plant-parasitic nematodes is used in plant nematology literature to denote predisposition of plants resistant to a given pathogen. Its precise meaning is explained in Chapter 5 of this book. The gene(s) for resistance for a pathogen is rendered ineffective through physiological alterations in the plant caused by the nematode even though the gene(s) remains operational. Resistance and susceptibility of plants to nematodes are related to plant damage and nematode reproduction. Resistant plants therefore tolerate nematode damage and limit nematode reproduction.

Fassuliotis and Rau (1969) postulated two types of resistance of *Fusarium* wilt – qualitative and quantitative. Qualitative resistance is based on incompatibility between the host and the fungal pathogen. Quantitative (polygenic) resistance is influenced by nematode infection because of improved nutrition for the fungus resulting from nematode attack (Pitcher, 1965). Sidhu and Webster (1974) also proclaimed that the physiological predisposition phenomenon has a genetic basis.

Root-knot nematodes are reported to break the monogenic *Fusarium* wilt resistance in tomato cultivars (Jenkins and Coursen, 1957; Sidhu and Webster, 1974, 1977; Liburd, 1977; Webster, 1985). Sidhu and Webster (1974, 1977), based on a genetic study, suggested that this breakdown of resistance is genetically controlled and one parasite induced susceptibility to another parasite. In contrast, there are reports which indicate that *Fusarium* wilt resistance in cultivars of cabbage, peas and other crops, possessing a single dominant gene for resistance against wilt fungi, is not altered by root-knot nematode infection (Fassuliotis and Rau, 1969; Walter, 1965). In tomato as well, some reports demonstrate that monogenic resistance to the wilt fungus is not broken by root-knot nematodes (Binder and Hutchinson, 1959; Jones *et al.*, 1976; Hirano *et al.*, 1979; Abawi and Barker, 1984). Mai and Abawi (1987) acknowledge that the role of root-knot nematodes in the modification or breakdown of the monogenic resistance to *Fusarium* wilt fungi is controversial, and in-depth studies are needed to elucidate the effect and mechanism. In overall analysis, most apparently histophysiological changes in the host caused by nematode infection are responsible for rendering the gene(s) for resistance ineffective and as a result the host is not able to express the resistant reaction.

4.4 NEMATODE–NEMATODE INTERACTIONS

Plant-parasitic nematodes interact with each other ecologically and aetiologically. These two aspects of nematode–nematode interactions are interrelated. Ecological interactions affect the reproduction capacity of the individual nematode populations. Aetiological interactions alter the development of the disease (Eisenback, 1985; Eisenback and Griffin, 1987). Ecological interactions may be beneficial or detrimental to one or all interacting nematodes or there may be no effect on either. Most investigations indicate that nematode–nematode interactions are often antagonistic to at least one of the species.

Competition is the key factor of all the ecological interactions. The interacting nematodes compete mainly for food and space and consequently the nature of competition is either physical or physiological or both. Host suitability, pathogenicity, mode of parasitism, nematode population density and time are important determinants of the competition. These aspects have been adequately dealt with by Eisenback (1985) and Eisenback and Griffin (1987). The importance of host suitability in determining the population build-up of the nematodes in cohabitance has been demonstrated in a number of studies (Johnson and Nusbaum, 1970; Turner and Chapman, 1972; Gay and Bird, 1973; Khan *et al.*, 1986a, 1986b, 1987). The mode of parasitism of the cohabiting nematodes is another important determinant of the nature of competition and resultant effects on the nematode populations. The antagonistic interaction among surface feeders is more intense than between a surface feeder and a subsurface feeder, probably because of their spatially separated feeding sites. Destruction of penetration sites of infective stages of sedentary endoparasites by feeding of ectoparasites leads to suppression of the endoparasites. Sedentary endoparasites suppress ectoparasites by physiological mechanisms (Norton, 1969). In general, competition between nematodes of similar feeding habits is more intense because of similar physical and physiological demands. Mechanical destruction of penetration sites and root tissue, physical occupation of feeding sites or altered host physiology are attributed as mechanisms of ecological interactions between nematodes in cohabitance.

Aetiological interactions between nematode species alter the course of the disease and consequently plant damage (Duncan and Ferris, 1983). In relation to disease intensity this interaction may be negative (antagonistic) or positive (synergistic) or neutral (additive) (Burrows, 1987; Wallace, 1983). Synergistic interaction is rather rare. Intense competition between nematode species is generally antagonistic in relation to disease intensity. Physical damage to the infection sites and physiological changes have been suggested as possible mechanisms of such interactions (Duncan and Ferris, 1983; Eisenback, 1985; Eisenback and Griffin, 1987). Changes in host physiology that facilitate penetration and establishment of the other nematode species is a possible mechanism of positive interactions (Griffin and Waite, 1982). A coherent view on the mechanisms that may explain various conflicting observations is difficult to offer at the moment, though some generalizations are made.

4.5 NEMATODES AND ROOT SYMBIONTS

Interactions of nematodes have been examined with two groups of symbiotic root associates: root nodule bacteria, and mycorrhizae and mycorrhizal fungi. In general, the benefits obtained by the plants through symbiotic associations are adversely affected by the involvement of the nematode as a parasite in this system.

4.5.1 Root nodule bacteria

Root nodule bacteria (*Rhizobium* and *Bradyrhizobium*) and leguminous plants show a mutually beneficial symbiotic relationship presenting a unique system of biological nitrogen fixation. Plant-parasitic nematodes affect this system at various stages from its establishment to efficient functioning. Survival of root nodule bacteria in the rhizosphere and colonization in the rhizoplane are influenced by root exudates. Alteration in root exudates in nematode-infected plants may influence the establishment of rhizobia on legumes (Huang, 1987).This is, however, not always true.

Most investigators have reported that plant nematodes irrespective of their mode of parasitism cause reduced nodulation on leguminous plants, reduction in nodulation being measured by nodule number or nodule mass. Nutrient depletion by the nematodes (Masefield, 1958), competition between nematode juveniles and root nodule bacteria (Epps and Chambers, 1962), devitalization of root tips as in the case of *Trichodorus christiei* (Malek and Jenkins, 1964) and suppression of lateral root formation as in *Rotylenchulus reniformis* infection (Oteifa and Salem, 1972) were suggested as possible causes of reduced nodulation. Huang *et al.* (1984) suggested that reduced nodulation on soya bean resulted from interference of the nematode *Heterodera glycines* with soya bean lectin metabolism. The nematodes reduced the binding of rhizobia to infected roots.

Nematodes also damage root nodules by direct invasion. Species of *Meloidogyne*, *Heterodera*, *Pratylenchus* and *Rotylenchulus reniformis* and *Belonolaimus longicaudatus* invade root nodules on legumes. The damage caused to nodules depends upon the nematode–host combination, and nitrogen-fixing capability varies accordingly. Invasion of nodules by root-knot nematodes is of special interest. *Meloidogyne* spp. induce histological changes in nodular tissue and giant cells develop inside the nodules. Nodules also develop on root galls induced by the nematode. Cyst nematodes *H. trifolii* and *H. glycines* induce development of syncytia in nodules. Infected nodules show abnormal histology and degenerate earlier than healthy ones. Nodular tissue volume and lifespan of nodules are directly correlated with nitrogen-fixing efficiency of the roots. Reduction in size and number of nodules and early degeneration of nodules on nematode-infected leguminous plants are considered as two possible reasons for adverse effects on the nitrogen-fixing capability of the roots. Huang and Barker (1983) found a lower leghaemoglobin (Lb) content of nodules in soya bean plants infected with *H. glycines* race 1 than uninfected plants.

Nodules from nematode-infected plants showed a higher Lbc/Lba ratio, which indicated that nodule development was impaired. Significant reduction in overall Lb content, however, suggested that nodules were senescent. Reduction in nitrogen-fixing efficiency results, therefore, either from reduced number and size of nodules, or their impaired development or early senescence on soya bean infected with *Heterodera glycines*, or all act jointly, and the overall effect is deleterious for the crop.

Chahal and Chahal (1988) found significant reduction in Lb and bacteroid content and nitrogenase activity in chickpea nodules on roots infected with *Meloidogyne incognita*. Verdejo *et al.* (1988), however, did not find any alteration in functioning of nodules on black bean and pea. In the interaction mechanism, root nodule bacterium, legume host and nematode are three important determinants and interaction occurs at the histophysiological level.

Stimulation of nodule formation by nematode infection of legume roots has been recognized. *Meloidogyne incognita*, *M. hapla*, *Pratylenchus penetrans* and *Belonolaimus longicaudatus* stimulated nodulation by *Bradyrhizobium japonicum* on soya bean (Huang, 1987). Root nodulation on pea and black bean was enhanced by *M. incognita* (Verdejo *et al.*, 1988).

Some reports show no apparent effect of nematode infection on root nodulation by rhizobia. Infection of *Meloidogyne javanica* on cowpea (Taha and Kassab, 1980) and of *M. hapla*, *Pratylenchus penetrans* and *Belonolaimus longicaudatus* on peanut (Barker and Hussey, 1976) did not affect root nodulation. Races of *Heterodera glycines* differ in their influence on soya bean root nodulation. The influence of the nematode is also determined by the soya bean cultivar (Hussey and Barker, 1976).

4.5.2 Mycorrhizal fungi

Information on mechanisms of interaction of mycorrhizal fungi and plant-parasitic nematodes is rather fragmentary and suggested mechanisms are speculative, lacking experimental evidence. Interactions of nematodes with mycorrhizal fungi have been determined on the basis of host efficiency (nematode population densities) for the nematode and host sensitivity (growth suppression or yield reduction) to nematodes in the presence or absence of vesicular–arbuscular mycorrhizal (VAM) fungi. Root colonization by mycorrhizal fungi and their sporulation are affected by plant-parasitic nematodes. At the same time, development and reproduction of the nematodes are influenced by VAM fungi. Direct interaction also occurs between sedentary females of endoparasitic nematodes and VAM fungi. Some nematodes feed upon mycorrhizal fungi. As a result of such interactions, plant growth may be increased or decreased, or there may not be any discernible effect on it. The literature, however, shows conflicting results attributable to various factors. Hussey and Roncadori (1982) and Smith (1987) have reviewed well the interactions of VAM fungi with plant-parasitic nematodes.

Destruction of root tissue by migratory endoparasites, nematodes and various transformations induced by sedentary endoparasitic nematodes greatly reduce the root colonization by VAM fungi. Mechanisms of inhibition of root colonization by nematodes are apparently physical alterations of root tissues and biochemical changes within the roots. VAM fungi increase host tolerance to plant-parasitic nematodes. Mechanisms of adverse effects of VAM fungi on nematode parasitism are also physical and biochemical. VAM fungi indirectly affect the host–parasite relationship of nematodes. VAM fungi influence root growth, root exudation, nutrient absorption and physiological response of the hosts (Gerdemann, 1968; Hayman, 1982; Harley and Smith, 1983). Changes in root exudation alter the chemotactic attraction of nematodes to roots, and affect hatching stimulus emanation from roots or initial nematode development within root tissues.

VAM fungi reduce nematode infection and development on a number of hosts. Changes in hormones, amino acids and permeability of cell membranes occur in roots due to mycorrhizal symbiosis (Hayman, 1982). Most likely, VAM fungi inhibit nematodes by altering root physiology (Smith, 1987). According to Smith possibly a factor(s) inherent to mycorrhizal symbiosis transforms mycorrhizal roots unfavourable to nematodes. He speculated about the possibility of production or increase of compounds with nematostatic properties in mycorrhizal roots. Some reports indicate that formation of phenolic compounds and phytoalexins in mycorrhizal roots adversely affect nematode pathogenesis (Sylvia and Sinclair, 1983; Morandi, 1987; Smith, 1987). Some mycorrhizal fungi like *Glomus* sp. occur within the cysts of *Heterodera* and *Globodera* species (Tribe, 1977). The invaded cysts contain chlamydospores of the fungus instead of eggs (Tribe, 1977; Francl and Dropkin, 1985). These capabilities of the VAM fungi may be additionally responsible for inhibition of such nematodes.

Direct competition between VAM fungi and endoparasitic nematodes for space may result in inhibition of the nematode. However, there is insufficient evidence to support this mechanism of nematode inhibition. Root coloniza-tion by VAM fungi leaves sufficient space for nematode infection. More-over, VAM fungi do not colonize the zone of elongation near the root tips, the penetration site of several nematode species. Evidence for competition between VAM fungi and nematodes for host photosynthates is lacking but it may be important in this respect. Both organisms utilize host photosyn-thates and in their combined demands nematodes may be inhibited because nematode activities are negatively affected by overcrowding or lack of infec-tion sites on jointly infected plants (Wallace, 1973; Smith, 1987).

Plant growth and yield of mycorrhizal plants infected with nematodes are increased over non-mycorrhizal plants with nematodes. Mycorrhizal fungi reduce host sensitivity to nematode infection, and nematode development and reproduction are suppressed. Therefore, plant growth is increased because of increased tolerance of plants to nematodes. Mycorrhizal-induced tolerance of plants allows them to grow and yield more even in the presence of nematodes. Improved nutritional status of mycorrhizal plants adds this benefit to the plants in relation to nematode parasitsm. Addition of phos-phate fertilizers did not offer such benefits to the plants (Tylka *et al.*, 1991). The improved nutrition through VAM association indirectly and the antagonistic relationship between nematodes and VAM together perhaps improve plant vigour and yield. In some instances this proposition was not found to be true (Smith and Kaplan, 1988).

4.6 CONCLUSIONS

An understanding of the mechanisms of interactions of plant-parasitic nema-todes with other plant pathogens is of fundamental importance in order to develop an effective management strategy for disease complexes. The plant-parasitic nematode and the other pathogen(s) involved in a given interaction are primary determinants of the operative mechanism. The host plant is another key factor and contributes to it substantially. The mechanism(s) operative in the vector relationship of the nematodes in transmitting plant viruses or some bacterial plant pathogens is basically different from the mechanism(s) in synergistic interactions resulting from association of the nematodes and fungal plant pathogens. Substantial progress has been made to understand the mechanism of interactions of nematodes in their vector relationship with bacteria and viruses. But answers to a host of questions are still lacking. Since most of the studies on synergistic interactions between a fungus and a nematode did not examine the mechanism of interaction, the evidence for mechanisms of such interactions is indirect and speculative. Wilt and root-rot fungi derive benefit from nematode-modified host physiology. This appears to be a general feature of such interactions. The proclaimed translocatable metabolites in some instances of *Meloidogyne-Fusarium* or *Pratylenchus–Verticillium* interactions have not been characterized. Their biochemical nature, mode of action and specificity, if any, need to be determined. The basis of interaction needs to be examined in all such claims. For practical reasons, further research is required on synergistic interactions of root-knot nematodes with wilt or root-rot fungi.

Our knowledge on saprophytic fungi becoming parasites on nematode-infected roots has not advanced further. Research on this significant aspect should elucidate the mechanism of induction of pathogenicity in otherwise saprophytic soil fungi.

Modification of monogenic or polygenic resistance in crop cultivars by nematodes is the most significant aspect of nematode–fungus interactions. The genetic basis of breakdown of resistance by nematodes and the operative mechanism(s) need further research. A significant breakthrough in this area of study is needed because of the considerable damage caused by disease complexes. Concerted and continued research in this fascinating field of nematode interactions should receive priority. Exciting developments may result from future research which may elucidate the essential mechanism(s) of the interactions as well. With this awareness and availability of sophisticated tools and technology, future prospects now look brighter than ever before.

REFERENCES

Abawi, G.S. and Barker, K.R. (1984) Effect of cultivar, soil temperature, and population levels of *Meloidogyne incognita* on root necrosis and Fusarium wilt of tomatoes. *Phytopathology*, **74**, 433–8.

Armstrong, G.M. and Armstrong, J.K. (1981) Formae speciales and races of *Fusarium oxysporum* causing wilt diseases, in *Fusarium: Diseases, Biology and Taxonomy* (eds P.E. Nelson, T.A. Toussoun and R.J. Cook), Pennsylvania State University Press, University Park, pp. 391–9.

Atanasoff, D. (1925) The Dilophospora disease of cereals. *Phytopathology*, **15**, 11–40.

Atkinson, G.F. (1892) Some diseases of cotton. *Alabama Agricultural Experiment Station Bulletin*, **41**, 65 pp.

Barker, K.R. and Hussey, R.S. (1976) Histopathology of nodule tissues of legumes infected with certain nematodes. *Phytopathology*, **66**, 851–5.

Bell, A.A. and Mace, M.E. (1981) Biochemistry and physiology of resistance, in *Fungal Wilt Diseases of Plants* (eds M.E. Mace, A.A. Bell and C.H. Beckman), Academic Press, New York, pp. 431–86.

Bergeson, G.S. (1963) Influence of *Pratylenchus penetrans* alone and in combination with *Verticillium albo-atrum* on growth of peppermint. *Phytopathology*, **53**, 1164–6.

Bergeson, G.B. (1972) Concepts of nematode-fungus associations in plant disease complexes: A review. *Experimental Parasitology*, **32**, 301–14.

Binder, E. and Hutchinson, M.T. (1959) Further studies concerning the effect of the root-knot nematode *Meloidogyne incognita acrita* on the susceptibility of the Chesapeake tomato to Fusarium wilt. *Plant Disease Reporter*, **43**, 972–8.

Bird, A.F. (1972) Quantitative studies on the growth of syncytia induced in plants by root-knot nematodes. *International Journal of Parasitology*, **2**, 157–70.

Bird, A.F. (1981) The *Anguina-Corynebacterium* association, in *Plant Parasitic Nematodes*, Vol. III (eds B.M. Zuckerman and R.A. Rohde), Academic Press, New York, pp. 303–22.

Bird, A.F., Stynes, B.A. and Thomson, W.W. (1980) A comparison of nematode and bacteria-colonized galls induced by *Anguina agrostis* in *Lolium rigidum*. *Phytopathology*, **70**, 1104–9.

Booth, C. (1971) *The Genus Fusarium*, Commonwealth Mycological Institute, Kew.

Booth, C. and Stover, R.H. (1974) *Cylindrocarpon musae* sp. nov., commonly associated with burrowing nematode (*Radopholus similis*) lesions on bananas. *Transactions of the British Mycological Society*, **63**, 503–7.

Bowman, P. and Bloom, J.R. (1966) Breaking of resistance of tomato varieties to Fusarium wilt by *Meloidogyne incognita*. *Phytopathology*, **56**, 87.

Brown, D.J.F., Ploeg, A.T. and Robinson, D.J. (1989) A review of reported associations between *Trichodorus* and *Paratrichodorus* species (Nematoda: Trichodoridae) and tobraviruses with a description of laboratory methods for examining virus transmission by Trichodorids. *Revue de Nematologie*, **12**, 235–41.

Burrows, P.M. (1987) Interaction concepts for analysis of responses of mixtures of nematode populations, in *Vistas on Nematology* (eds J.A. Veech and D.W. Dickson), Society of Nematologists, Hyattsville, pp. 82–93.

Chahal, P.P.K. and Chahal, V.P.S. (1988) Effects of different population levels of *Meloidogyne incognita* on nitrogenase activity, leghaemoglobin and bacteroid contents of chickpea (*Cicer arietinum*) nodules formed by *Rhizobium* spp. *Zentralblatt für Mikrobiologie*, **143**, 63–5.

Conroy, J.J., Green, R.J., Jr and Ferris, J.M. (1972) Interaction of *Verticillium albo-atrum* and the root-lesion nematode *Pratylenchus penetrans* in tomato roots at controlled inoculum densities. *Phytopathology*, **62**, 362–6.

Cook, R.J. and Baker, K.F. (1983) *The Nature and Practice of Biological Control of Plant Pathogens*, American Phytopathological Society, St Paul.

Crosse, J.E. and Pitcher, R.S. (1952) Studies in the relationship of eelworms and bacteria to certain plant diseases 1. The etiology of strawberry cauliflower disease. *Annals of Applied Biology*, **39**, 475–86.

Duncan, L. and Ferris, H. (1983) Validation of a model for prediction of host damage by two nematode species. *Journal of Nematology*, **15**, 227–34.

Eisenback, J.D. (1985) Interactions among concomitant populations of nematodes, in *An Advanced Treatise on Meloidogyne, Vol. I: Biology and Control* (eds J.N. Sasser and C.C. Carter), North Carolina State University Graphics, Raleigh, pp. 193–213.

Eisenback, J.D. and Griffin, G.D. (1987) Interactions with other nematodes, in *Vistas on Nematology* (eds J.A. Veech and D.W. Dickson), Society of Nematologists, Hyattsville, pp. 313–20.

Epps, J.M. and Chambers, A.Y. (1962) Effect of seed inoculation, soil fumigation and cropping sequences on soybean nodulation in soybean-cyst nematode infested soil. *Plant Disease Reporter*, **46**, 48–51.

Evans, K. (1987) The interactions of potato cyst nematodes and *Verticillium dahliae* on early and main crop potato cultivars. *Annals of Applied Biology*, **110**, 329–39.

Fassuliotis, G. and Rau, G.J. (1969) The relationship of *Meloidogyne incognita* to the incidence of cabbage yellows. *Journal of Nematology*, **1**, 219–22.

Faulkner, L.R. and Bolander, W.J. (1969) Interaction of *Verticillium dahliae* and *Pratylenchus minyus* in Verticillium wilt of peppermint: Effects of soil temperature. *Phytopathology*, 59, 868–70.

Faulkner, L.R., Bolander, W.J. and Skotland, C.B. (1970) Interaction of *Verticillium dahliae* and *Pratylenchus minyus* in Verticillium wilt of peppermint: Influence of the nematode as determined by a double root technique. *Phytopathology*, 60, 100–3.

Francl, L.J. and Dropkin, V.H. (1985) *Glomus fasciculatus*, a weak pathogen of *Heterodera glycines*. *Journal of Nematology*, 17, 470–5.

Fukudome, N. and Sakasegawa, Y. (1972) Interaction between root-knot nematode (*Meloidogyne incognita*) on the occurrence of Granville wilt of tobacco. *Proceedings of the Association of Plant Protection, Kyushu*, 18, 100–2.

Gay, C.M. and Bird, G.W. (1973) Influence of concomitant *Pratylenchus branchyurus* and *Meloidogyne* spp. on root penetration and population dynamics. *Journal of Nematology*, 5, 212–17.

Gerdemann, J.W. (1968) Vesicular–arbuscular mycorrhizae and plant growth. *Annual Review of Phytopathology*, 6, 397–418.

Gokte, N. and Swarup, G. (1988) On the association of bacteria with larvae and galls of *Anguina tritici*. *Indian Journal of Nematology*, 18, 313–18.

Golden, J.K. and Van Gundy, S.D. (1975) A disease complex of okra and tomato involving the nematode, *Meloidogyne incognita* and the soil inhabiting fungus, *Rhizoctonia solani*. *Phytopathology*, 65, 265–73.

Griffin, G.D. and Waite, M.W. (1982) Pathological interaction of a combination of *Heterodera schachtii* and *Meloidogyne hapla* on tomato. *Journal of Nematology*, 14, 182–7.

Gupta, P. (1966) Studies on earcockle and 'tundu' disease of wheat. PhD thesis, Indian Agricultural Research Institute, New Delhi.

Gupta, P. and Swarup, G. (1968) On the ear-cockle and yellow ear-rot diseases of wheat. 1. Symptoms and histopathology. *Indian Phytopathology*, 21, 318–23.

Gupta, P. and Swarup, G. (1972) Ear-cockle and yellow ear-rot diseases of wheat. II. Nematode bacterial association. *Nematologica*, 18, 320–4.

Harley, J.L. and Smith, S.E. (1983) *Mycorrhizal Symbiosis*, Academic Press, London.

Harris, K.F. (1981) Arthropod and nematode vectors of plant viruses. *Annual Review of Phytopathology*, 19, 391–426.

Hawn, E.J. (1963) Transmission of bacterial wilt of alfalfa by *Ditylenchus dipsaci* (Kuhn). *Nematologica*, 9, 65–8.

Hawn, E.J (1971) Mode of transmission of *Corynebacterium insidiosum* by *Ditylenchus dipsaci*. *Journal of Nematology*, 3, 420–1.

Hayman, D.S. (1982) The physiology of vesicular–arbuscular endomycorrhizal symbiosis. *Canadian Journal of Botany*, 61, 944–63.

Hide, G.A. and Corbett, D.C.M. (1973) Controlling early death of potatoes caused by *Heterodera rostochiensis* and *Verticillium dahliae*. *Annals of Applied Biology*, 75, 461–2.

Hide, G.A., Corbett, D.C.M. and Evans, K. (1984) Effects of soil treatments and cultivars on 'early dying' disease of potatoes caused by *Globodera rostochiensis* and *Verticillium dahliae*. *Annals of Applied Biology*, 104, 277–89.

Hillocks, R.J. (1986) Localised and systemic effects of root-knot nematode on incidence and severity of Fusarium wilt in cotton. *Nematologica*, 32, 202–8.

Hirano, K., Sugiyama, S. and Iida, W. (1979) Relation of the rhizosphere microflora to the occurrence of Fusarium wilt of tomato under presence of the root-knot nematode, *Meloidogyne incognita* (Kofoid & White) Chitwood. *Japanese Journal of Nematology*, 9, 60–8.

Huang, J.S. (1987) Interactions of nematodes with rhizobia, in *Vistas in Nematology* (eds J.A. Veech and D.W. Dickson), Society of Nematologists, Hyattsville, pp. 301–6.

Huang, J.S. and Barker, K R. (1983) Influence of *Heterodera glycines* on leghemoglobins of soybean nodule. *Phytopathology*, **73**, 1002–4.

Huang, J.S., Barker, K.R. and Van Dyke, C.G. (1984) Suppression of binding between rhizobia and soybean roots by *Heterodera glycines*. *Phytopathology*, **74**, 1381–4.

Hussey, R.S. and Barker, K.R. (1976) Influence of nematodes and light sources on growth and nodulation of soybean. *Journal of Nematology*, **8**, 48–52.

Hussey, R.S. and Roncadori, R.W. (1982) Vesicular–arbuscular mycorrhizae may limit nematode activity and improve plant growth. *Plant Disease*, **66**, 9–14.

Inagaki, H. and Powell, N.T. (1969) Influence of root-lesion nematode on black shank symptom development in flue-cured tobacco. *Phytopathology*, **59**, 1350–5.

Jenkins, W.R. and Coursen, B.W. (1957) The effect of root-knot nematodes, *Meloidogyne incognita acrita* and *M. hapla*, on Fusarium wilt of tomato. *Plant Disease Reporter*, **41**, 182–6.

Jensen, H.J. and Siemer, S.R. (1971) Protection of *Fusarium* and *Verticillium* propagules from selected biocider following ingestion by *Pristionchus lheritieri*. *Journal of Nematology*, **3**, 23–7.

Johnson, A.W. and Nusbaum, C.J. (1970) Interactions between *Meloidogyne incognita*, *M. hapla* and *Pratylenchus brachyurus* in tobacco. *Journal of Nematology*, **2**, 334–40.

Johnson, H.A. and Powell, N.T. (1969) Influence of root-knot nematodes on bacterial wilt development in flue-cured tobacco. *Phytopathology*, **59**, 486–91.

Jones, J.P., Overman, A.J. and Crill, P. (1976) Failure of root-knot nematode to affect Fusarium wilt resistance of tomato. *Phytopathology*, **66**, 1339–41.

Jorgenson, E.C. (1970) Antagonistic interaction of *Heterodera schachtii* Schmidt and *Fusarium oxysporum* (Wall.) on sugarbeets. *Journal of Nematology*, **2**, 393–8.

Kalinenko, V.O. (1936) The inoculation of phytopathogenic microbes into rubber-bearing plants by nematodes. *Phytopathologische Zeitschrift*, **9**, 407–16.

Khan, M.W. and Müller, J. (1982) Interaction between *Rhizoctonia solani* and *Meloidogyne hapla* on radish in gnotobiotic culture. *Libyan Journal of Agriculture*, **11**, 133–40.

Khan, R.M., Khan, A.M. and Khan, M.W. (1986a) Interaction between *Meloidogyne incognita*, *Rotylenchulus reniformis* and *Tylenchorhynchus brassicae* on tomato. *Revue de Nematologie*, **9**, 245–50.

Khan, R.M., Khan, M.W. and Khan, A.M. (1986b) Interactions between *Meloidogyne incognita*, *Rotylenchulus reniformis* and *Tylenchorynchus brassicae* as cohabitants on eggplant. *Nematologia Mediterranea*, **14**, 201–6.

Khan, R.M., Khan, M.W. and Khan, A.M (1987) Competitive interactions between *Meloidogyne incognita*, *Rotylenchulus reniformis* and *Tylenchorhynchus brassicae* on cauliflower. *Pakistan Journal of Nematology*, **5**, 103–6.

Libman, G. and Leach, J.G. (1962) A study of the role of some nematodes in the incidence and severity of southern bacterial wilt of tomato. *Phytopathology*, **52**, 1219.

Liburd, O.W. (1977) Interactions involving root-knot nematodes, *Fusarium oxysporum* f. sp. *lycopersici* and tomatoes. PhD Thesis, Cornell University, Ithaca, New York.

Lucas, G.B., Sasser, J.N. and Kelman, A. (1955) The relationship of root-knot nematodes to Granville wilt resistance in tobacco. *Phytopathology*, **45**, 537–40.

MacHardy, W.E. and Beckman, C.H. (1981) Vascular wilt fusaria: Infection and pathogenesis, in *Fusarium: Diseases, Biology and Taxonomy* (eds P.E. Nelson, T.A. Toussoun and R.J. Cook), Pennsylvania State University Press, University Park, pp. 365–90.

Mai, W.F. and Abawi, G.S. (1987) Interactions among root-knot nematodes and Fusarium wilt fungi on host plants. *Annual Review of Phytopathology*, 25, 317–38.

Malek, R.B. and Jenkins, W.R. (1964) Aspects of the host–parasite relationships of nematodes and hairy vetch. *New Jersey Agricultural Experiment Station Bulletin*, 813, 31 pp.

Masefield, G.B. (1958) Some factors affecting nodulation in the tropics, in *Nutrition of the Legumes* (ed. E.G. Hallsworth), Butterworths Scientific Publications, London, pp. 202–15.

Melendez, P.L. and Powell, N.T. (1967) Histological aspects of the Fusarium wilt–root-knot complex in flue-cured tobacco. *Phytopathology*, 57, 286–92.

Melendez, P.L. and Powell, N.T. (1970) The *Pythium*–root-knot nematode complex in flue-cured tobacco. *Phytopathology*, 60, 1303.

Morandi, D. (1987). VA mycorrhizae, nematodes, phosphorus and phytoalexins, in *Mycorrhizae in the Next Decade: Practical Applications and Research Priorities*, Proceedings of the 7th North American Conference on Mycorrhizae (eds D.M. Sylvia, L.L. Hung and J.H. Graham), Institute of Food and Agricultural Sciences, University of Florida, Gainesville, p. 212.

Moura, R.M. de, Echandi, E. and Powell, N.T. (1975) Interaction of *Corynebacterium michiganense* and *Meloidogyne incognita* on tomato. *Phytopathology*, 65, 1332–5.

Nelson, P.E. (1981) Life cycle and epidemiology of *Fusarium oxysporum*, in *Fungal Wilt Diseases of Plants* (eds M.E. Mace, A.A. Bell and C.H. Beckman), Academic Press, New York, pp. 51–80.

Noguera, G.R. (1982) Alterations in the production of rishistin in roots and tyloses in stems and the interaction of *Meloidogyne–Fusarium* in tomato plants. *Agronomia Tropical*, 32, 303–8.

Norton, D.C. (1969) *Meloidogyne hapla* as a factor in alfalfa decline in Iowa. *Phytopathology*, 59, 1824–8.

Oteifa, B.A. and Salem, A.A. (1972) Biology and histopathogenesis of the reniform nematode, *Rotylenchulus reniformis*, on Egyptian cotton, *Gossypium barbadense*. *Proceedings of the Third Congress of the Mediterranean Phytopathological Union*, Oeiras, Portugal, pp. 299–304.

Pathak, K.N. and Swarup, G. (1984) Incidence of *Corynebacterium michiganense* pv. *tritici* in the ear-cockle nematode (*Anguina tritici*) galls and pathogenicity. *Indian Phytopathology*, 37, 267–70.

Pegg, G.F. (1981) Biochemistry and physiology of pathogenesis, in *Fungal Wilt Disease of Plants* (eds M.E. Mace, A.A. Bell and C.H. Beckman), Academic Press, New York, pp. 193–253.

Perry, D.A. (1963) Interaction of root-knot and Fusarium wilt of cotton. *Empire Cotton Growers Review*, 40, 41–7.

Pitcher, R.S. (1963) The role of plant-parasitic nematodes in bacterial diseases. *Phytopathology*, 53, 35–9.

Pitcher, R.S. (1965) Interrelationships of nematodes and other pathogens of plants. *Helminthological Abstract*, 34, 1–17.

Pitcher, R.S. (1978) Interactions of nematodes with other pathogens, in *Plant Nematology* (ed. J.F. Southey), Her Majesty's Stationery Office, London, pp. 63–77.

Polychronopoulos, A.G., Houston, B.R. and Lownsbery, B.F. (1969) Penetration and development of *Rhizoctonia solani* in sugarbeet seedlings infected with *Heterodera schachtii*. *Phytopathology*, 59, 482–5.

Powell, N.T. (1963) The role of plant-parasitic nematodes in fungus diseases. *Phytopathology*, 53, 28–35.

Powell, N.T. (1971a) Interactions between nematodes and fungi in disease complexes. *Annual Review of Phytopathology*, 9, 253–74.

Powell, N.T. (1971b) Interaction of plant-parasitic nematodes with other disease causing agents, in *Plant Parasitic Nematodes*, Vol. II (eds B.M. Zuckerman, W.F. Mai and R.A. Rohde), Academic Press, New York, pp. 119–36.

Powell, N.T. (1979) Internal synergisms among organisms inducing disease, in *Plant Disease: An Advanced Treatise*, Vol. IV (eds J.G. Horsfall and E.B. Cowling), Academic Press, New York, pp. 113–33.

Powell, N.T. and Batten, C.K. (1967) The influence of *Meloidogyne incognita* on *Rhizoctonia* root-rot of tobacco. *Phytopathology*, 57, 826.

Powell, N.T. and Nusbaum, C.J. (1960) The black shank–root-knot complex in flue-cured tobacco. *Phytopathology*, 50, 899–906.

Powell, N.T., Melendez, P.L. and Batten, C.K. (1971) Disease complexes in tobacco involving *Meloidogyne incognita* and certain soil-borne fungi. *Phytopathology*, 61, 1332–7.

Ray, S.N. (1987) Studies on some aspects of *Anguina tritici* nematode infesting wheat. MSc thesis, Rajendra Agricultural University, Pusa.

Riley, I.T. and McKay, A.C. (1990) Specificity of the adhesion of some plant pathogenic microorganisms to the cuticle of nematodes in the genus *Anguina* (Nematoda: Anguinidae). *Nematologica*, 35, 90–103.

Ross, J.P. (1965) Predisposition of soybeans to *Fusarium* wilt by *Heterodera glycines* and *Meloidogyne incognita*. *Phytopathology*, 55, 361–4.

Schuster, M.L. (1959) Relation of root-knot nematodes and irrigation water to the incidence and dissemination of bacterial wilt of bean. *Plant Disease Reporter*, 43, 27–32.

Sidhu, G.S. and Webster, J.M. (1974) Genetics of resistance in the tomato to root-knot nematode–wilt fungus complex. *Journal of Heredity*, 65, 153–6.

Sidhu, G.S. and Webster, J.M. (1977) Predisposition of tomato to the wilt fungus (*Fusarium oxysporum lycopersici*) by the root-knot nematode (*Meloidogyne incognita*). *Nematologica*, 23, 433–42.

Smith, G.S. (1987) Interactions of nematodes with mycorrhizal fungi, in *Vistas on Nematology* (eds J.A. Veech and D.W. Dickson), Society of Nematologists, Hyattsville, pp. 292–300.

Smith, G.S. and Kaplan, D.T. (1988) Influence of mycorrhizal fungus, phosphorus, and burrowing nematode interactions on growth of rough lemon citrus seedlings. *Journal of Nematology*, 20, 539–44.

Stewart, R.N. and Schindler, A.F. (1956) The effect of some ectoparasitic and endoparasitic nematodes on the expression of bacterial wilt in carnations. *Phytopathology*, 46, 219–22.

Storey, G.W. and Evans, K. (1987) Interactions between *Globodera pallida* juveniles and *Verticillium dahliae* and three potato cultivars, with descriptions of associated histopathologies. *Plant Pathology*, 36, 192–200.

Sylvia, D.M. and Sinclair, W.A. (1983) Phenolic compounds and resistance to fungal pathogens induced in primary roots of Douglas-fir seedlings by the ecto-mycorrhizal fungus *Laccaria laccata*. *Phytopathology*, 73, 390–7.

Taha, A.H.Y. and Kassab, A.S. (1980) Interactions between *Meloidogyne javanica*, *Rotylenchulus reniformis*, and *Rhizobium* sp. on *Vigna sinensis*. *Journal of Nematology*, 12, 57–62.

Taylor, C.E. (1979) *Meloidogyne* interrelationships with microorganisms, in *Root-Knot Nematodes (Meloidogyne Species) Systematics, Biology and Control* (eds F. Lamberti and C.E. Taylor), Academic Press, London, pp. 375–98.

Taylor, C.E. (1990) Nematode interactions with other pathogens. *Annals of Applied Biology*, 116, 405–16.

Taylor, C.E. and Brown, B.J.F. (1981) Nematode–virus interactions, in *Plant Parasitic Nematodes*, Vol. III (eds B.M. Zuckerman and R.A. Rohde), Academic Press, New York, pp. 281–301.

Taylor, C.E. and Robertson, W.M. (1969) The location of raspberry ringspot and

78 *Nematodes and other plant pathogens*

tomato black ring virus in the nematode vector *Longidorus elongatus* (de Man.). *Annals of Applied Biology*, **64**, 43–8.

Taylor, C.E. and Robertson, W.M. (1970) Sites for virus retention in the alimentary tract of the nematode vectors *Xiphinema diversicaudatum* (Micol.) and *X. index* (Thorne & Allen). *Annals of Applied Biology*, **66**, 373–80.

Taylor, C.E. and Robertson, W.M. (1975) Acquisition, retention and transmission of viruses by nematodes, in *Nematode Vectors of Plant Viruses* (eds F. Lamberti, C.E. Taylor and J.W. Seinhorst), Plenum Press, New York, pp. 253–76.

Taylor, C.E. and Robertson, W.M. (1977) Virus vector relationships and mechanisms of transmission. *Proceedings of the American Phytopathological Society*, **4**, 20–9.

Tribe, H.T. (1977) Pathology of cyst-nematodes. *Biological Review*, **52**, 477–507.

Trudgill, D.L., Brown, D.J.F. and Robertson, W.M. (1981) A comparison of the four British vector species of *Longidorus* and *Xiphinema*. *Annals of Applied Biology*, **99**, 63–7.

Turner, D.R. and Chapman, R.A. (1972) Infection of seedling of alfalfa and red clover by concomitant populations of *Meloidogyne incognita* and *Pratylenchus penetrans*. *Journal of Nematology*, **4**, 280–6.

Tylka, G.L., Hussey, R.S. and Roncadori, R.W. (1991) Interactions of vesicular–arbuscular mycorrhizal fungi, phosphorus and *Heterodera glycines* on soybean. *Journal of Nematology*, **23**, 122–33.

Verdejo, S., Green, C.D. and Podder, A. K. (1988) Influence of *Meloidogyne incognita* on nodulation and growth of pea and black bean. *Nematologica*, **34**, 88–97.

Wallace, H.R. (1973) *Nematode Ecology and Plant Disease*, Edward Arnold, London.

Wallace, H.R. (1983) Interactions between nematodes and other factors on plants. *Journal of Nematology*, **15**, 221–7.

Walter, I.C. (1965) Host resistance as it relates to root pathogens and soil microorganisms, in *Ecology of Soil-Borne Plant Pathogens* (eds K.F. Barker and W.C. Snyder), University of California Press, Berkeley, pp. 314–20.

Wang, L.H. (1973) Biochemical and physiological changes in root exudates, xylem sap and cell permeability of tomato plants infected with *Meloidogyne incognita*. PhD thesis, Purdue University, Indiana.

Webster, J.M. (1975) Aspects of the host parasite relationship of plant parasitic nematodes. *Advanced Parasitology*, **13**, 225–50.

Webster, J.M. (1985) Interaction of *Meloidogyne* with fungi on crop plants, in *An Advanced Treatise on Meloidogyne, Vol. I: Biology and Control* (eds J.N. Sasser and C.C. Carter), North Carolina State University Graphics, Raleigh, pp. 183–92.

Westerlund, F.U., Jr, Campbell, R.N. and Simble, K.A. (1974) Fungal root-rot and wilt of chick-pea in California. *Phytopathology*, **64**, 432–6.

5

Interaction of plant-parasitic nematodes with wilt-inducing fungi

L.J. Francl and T.A. Wheeler

Of all the interactions of pathogens with nematodes that are presented in this book, it can be persuasively argued that none are more damaging to crops worldwide as the combined effects of wilt-inducing fungi and plant-parasitic nematodes. The combination of nematode and fungus often results in a synergistic interaction wherein the crop loss is greater than expected from either pathogen alone or an additive effect of the two together. The result for a cultivar sensitive to the interaction can be total crop failure. The problem for the grower is exacerbated by such factors as saprophytic capability, wide host range and long-term survival of the pathogens; thus, productivity of land for what may be a highly valuable crop is impaired for many years.

We detail in this chapter, after a brief introduction to the phenomenon of wilt disease, interactions of *Fusarium oxysporum* with plant-parasitic species of nematodes, primarily *Meloidogyne* spp., and *Verticillium* spp. with *Pratylenchus* spp. and *Globodera* spp. Aspects of these subjects have been reviewed previously (Powell, 1971; Bergeson, 1972; Webster, 1985; Mai and Abawi, 1987; Rowe *et al.*, 1987). Beyond the scope that we have delimited are interactions of wilt-inducing fungi and non-parasitic nematodes. There is also apparently no information available concerning the potential interaction between nematodes and *Ophiostoma* (*Ceratocystis*) *ulmi* and *Ceratocystis fagacearum*, fungi that cause Dutch elm disease and oak wilt. A similar lack of information holds for other species, such as *Phialophora cinerescens*, *Cephalosporium gramineum* and *Fusicoccum amygdali*, which are considered as wilt-inducing fungi.

5.1 WILTING PATHOGENESIS

We follow here the conceptual model of Beckman (1987). Talboys (1964) originally divided the aetiology of wilt disease into primary determinants or prevascular infection stages, and secondary determinants which operate after vascular infection occurs. The determinative phase may now be conceptualized as having rhizosphere, extravascular and xylem components (Fig. 5.1). Microorganisms responsible for wilt-conducive or suppressive soils operate in the rhizosphere and help to determine how extensively the plant becomes colonized. Competing microorganisms and edaphic factors also help to determine the length of time resting structures remain viable, and the saprophytic competence of the pathogen. Fungal propagules germinate in response to root exudates, commonly ingress near the growing point of a root, and gain access to the vascular system via the protostele (MacHardy and Beckman, 1981) or form small colonies in the cortex and traverse the endodermis through natural breaks, wounds or by direct penetration (Huisman and Gerik, 1989). Entrance as well into wounded and exposed mature xylem vessels is probably not uncommon (Pegg, 1989). Plant reaction to invasion occurs within hours and involves a number of physiological and structural alterations. One of the earliest defence mechanisms observed is the production of phytoalexins that coat the surface of the fungi and tylosis that prevents movement of conidia (Mace *et al.*, 1976). In a resistant reaction, the fungus is compartmentalized within a short distance of the site of infection. Other plant mechanisms responsible for lateral isolation, vascular occlusion and fungitoxicity include callose deposition, phenolic syntheses leading to lignification of cell walls, gel and gum formation in the xylem vessel, and possibly enzymes and enzyme inhibitors. That compartmentalization results in reduced vascular functionality is implicit in this model of resistance. Wilting is exhibited when the fungus avoids or overcomes these defences or when the plant ineffectually prepares its defensive posture. Beckman (1987) emphasized differences between resistant and susceptible plants in their rate of forming recognition reactions such as callose deposition. In a susceptible host, fungal conidia passively move in the vascular stream and germinate to pass through vessel perforation plates. The fungus systemically invades the shoot's xylem tissue but is otherwise restricted in its spread. The plant eventually expresses wilting some weeks or months after fungal infection. The actual cause of wilting is still subject to debate. Fungal toxins that affect cell membranes, mechanical plugging of perforation plates or air embolisms that cause vascular occlusion and consequent water deficits, hydrolytic enzymes and plant growth regulator imbalances are working models (Beckman, 1987; Van Alfen, 1989; Pegg, 1989). Based upon the available evidence, not all hypotheses are equally tenable (Puhalla and Howell, 1975) but these models also do not seem to be mutually exclusive. When the disease has run its course, the fungus freely colonizes senescing or dead plant tissue and forms reproductive structures. For *Fusarium oxysporum*, these are chlamydospores or conidia that are capable of colonizing a variety of substrates (Fig. 5.1). *Verticillium dahliae* produces long-lived microsclerotia.

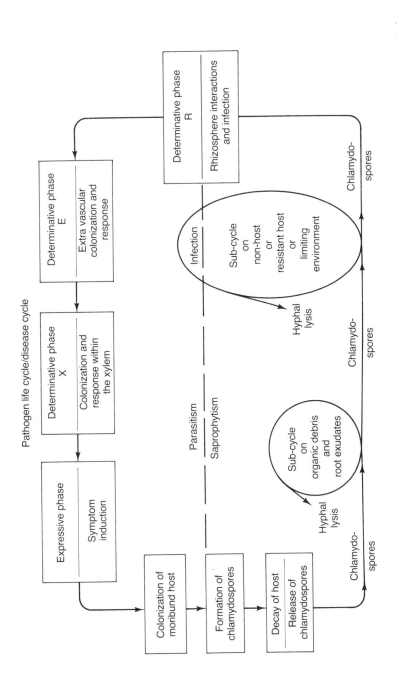

Figure 5.1 Model of disease and life cycles of *Fusarium oxysporum*, courtesy of C.H. Beckman. Determinative phases are divided into rhizosphere (R), extravascular (E) and xylem (X) components.

5.2 STUDIES OF PLANT-PARASITIC NEMATODES AND WILT-INDUCING FUNGI

Research on plant-parasitic nematodes and wilt-inducing fungi has been aimed at: (1) proving an interaction exists; (2) investigating control measures which relate to one or both organisms involved in the interaction; (3) understanding the mechanisms which cause the disease complex; (4) exploring the effects of environmental factors on expression of the interaction; and (5) developing models to predict yield loss primarily as a function of fungal and nematode preplant population density. We discuss these in turn.

5.2.1 Existence of an interaction

Fusarium–*nematodes*

Disease complexes consisting of *Fusarium oxysporum* and plant-parasitic nematodes have had a long history of investigation (Atkinson, 1892). A considerable number of the approximately 70 described *F. oxysporum* formae speciales (Armstrong and Armstrong, 1981) are known to interact with nematodes (Table 5.1); the most studied, though, are *F. oxysporum* f. sp.

Table 5.1 Interactions of *Fusarium oxysporum* and plant-parasitic nematodes that have been reported to increase the intensity of *Fusarium* wilt

Plant host	Nematode	Source
Alfalfa (*Medicago sativa*)	*Ditylenchus dipsaci* *Meloidogyne arenaria* *M. hapla* *M. incognita* *M. javanica*	Griffin (1990) McGuire *et al.* (1958)
Banana (*Musa paradisiaca*)	*M. acrita* *Radopholus similis*	Loos (1959)
Bean (*Phaseolus vulgaris*)	*M. incognita* *M. javanica*	Singh *et al.* (1981) Ribeiro and Ferraz (1983)
Cabbage (*Brassica oleracea*)	*M. acrita*	Fassuliotis and Rau (1969)
Carnation (*Dianthus carophyllus*)	*Meloidogyne* spp. *Criconemella curvata* *M. hapla* *Paratylenchus dianthus*	Schindler *et al.* (1961) Ferraz and Lear (1976)
Chickpea (*Cicer arietinum*)	*M. incognita*	Mani and Sethi (1987)
Chrysanthemum (*Chrysanthemum morifolium*)	*Meloidogyne* spp.	Johnson and Littrell (1969)

Table 5.1   *(cont.)*

Plant host	Nematode	Source
Coffee (*Coffea arabica*)	M. incognita	Negron and Acosta (1989)
Cotton (*Gossypium hirsutum*)	Belonolaimus longicaudatus	Holdeman and Graham (1954)
	Hoplolaimus seinhorsti	Shanmugam et al. (1977)
	M. incognita	Martin et al. (1956)
	Pratylenchus brachyurus	Michell and Powell (1972)
	Rotylenchulus reniformis	Neal (1954)
Cotton (*Gossypium barbadense*)	P. sudanensis	Yassin (1974)
Cowpea (*Vigna sinensis*)	M. javanica	Thomason et al. (1959)
Cucumber (*Cucumis sativus*)	M. incognita	Fritzsche et al. (1983)
Eggplant (*Solanum melongena*)	M. incognita	Smits and Noguera (1982)
Mimosa (*Albizzia julibrissin*)	M. incognita	Gill (1958)
	M. javanica	
Muskmelon (*Cucumis melo*)	M. incognita	Bergeson (1975)
Pea (*Pisum sativum*)	M. acrita	Davis and Jenkins (1963)
	M. incognita	
	P. penetrans	Oyekan and Mitchell (1971)
Soya bean (*Glycine max*)	Heterodera glycines	Ross (1965)
	M. incognita	
Squash (*Cucurbita pepo*)	M. incognita	Caperton et al. (1986)
Tobacco (*Nicotiania tabacum*)	Globodera tabacum	LaMondia and Taylor (1987)
	Meloidogyne spp.	Porter and Powell (1967)
	Tylenchorhynchus claytoni	Holdeman (1956)
Tomato (*Lycopersicon esculentum*)	M. acrita	Jenkins and Coursen (1957)
	M. hapla	
	M. incognita	Harrison and Young (1941)
	M. javanica	
	P. coffeae	Hirano and Kawamura (1972)
	P. penetrans	
Watermelon (*Citrullus lanatus*)	M. arenaria	Sumner and Johnson (1973)
	M. incognita	

lycopersici and *F. oxysporum* f. sp. *vasinfectum*, causative agents of wilt in tomato and cotton.

Verticillium–*nematodes*

Synergistic and additive responses between *Verticillium* spp. and several species of plant-parasitic nematodes have been reported over the years (Table 5.2). The majority of these interactions occur with the root-lesion nematodes, *Pratylenchus* spp., and potato cyst nematodes, *Globodera rostochiensis* and *G. pallida*. Synergistic interactions have been defined in terms of earlier symptom development (Bergeson, 1963; Burpee and Bloom, 1978; Faulkner and Skotland, 1965), higher disease incidence and/or severity (Conroy *et al.*, 1972; Faulkner and Skotland, 1965; Harrison, 1971; McKeen and Mountain, 1960), and lower yields (Francl *et al.*, 1987; MacGuidwin and Rouse, 1990b; Rowe *et al.*, 1985), as compared with the sum effects of the two pathogens separately.

Because of inconsistencies in the data, some scepticism has been expressed concerning the validity of many reports of interactions (Sikora and Carter, 1987). Inconsistencies can in part be explained by an understanding of the infection processes of *Verticillium dahliae*. The primary survival propagule of the fungus is a multicellular microsclerotia which can germinate repeatedly in response to moisture or other environmental stimuli (Farley *et al.*, 1971). Infection of roots by *V. dahliae* requires contact or very close (c.100 μm) placement between a root and fungal microsclerotia (Fitzell *et al.*, 1980). Experimental error when using microsclerotial inoculum of *V. dahliae* is high because the chance of a root making close contact with a microsclerotium follows a probability distribution. Even with high densities of microsclerotia, in some plants the necessary close contact (within a pathozone *sensu* Gilligan, 1985) will not occur at a high frequency. Large numbers of replicate observations are necessary to evince the underlying probability distribution and subsequently quantify an interaction between *V. dahliae* and specific nematode species. Additional experimental error is related to the precision in adding the same dose of viable nematodes to plots and uniform dispersal of nematodes in plots. Many of these same comments apply to research on interactions between *Fusarium oxysporum* and nematodes as well.

5.2.2 Control measures

Resistance

Plant resistance conceivably could have two means of disrupting the fungus–nematode interaction: through its action on the fungus or on the nematode. It is unlikely that a genetic mechanism exists that is specifically directed to resist the interaction. However, since different mechanisms of host resistance are known (Talboys, 1964; Beckman, 1987), it may be

Table 5.2 Greenhouse, field or microplot studies with *Verticillium* spp. and plant-parasitic nematode species resulting in claims of interactions

Plant host	Nematode	Source
Potato (*Solanum tuberosum*)	*Pratylenchus penetrans*	Gould (1973) MacGuidwin and Rouse (1990b) Martin *et al.* (1982) Morsink and Rich (1968) Rowe *et al.* (1985) Kotcon *et al.* (1985)[*]
	P. thornei	Krikun and Orion (1977) Seti *et al.* (1979)
	Rhizoctonia solani + *P. neglectus* or + *Meloidogyne* spp.	Scholte and s'Jacob (1989)
	M. hapla	Jacobsen *et al.* (1979) MacGuidwin and Rouse (1990a)[*]
	Globodera rostochiensis	Corbett and Hide (1971) Evans (1987) Harrison (1971)
	G. pallida	Farley *et al.* (1971)
Tomato (*Lycopersicon esculentum*)	*P. penetrans*	Conroy *et al.* (1972)
	Trichodorus christiei	Conroy and Green (1974)
	G. tabacum	Miller (1975)
	Tylenchorhynchus capitatus	Overman and Jones (1970)
	M. incognita	Overman and Jones (1970) Conroy and Green (1974)[*] Price *et al.* (1980)[*]
American elm (*Ulmus americana*)	*P. penetrans*	Dwinell and Sinclair (1967)
Cotton (*Gossypium hirsutum*)	*M. acrita*	Bazán de Segura and Aguilar (1955) Khoury and Alcorn (1973)
	Rotylenchulus reniformis	Tchatchova and Sikora (1983)
Eggplant (*S. melongena*)	*P. penetrans*	McKeen and Mountain (1960)
Impatiens (*Impatiens balsamina*) (garden balsam)	*P. penetrans*	Müller (1973)
Peppermint (*Mentha piperita*)	*P. neglectus*	Faulkner and Skotland (1965)
	P. penetrans	Bergeson (1963)
Strawberry (*Fragaria ananassa*)	*P. penetrans*	McKinley and Talboys (1979)
Sugar maple (*Acer saccharum*)	*P. penetrans*	Dwinell and Sinclair (1967)

[*] Reports of additive effects.

possible to select for a type of resistance which is effective in disrupting the mechanism(s) responsible for the fungus–nematode interaction. Still, selection of resistance to one organism without the other necessary pathogen for the interaction may lead to a susceptible reaction in a host possessing genes for resistance.

An imprecise term, namely 'resistance breaking' or variants thereof, has been used in the literature to describe the predisposition to wilting of plants that possess gene(s) for wilt resistance when plant-parasitic nematodes interact with wilt-inducing fungi. The term is unfortunate in that some readers may infer that the nematode somehow interferes with the transcription of genes responsible for the defence of the plant; whereas, users of the term seem to imply that the nematode affects the physiology of the plant in some way, rendering it totally or partially incapable of expressing the resistant reaction (i.e. the gene action is rendered ineffective due to disruption at the cellular or tissue level; the gene is *not* made inoperative). We will endeavour to avoid confusion in this respect.

Single dominant genes for resistance to formae speciales of *Fusarium oxysporum* have been identified in many crops but recessiveness and polygenic resistance are also known (Wilhelm, 1981; Goth and Webb, 1981). It is somewhat common for plants possessing polygenic resistance to have that expression overcome by nematodes interacting with fungi (Mai and Abawi, 1987). Those plants having single dominant genes, however, often retain their resistant reaction in the presence of nematodes. Yet there are conflicting results about whether such genes in tomato are defeated in their effects by nematodes. Workers testing different tomato cultivars possessing the *I* gene for *F. oxysporum* resistance reported either no interaction (Jones *et al.*, 1976; Hirano *et al.*, 1979; Binder and Hutchinson, 1959) or increased wilting in the presence of *Meloidogyne* spp. (Jenkins and Coursen, 1957; Sidhu and Webster, 1977a). The reason for these conflicting results may not lie entirely with differences in experimental methodology but may reflect fundamental mechanisms that affect the interaction. Experimentation under identical conditions with the different cultivars used as sources of resistance in these aforementioned studies would do much to explain these conflicting accounts.

Plant breeders typically test germplasm in field soil infested with both *Fusarium oxysporum* and nematodes. However, Starr and Veech (1986) issued a cautionary note showing that field resistance (or possibly field tolerance since soil populations are usually not monitored) to the disease complex may not necessarily mean resistance to *Meloidogyne incognita*. Also, greenhouse tests designed to determine vascular resistance (Fig. 5.1) failed to discriminate cultivars with field resistance to *M. incognita* and *F. oxysporum* f. sp. *vasinfectum*, possibly due to different mechanisms of pathogenesis (Shepherd and Kappelman, 1986).

Resistance in potato to the cyst nematode has been better characterized than resistance to *Pratylenchus* spp. In a pot study, *Globodera rostochiensis*,

G. pallida and *Verticillium dahliae* were placed separately and combined on early and late-maturing potato cultivars with or without the *H1* gene for resistance to the cyst nematode (Evans, 1987). The *H1* gene was not effective in stopping the interaction between the two pathogens, which was defined by earlier symptom development and plant death when both pathogens were present in early-maturing cultivars.

Partial plant resistance to *Verticillium* spp. has been found in some potato cultivars (Akeley *et al.*, 1971). In a microplot study, two resistant and susceptible cultivars were grown in microplots infested with *V. dahliae* and/or *Pratylenchus penetrans* (Rowe, 1989). Yield loss was modelled as a function of *V. dahliae* alone and as the transformed product of the preplant density of both pathogens. Parameter estimates for *V. dahliae* alone were significantly lower for resistant cultivars than susceptible cultivars, indicating that resistance was effective. Parameter estimates for the interaction term were not significantly different between susceptible and *Verticillium*-resistant cultivars (Wheeler and Rowe, unpublished). Again, host resistance to one organism therefore was judged to be ineffective in disrupting the mechanism leading to the synergistic response.

Cultural control

The limited applicability of cultural control has led to shifting agriculture on land infested by wilt organisms. Rotation has had marginal success in controlling wilt disease complexes. There are several aspects of the problem: the slow attrition rate for *Verticillium* and *Fusarium* on fallow land and reproduction on plants considered as non-hosts (Wilhelm, 1955; Huisman and Ashworth, 1976; Huisman and Gerik, 1989); the saprophytic ability of *F. oxysporum*; the survivability of nematodes lacking a host; and host range characteristics of the nematode. *Pratylenchus* spp. have a wide host range, and include many of the grain crops which are commonly used in rotations. The potato cyst nematode has long-term survivability (Van Gundy, 1985), similar to *Verticillium* spp., and has a host range restricted to Solanaceae, in particular the genus *Solanum* (Southey, 1965). *Meloidogyne hapla* and *M. incognita* both have a wide host range and moderate survival characteristics (Van Gundy, 1985). In fields infested with root-knot nematodes and *V. dahliae*, Scholte (1990) found that a five-year rotation of maize–sugar-beet–barley–barley–potato had significantly higher potato yields than continuous potato. Yields, however, were not as high as those found in the five-year rotation.

Chemical control

Fumigation, while not a viable alternative for many parts of the world, is still a cost-effective control measure for soil-borne organisms in high-value cropping systems. Nematicides and soil-applied fungicides can be effective pesticides for reducing the population of one of the pathogens in the system. For example, methylisothiocyanate has been used successfully to reduce populations of soil-borne fungi and alleviate the effects of disease complexes of wilt-inducing fungi and nematodes (Rowe *et al.*, 1987) Beneficial effects rarely last more than one growing season, however. As a case in point, fumigation with methyl bromide resulted in yield increases of 68% for potato fields infested with *Globodera* spp. and *Verticillium dahliae*, while no increases were seen in uninfested fields (Hide and Corbett, 1973). In a field experiment comparing the effects of a fumigant (methyl bromide), nematicide (aldicarb) and fungicide (benomyl) on plots infested with *G. rostochiensis* and *V. dahliae*, yield increased and symptom development was delayed in all treatments for one year (Hide *et al.*, 1984). Methyl bromide reduced population densities of *V. dahliae* for the first year, but benomyl had little influence on microsclerotia density. All the chemical treatments reduced nematode populations for the first year. Control beyond the first year was not achieved with any of the chemical treatments in the study.

Biological and physical control

Little work on joint effects of nematodes and wilt-inducing fungi has been done in this important area. Probably, the reason for this lies in the reductionist scientific approach that focuses on a single target organism. Perhaps investigations are warranted into multiple biocontrol tactics; that is, attempts to reduce simultaneously both wilt-inducing fungus and plant-parasitic nematode to population levels that are economically inconsequential.

5.2.3 Mechanisms of the interactions

Beckman (1989) logically deduced the progression of wilt disease into well-defined spatial and temporal components and his model of fungal and plant processes was used to illustrate some of the possible fungus–nematode interaction mechanisms (Fig. 5.2). The mechanisms responsible for interactions between wilt-inducing fungi and plant-parasitic nematodes, not entirely understood, may occur at more than one time and location, and may differ between migratory and sedentary modes of parasitism and among host–parasite interactions. Research results and suppositions on potential interactive mechanisms are presented in a spatial and temporal sequence.

Rhizosphere

Nematode feeding can slow or temporarily halt root extension (Pitcher, 1967) and increase the quantity and alter the components of plant exudates (Wang and Bergeson, 1974). There is an inverse relationship between distance and spore germination and there is a distance beyond which the

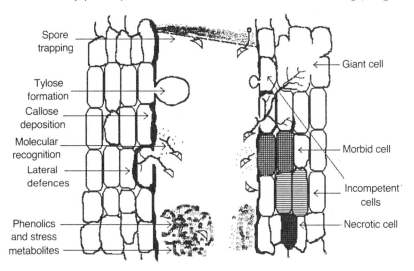

Spore trapping

Tylose formation

Callose deposition

Molecular recognition

Lateral defences

Phenolics and stress metabolites

Giant cell

Morbid cell

Incompetent cells

Necrotic cell

Figure 5.2 Model of host–parasite interactions for wilt-resistant plant (left side) and nematode pathologies that might interfere with resistant responses (right side). Multiple cellular reactions to the presence of a wilt-causing fungus can occur within a three-day period and serve to quickly localize the invasion. Nematodes might cause a number of physiological changes that limit structural responses, reduce the quantity and timing of biochemicals released, and provide additional substrate for the invader so that a systemic invasion proceeds.

propagule is not stimulated to germinate. According to Huisman and Gerik (1989), a propagule has only a few hours to germinate in response to a passing root. If a fungal propagule is stimulated to germinate and the root tip is still in the immediate vicinity due to stunted growth from nematode feeding, then conceivably infection is more likely than if the root tip is able to escape through extension. One can infer from this model a lack of specificity in host–nematode parasitism. Most observational evidence and subsequent research points to a higher degree of specificity than proposed here, a fact that reduces the probable importance of this phenomenon in some interactions. It has been reported, however, that *Belonolaimus* spp., by feeding on root tips, interacted with *Fusarium* to cause wilting of cotton as great as that of *Meloidogyne–Fusarium* (Minton and Minton, 1966).

Propagules of *Fusarium oxysporum* germinate in response to exudates from plant roots, with the root tip as the main source of stimuli (Griffin, 1981). Infective second-stage juveniles of *Meloidogyne* are attracted to the same area. Nematode feeding that causes more exudation would increase the probability of propagule germination and thus increase wilt disease pressure. *M. javanica* infection of tomato was associated with reduced actinomycete population, increased infection and more *Fusarium* spp. propagules in the rhizosphere (Bergeson *et al.*, 1970). The authors suggested that increased root exudation could have stimulated more colony-forming units of *Fusarium* to germinate or that accelerated root decomposition could have released more propagules. Another possibility recognized by Bergeson (1972) is the inhibition of antagonism of *Fusarium* by actinomycetes. This research demonstrates the importance of considering other organisms that may influence the disease complex.

The interaction between *Verticillium* and nematodes has been assumed to include a rhizosphere component but little or no research has been done on the subject. We have found no indication in the literature of a direct linkage between nematode and fungus, such as a species-specific sex pheromone produced by the nematode that could possibly enhance germination of microsclerotia. *V. dahliae* produces microsclerotia that germinate in response to plant exudates (Schreiber and Green, 1963), so we can speculate that similar events occur here as in the case of *Fusarium*–nematode interactions.

Prevascular

Attention on cortical interactions has focused on root wounding by nematodes that may allow entrance for wilt-inducing fungi. Indeed, the wounding hypothesis was often invoked when interactions between nematodes and wilt fungi were first being described. A second possible circumstance in cortical interactions, though, is an improved habitat for fungal extravascular colonization due to nematode injury of cortical tissue. Improved sustenance for the fungus at this stage might allow it to bridge the endodermis layer and gain access to the vascular system. These effects might be expected to be localized in their influence on the plant's physiology, although there is no assurance that this is strictly true.

Wounding by *Meloidogyne* spp. was shown to be unimportant for penetration of cotton roots by *Fusarium oxysporum* f. sp. *vasinfectum* (Perry, 1963) but was thought to be of primary importance in the association of *F. oxysporum* f. sp. *cubense* and the burrowing nematode *Radopholus similis* in banana wilt (Newhall, 1958). *Meloidogyne* juveniles penetrate roots without causing much damage to cortical cells, while *R. similis* individuals cause extensive damage to cortical tissue. Studies of *Fusarium*–*Meloidogyne* interactions that have used the split-root technique and sequential inoculations (Bowman and Bloom, 1966; Sidhu and Webster, 1977a) suggest that wounding, and therefore prevascular interactions, cannot fully explain the increased wilting often experienced. Furthermore, the expression of wilt resistance in the nematode and wilt-resistant cultivar 'Nematex' was unaltered when plants were challenged by *F. oxysporum* f. sp. *lycopersici* and *M. incognita* (Sidhu and Webster, 1977b). Second-stage juveniles are not deterred from entering root tips and therefore wounding resistant plants (Huang, 1986; Herman *et al.*, 1991). Localized damage of root tips does not seem to alter the expression of resistance by tomato to *F. oxysporum*.

Histological studies have been conducted with *Pratylenchus penetrans* and *Globodera pallida*, where wounds in the roots caused by the nematodes were examined for presence of *Verticillium dahliae*. Storey and Evans (1987) found that *G. pallida* assisted the fungus to evade defence mechanisms in the root by opening an invasion channel for the fungus to the vascular system. McKinley and Talboys (1979) also stated that *P. penetrans* appeared to cause local changes in the root cortex which led to hyphal penetration to the vascular system. These two studies have not shown conclusively that a wounding mechanism exists and a physiological one does not. Talboys (1964) emphasized the multiple ways in which plants have developed resistance mechanisms to invasion by vascular fungi, and more than one mechanism probably exists in *Verticillium*–nematode pathosystems as well.

The specificity between *Verticillium dahliae* and *Pratylenchus* spp. potentially can be exploited in determining possible mechanisms of the interaction. Riedel *et al.* (1985) have shown that of three common species of *Pratylenchus* found in Ohio potato fields (*P. crenatus*, *P. penetrans* and *P. scribneri*) only *P. penetrans* consistently interacted with *V. dahliae*. Detailed observations on *P. penetrans* have shown that the nematode enters roots primarily in the zone of root elongation, and migration and feeding seem to be restricted to the cortical cells (Zunke, 1990). Similar studies on *P. crenatus* with video-enhanced contrast microscopy may show important differences in behaviour. Furthermore, if *V. dahliae* and *P. penetrans* or *P. crenatus* were jointly utilized in an experiment, it may be possible to distinguish nematode behaviour that contributes to the interaction from behaviour which occurs generally in species of *Pratylenchus*.

Xylem

The vascular phase of infection by wilt-inducing fungi determines the overall expression of wilting by the plant. Therefore, purported systemic effects are discussed in this section. A select group of hypotheses has guided researchers to this point (Powell, 1971; Mai and Abawi, 1987). First, it has been argued that nematodes increase wilting through augmentation of nutritional requirements of wilt-inducing fungi so that the fungal growth rate is higher and more fungitoxins may be produced. Second, plant defences might be evaded if they are perturbed by nematodes and this effect could be felt systemically if the factors responsible are translocatable. In addition, it can be conjectured that nematode infection combined with compartmentalization of wilt-inducing fungi in the resistant reaction may reduce the capacity of the vascular system sufficiently to cause symptom expression that may be absent if either pathogen is present alone.

Giant cells induced by *Meloidogyne* spp. were shown to be a suitable substrate for *Fusarium oxysporum* growth in several systems (Minton and Minton, 1963; Melendez and Powell, 1967). A nutritional hypothesis is supported by the finding of up to a 700% increase in the amount of free amino acids in a *Meloidogyne*-infected tomato (Sidhu and Webster, 1977b). Jones and Waltz (1972) had previously found that supplemental amino acids increased the wilt incidence in a resistant tomato cultivar. It therefore seems likely that *F. oxysporum* grows significantly better in nutrient-enriched, *Meloidogyne*-infected roots.

Studies have shown that the wilt expression is greatest when *Meloidogyne* inoculation precedes fungal inoculation by two to four weeks (Porter and Powell, 1967; Powell, 1971). The biochemical status of giant cells changes over time after infection and differs among hosts (Bird, 1979), and the two to four-week stage of development coincides with the time when giant cells are highly active metabolically. By this time, the nematode site can be some distance from the root tip, suggesting that a physiological disturbance is responsible for the successful establishment of *Fusarium oxysporum*.

Ultrastructural studies have shown that the organization of giant cells induced by *Meloidogyne javanica* was affected adversely by toxic metabolites of *Fusarium oxysporum* f. sp. *lycopersici* about three weeks after inoculation (Fattah and Webster, 1983; Fattah and Webster, 1989). Cell walls and nuclei degenerated when *Fusarium*-resistant and *Fusarium*-susceptible tomato plants were co-inoculated with both organisms and when *F. oxysporum* culture filtrates were introduced. The capability of fungal filtrates to cause similar changes in giant cells as a live culture supports the notion that a translocatable metabolite deteriorated the giant cells. The nematodes themselves were not visibly affected in this case but other studies have shown that reproduction is reduced (El-Sherif and Elwakil, 1991).

Several questions about the system can be raised. For example, could the giant cell environment provide a refuge from fungitoxins produced in a wilt-resistant plant? Also, would not the enormous anatomical and physiological distortions manifested by giant cells respresent a region where resistance (e.g. tylose formation and lignification) cannot be expressed? The fungus, by marshalling its resources in this area, could invade more of the plant than would otherwise be the case. Finally, is there a difference in plant response between tissues distal versus proximal to the nematode site?

Several researchers have used a double-root (Faulkner *et al.*, 1970) or split-root (Conroy *et al.*, 1972) technique to test whether the interaction was of similar magnitude when *Verticillium* spp. and *Pratylenchus* spp. were placed in the same pot, connected to one-half of a root system (VP/0), or when the two organisms were in separate pots, each attached to one-half of the root system (V/P); i.e. whether the response was systemic or local. The control treatment for the system was a pot with *Verticillium* spp. alone attached to one-half of the root system (V/0). In both examples, the VP/0 combination showed more severe disease incidence and/or symptoms than the V/P combination. However, both VP/0 and V/P showed higher incidence and/or severity of symptoms than the V/0 combination. Whether the differences were statistically significant was difficult to determine. Since there was a synergistic response with both V/P and VP/0 with respect to V/0, it is difficult to proclaim that a physiological mechanism was more important than a wounding mechanism. Rather, the studies point out that more than one mechanism is probably responsible for the synergistic disease response.

5.2.4 Environmental factors

The activities of soil-borne organisms are influenced by abiotic factors in their environment; that is, physicochemical conditions are not neutral in their effect on pathogens and the organisms that interact with them. Much has been done to investigate which factors predominate in their effect on the separate species that interact, but much less has been accomplished to detail conditions when nematodes and wilt fungi occur together. The factors may affect either fungus, nematode or both organisms. A major goal in exploring the effect of environmental factors must be their manipulation to reduce disease incidence and severity.

Fusarium wilt of cotton in North America is associated with sandy, acidic soils, a soil environment favourable for root-knot nematodes. Reduced wilt incidence has been noted after liming. Other races of *F. oxysporum* f. sp. *vasinfectum* occur in the heavy clay, alkaline soils of the Nile valley of Egypt, an environment more suited to *Rotylenchulus reniformis* (Bell, 1989).

A resistant cultivar of Egyptian cotton (*Gossypium barbadense*) had 12% *Fusarium* wilt at 50 mg l^{-1} K, but at 100 mg l^{-1} K wilt was suppressed (Oteifa and Elgindi, 1976). When the reniform nematode was added to the system, wilting was not suppressed until 500 mg l^{-1} K was added. Wilting in a susceptible cultivar was appreciably reduced at 500 mg l^{-1} K without the nematode and 1000 mg l^{-1} with the nematode. Field research from the 1930s (Tisdale and Dick, 1939) supports the results of this controlled study in the case of root-knot nematode and upland cotton (*G. hirsutum*). Apparently, K cannot compensate for the damage caused by very high populations of nematodes. The mechanism of K alleviation of wilting is not known but cotton growers have been successfully using potassium fertilizer and lime along with resistant cultivars to combat *Fusarium* wilt for many years (Bell, 1989).

Nitrate nitrogen has been shown to decrease *Fusarium* wilt, while ammonia nitrogen increases wilt (Spiegel and Netzer, 1984). The former increases soil pH while the latter decreases pH, so this may be inferred to be the pertinent relationship. Soil pH is thought to act indirectly, through its influence of micronutrient availability and/or siderophore bacteria (Louvet, 1989). *F. oxysporum* is also sensitive, however, to the nitrite form of nitrogen and N uptake is inversely related to K uptake. The effect of nematodes on this system has not been determined.

The effects of soil texture on potato early dying was examined for three widely differing soils (Wooster silt loam, Rifle peat and Spinks fine sand) (Francl *et al.*, 1988). No differences were observed in yield losses between the different textures when both *Verticillium dahliae* and *Pratylenchus penetrans* were added to microplots in two experiments.

The effects of soil temperature on incidence and severity of wilt development in peppermint was examined in greenhouse temperature tanks (Faulkner and Bolander, 1969). It was found that optimal soil temperature for symptom development was 24°C for plants inoculated with *Verticillium dahliae* alone and 27°C when both *V. dahliae* and *Pratylenchus neglectus* were present. Interactions between the organisms were observed at all soil temperatures tested (18–30°C).

Soil moisture has the potential to influence the hatch and mobility of plant-parasitic nematodes (Kable and Mai, 1968; Townshend and Webber, 1971; Wallace, 1958; Wheeler, 1990). *Verticillium* has the capability to germinate and grow on roots over a wide range of soil matric tensions (Besri, 1980). Relatively high rainfall (and resultant high soil moisture) early in the growing season, i.e. prior to symptom expression, and high temperatures late in the season were correlated with more severe yield losses in potato early dying over an eight-year period (Francl *et al.*, 1990). Using field microplots infested with *V. dahliae*, Cappaert *et al.* (1990) and Wheeler *et al.* (1991) found that frequent irrigation was associated with greater disease severity and higher early incidence of wilt symptoms than in moderate or low irrigation treatments. Presumably, high soil moisture tension is associated with transient low soil oxygen tension, increased nematode mobility and faster plant growth rates. The relationship of these and other factors to increased disease severity is open to further experimentation.

5.2.5 Yield-loss models

Interest in synergistic interactions between wilt-inducing fungi and nematodes has been stimulated because of severe yield losses which may result when both organisms are present (Rowe *et al.*, 1985; Francl *et al.*, 1987; MacGuidwin and Rouse, 1990b). Potato early dying (*Verticillium dahliae* and *Pratylenchus penetrans*) is an example where much work has been accomplished to characterize yield losses and develop predictive models which may be of use in selecting and implementing management strategies.

Francl *et al.* (1987) proposed a predictive model, which was derived from regression analysis, that related yield losses to preplant levels of *Verticillium dahliae* and the product of the densities of *V. dahliae* and *Pratylenchus penetrans*, with a natural logarithmic transformation. This model assumes that yield loss is proportional to densities of both *V. dahliae* and *P. penetrans*. Wheeler *et al.* (1992), using similar microplot data to Francl *et al.* (1987), proposed a model where yield loss was proportional only to the density of *V. dahliae*, and the addition of the nematode 'turned on' a more severe yield loss curve. Both models fit a collection of data from seven microplot experiments equally well. On average, *P. penetrans* without *V. dahliae* caused no yield reduction; thus the interaction was synergistic. Figure 5.3 demonstrates the average yield loss over five years for potato plants growing in microplots infested with *V. dahliae* and/or *P. penetrans*.

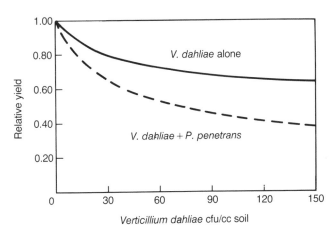

Figure 5.3 Yield–loss relationship for the potato cultivar 'Superior' grown in microplots having soil infested with *Verticillium dahliae* and/or *Pratylenchus penetrans*.

There is a danger of using microplot data for development of prediction relationships and applying them to grower fields. Studies of *Verticillium* wilt of cotton (Brody *et al.*, 1990; Ranney, 1973) have indicated that a negative relationship exists between inoculum density of *Verticillium* spp. and plant density; however, this inverse relationship may differ quantitatively from that found in microplots. In a preliminary test on potatoes, a higher rate of yield loss per colony-forming unit of *V. dahliae* was found when microplots were placed at 90 cm than with a row of microplots spaced at 30 cm (Wheeler, unpublished). Francl *et al.* (1990) added terms to their empirical yield-loss model to correct for a canopy effect.

There is no reported yield-loss model for potato early dying which has been validated under field conditions. However, a large body of data has been accumulated which supports the claim of a synergistic interaction between *Verticillium dahliae* and *Pratylenchus penetrans*. There are only limited data sets which have been published for other *Verticillium*–nematode combinations which can be utilized for yield-loss predictive models. This is an area of importance, particularly as more information is gained on yield increases due to control measures. Regev *et al.* (1990) used dynamic programming to determine the optimal, long-term control measures for *Verticillium* wilt of cotton. Similar methods could be employed for *Verticillium* and plant-parasitic nematodes using existing control strategies.

5.3 CONCLUSIONS

The importance of nematode interactions with wilt-inducing fungi has been acknowledged by the continuation of work to understand and quantify the phenomena. There is still an enormous amount of work that can be done in developing yield-loss models, breeding cultivars resistant to the interaction, understanding environmental factors that modify disease expression and optimizing control strategies. While there are endless possibilities for investigation with respect to field applications, more progress needs to be made on answering fundamental questions about the mechanisms of the interaction. The molecular biologists can produce plant isolines and fungal mutants, which assuredly differ in only one gene, and this opens up new avenues for a rigorous study of pathogenicity. Such a reductionist approach should retain an appreciation for the complexity of the system over time and space and among associated organisms, a view we have tried to promulgate here.

REFERENCES

Akeley, R.V., Murphy, H.J. and Cetas, R.C. (1971) Abnaki: A new high yielding potato variety resistant to Verticillium wilt and leaf roll. *American Potato Journal*, 48, 230–3.

Armstrong, G.M. and Armstrong, J.K. (1981) Formae speciales and races of *Fusarium oxysporum* causing wilt diseases, in *Fusarium: Diseases, Biology, and Taxonomy* (eds P.E. Nelson, T.A. Toussoun and R. J. Cook), Pennsylvania State University Press, University Park, pp. 391–9.

Atkinson, G.F. (1892) Some diseases of cotton. *Bulletin of Alabama Agricultural Experimental Station*, 41, 61–5.

Bazán de Segura, C. and Aguilar, F.P. (1955) Nematodes and root rot diseases of Peruvian cotton. *Plant Disease Reporter*, 39, 12.

Beckman, C.H. (1987) *The Nature of Wilt Diseases of Plants*, American Phytopathological Society Press, St Paul.

Beckman, C.H. (1989) Recognition and response between host and parasite as determinants in resistance and disease development, in *Vascular Wilt Diseases of Plants* (eds E.C. Tjamos and C.H. Beckman), Springer-Verlag, Berlin, pp. 153–62.

Bell, A.A. (1989) Role of nutrition in diseases of cotton, in *Soilborne Plant Pathogens: Management of Diseases with Macro- and Microelements* (ed. A.W. Englehard), American Phytopathological Society Press, St Paul, pp. 167–204.

Bergeson, G.B. (1963) Influence of *Pratylenchus penetrans* alone and in combination with *Verticillium albo-atrum* on growth of peppermint. *Phytopathology*, 53, 1164–6.

Bergeson, G.B. (1972) Concepts of nematode–fungus associations in plant disease complexes: A review. *Experimental Parasitology*, 32, 301–14.

Bergeson, G.B. (1975) The effect of *Meloidogyne incognita* on the resistance of four muskmelon varieties to Fusarium wilt. *Plant Disease Reporter*, 59, 410–13.

Bergeson, G.B., Van Gundy, S.D. and Thomason, I.J. (1970) Effect of *Meloidogyne javanica* on rhizosphere microflora and Fusarium wilt of tomato. *Phytopathology*, 60, 1245–9.

Besri, M. (1980) Effects of water potential changes on growth of *Fusarium oxysporum* f. sp. *lycopersici* and *Verticillium dahliae*. *Phytopathologische Zeitschrift*, 99, 1–9.

Binder, E. and Hutchinson, M.T. (1959) Further studies concerning the effect of the root-knot nematode *Meloidogyne incognita acrita* on the susceptibility of the Chesapeake tomato to Fusarium wilt. *Plant Disease Reporter*, 43, 972–8.

Bird, A.F. (1979) Histopathology and physiology of syncytia, in *Root-Knot-*

Nematodes (Meloidogyne Species): Systematics, Biology and Control (eds F. Lamberti and C.E. Taylor), Academic Press, New York, pp. 155-71.

Bowman, P. and Bloom, J.R. (1966) Breaking the resistance of tomato varieties to Fusarium wilt by *Meloidogyne incognita*. *Phytopathology*, 56, 871 (Abstract).

Brody, A.K., Karban, R. and Schnathorst, W.C. (1990) Inverse relationship between cotton plant density and Verticillium wilt incidence and severity: Evidence for an alternative hypothesis. *Crop Protection*, 9, 174-6.

Burpee, L.L. and Bloom, J.R. (1978) The influence of *Pratylenchus penetrans* on the incidence and severity of Verticillium wilt of potato. *Journal of Nematology*, 10, 95-9.

Caperton, C.M., Martyn, R.D. and Starr, J.L. (1986) Effects of *Fusarium* inoculum density and root-knot nematodes on wilt resistance in summer squash. *Plant Disease*, 70, 207-9.

Cappaert, M.R., Powelson, M.L. and Christensen, N.W. (1990) Soil moisture, genotype, and *Verticillium dahliae* interactions in potato early dying. *Phytopathology*, 80, 971 (Abstract).

Conroy, J. and Green, R.J., Jr (1974) Interactions of the root knot nematode *Meloidogyne incognita* and the stubby root nematode *Trichodorus christiei* with *Verticillium albo-atrum* on tomato at controlled inoculum densities. *Phytopathology*, 64, 1118-21.

Conroy, J.J., Green, R.J., Jr and Ferris, J.M. (1972) Interaction of *Verticillium albo-atrum* and the root lesion nematode, *Pratylenchus penetrans*, in tomato roots at controlled inoculum densities. *Phytopathology*, 62, 362-6.

Corbett, D.C.M. and Hide, G.A. (1971) Interactions between *Heterodera rostochiensis* Woll. and *Verticillium dahliae* Kleb. on potatoes and the effect of CCC on both. *Annals of Applied Biology*, 68, 71-80.

Davis, R.A. and Jenkins, W.R. (1963) Effects of *Meloidogyne* spp. and *Tylenchorhynchus claytoni* on pea wilt incited by *Fusarium oxysporum* f. *pisi* race 1. *Phytopathology*, 53, 745 (Abstract).

Dwinell, L.D. and Sinclair, W.A. (1967) Effects of N, P, K, and inoculum density of *Verticillium dahliae* on populations of *Pratylenchus penetrans* in roots of American elm and sugar maple. *Phytopathology*, 57, 810 (Abstract).

El-Sherif, A.G. and Elwakil, M.A. (1991) Interaction between *Meloidogyne incognita* and *Agrobacterium tumefaciens* or *Fusarium oxysporum* f. sp. *lycopersici* on tomato. *Journal of Nematology*, 23, 239-42.

Evans, K. (1987) The interactions of potato cyst nematodes and *Verticillium dahliae* on early and main crop potato cultivars. *Annals of Applied Biology*, 110, 329-39.

Farley, J.D., Wilhelm, S. and Snyder, W.C. (1971) Repeated germination and sporulation of microsclerotia of *Verticillium albo-atrum* in soil. *Phytopathology*, 61, 260-4.

Fassuliotis, G. and Rau, G.J. (1969) The relationship of *Meloidogyne incognita acrita* to the incident of cabbage yellows. *Journal of Nematology*, 1, 219-22.

Fattah, F.A. and Webster, J.M. (1983) Ultrastructural changes caused by *Fusarium oxysporum* f. sp. *lycopersici* in *Meloidogyne javanica* induced giant cells in *Fusarium* resistant and susceptible tomato cultivars. *Journal of Nematology*, 15, 128-35.

Fattah, F.A. and Webster, J.M. (1989) Ultrastructural modifications of *Meloidogyne javanica* induced giant cells caused by fungal culture filtrates. *Revue de Nematologie*, 12, 197-210.

Faulkner, L.R. and Bolander, W.J. (1969) Interaction of *Verticillium dahliae* and *Pratylenchus minyus* in Verticillium wilt of peppermint: Effect of soil temperature. *Phytopathology*, 59, 868-70.

Faulkner, L.R. and Skotland, C.B. (1965) Interactions of *Verticillium dahliae* and

Pratylenchus minyus in Verticillium wilt of peppermint. *Phytopathology*, 55, 583–6.

Faulkner, L.R., Bolander, W.J. and Skotland, C.B. (1970) Interactions of *Verticillium dahliae* and *Pratylenchus minyus* in Verticillium wilt of peppermint: Influence of the nematode as determined by a double root technique. *Phytopathology*, 60, 100–3.

Ferraz, S. and Lear, B. (1976) Interaction of four plant parasitic nematodes and *Fusarium oxysporum* f. sp. *dianthi* on carnation. *Experientiae*, 22, 272–7.

Fitzell, R., Evans, G. and Fahy, P.C. (1980) Studies on the colonization of plant roots of *Verticillium dahliae* Klebahn with use of immunofluorescent staining. *Australian Journal of Botany*, 28, 357–68.

Francl, L.J., Madden, L.V., Rowe, R.C. and Riedel, R.M. (1987) Potato yield loss prediction and discrimination using preplant population densities of *Verticillium dahliae* and *Pratylenchus penetrans*. *Phytopathology*, 77, 579–84.

Francl, L.J., Rowe, R.C., Riedel, R.M. and Madden, L.V. (1988) Effects of three soil-types on potato early dying disease and associated yield reduction. *Phytopathology*, 78, 159–66.

Francl, L.J., Madden, L.V., Rowe, R.C. and Riedel, R.M. (1990) Correlation of growing season environmental variables and the effect of early dying on potato yield. *Phytopathology*, 80, 425–32.

Fritzsche, R., Pelcz, J., Ottel, G. and Thiele, S. (1983) Interaction between *Meloidogyne incognita* and *Fusarium oxysporum* f. sp. *cucumerinum* in greenhouse cucumbers. *Tagungsbericht Akademie der Landwirschaftwissenschaften der Deutschen Demokratischen Republik*, 261, 685–90.

Gill, D.L. (1958) Effect of root-knot nematodes on Fusarium wilt of mimosa. *Plant Disease Reporter*, 42, 587–90.

Gilligan, C.A. (1985) Probability models for host infection by soilborne fungi. *Phytopathology*, 75, 61–7.

Goth, R.W. and Webb, R.E. (1981) Sources and genetics of host resistance in vegetable crops, in *Fungal Wilt Disease of Plants* (eds M.E. Mace, A.A. Bell and C.H. Beckman), Academic Press, New York, pp. 377–411.

Gould, M.D. (1973) The root lesion nematode–Verticillium wilt disease complex of potatoes and pathogenicity studies of the lesion nematodes on selected ornamental and cover crops. PhD thesis, University of Rhode Island, Kingston, 72 pp.

Griffin, G.D. (1990) Pathological relationship of *Ditylenchus dipsaci* and *Fusarium oxysporum* f. sp. *medicaginis* on alfalfa. *Journal of Nematology*, 22, 333–6.

Griffin, G.J. (1981) Physiology of conidium and chlamydospore germination in *Fusarium*, in *Fusarium: Diseases, Biology and Taxonomy* (eds P.E. Nelson, T.A. Toussoun and R.J. Cook), Pennsylvania State University Press, University Park, pp. 331–9.

Harrison, A.L. and Young, P.A. (1941) Effect of root-knot nematode on tomato wilt. *Phytopathology*, 31, 749–52.

Harrison, J.A.C. (1971) Association between the potato cyst nematode, *Heterodera rostochiensis* Woll., and *Verticillium dahliae* Kleb. in the early-dying disease of potatoes. *Annals of Applied Biology*, 67, 185–93.

Herman, M., Hussey, R.S. and Boerma, H.R. (1991) Penetration and development of *Meioidogyne incognita* on roots of resistant soybean genotypes. *Journal of Nematology*, 23, 155–61.

Hide, G.A. and Corbett, D.C.M. (1973) Controlling early death of potatoes caused by *Heterodera rostochiensis* and *Verticillium dahliae*. *Annals of Applied Biology*, 75, 461–2.

Hide, G.A., Corbett, D.C.M. and Evans, K. (1984) Effects of soil treatments and cultivars on 'early dying' disease of potatoes caused by *Globodera rostochiensis* and *Verticillium dahliae*. *Annals of Applied Biology*, 104, 277–89.

Hirano, K. and Kawamura, T. (1972) Incidence of complex disease caused by *Pratylenchus penetrans* or *P. coffeae* and Fusarium wilt fungus in tomato seedlings. *Technical Bulletin of the Faculty of Horticulture, Chiba University,* **20**, 37–43.

Hirano, K., Sugiyami, S. and Iida, W. (1979) Relation of the rhizosphere microflora to the occurrence of Fusarium wilt of tomato under presence of the root-knot nematode, *Meloidogyne incognita* (Kofoid and White) Chitwood. *Japanese Journal of Nematology,* **9**, 60–8.

Holdeman, Q.L. (1956) The effect of the tobacco stunt nematode on the incidence of Fusarium wilt in flue-cured tobacco. *Phytopathology,* **46**, 129.

Holdeman, Q.L. and Graham, T.W. (1954) Effect of the sting nematode on expression of Fusarium wilt of cotton. *Phytopathology,* **44**, 683–5.

Huang, S.P. (1986) Penetration, development, reproduction, and sex ratio of *Meloidogyne javanica* in three carrot cultivars. *Journal of Nematology,* **18**, 408–12.

Huisman, O.C. and Ashworth, L.J., Jr (1976) Influence of crop rotation on survival of *Verticillium albo-atrum* in soils. *Phytopathology,* **66**, 978–81.

Huisman, O.C. and Gerik, J.S. (1989) Dynamics of colonization of plant roots by *Verticillium dahliae* and other fungi, in *Vascular Wilt Disease of Plants* (eds E.C. Tjamos and C.H. Beckman), Springer-Verlag, Berlin, pp. 1–18.

Jacobsen, B.J., MacDonald, D.H. and Bissonnette, H.L. (1979) Interaction between *Meloidogyne hapla* and *Verticillium albo-atrum* in the Verticillium wilt of potato. *Phytopathology,* **69**, 288–92.

Jenkins, W.R. and Coursen, B.W. (1957) The effect of root-knot nematodes, *Meloidogyne incognita acrita* and *M. hapla,* on Fusarium wilt of tomato. *Plant Disease Reporter,* **41**, 182–6.

Johnson, A.W. and Littrell, R.H. (1969) Effect of *Meloidogyne incognita, M. hapla* and *M. javanica* on the severity of Fusarium wilt on chrysanthemum. *Journal of Nematology,* **1**, 122–5.

Jones, J.P. and Waltz, S.S. (1972) Effect of amino acids on development of Fusarium wilt of resistant and susceptible tomato cultivars. *Proceedings of the 85th Annual Meeting of the Florida State Horticulture Society,* **85**, 148–51.

Jones, J.P., Overman, A.J. and Crill, P. (1976) Failure of root-knot nematode to affect Fusarium wilt resistance of tomato. *Phytopathology,* **66**, 1339–41.

Kable, P.F. and Mai, W.F (1968) Influence of soil moisture on *Pratylenchus penetrans. Nematologica,* **14**, 101–22.

Khoury, F.Y. and Alcorn, S.M. (1973) Effect of *Meloidogyne incognita acrita* on the susceptibility of cotton plants to *Verticillium albo-atrum. Phytopathology,* **63**, 485–90.

Kotcon, J.B., Rouse, D.I. and Mitchell, J.E. (1985) Interactions of *Verticillium dahliae, Colletotrichum coccodes, Rhizoctonia solani,* and *Pratylenchus penetrans* in the early dying syndrome of Russet Burbank potatoes. *Phytopathology,* **75**, 68–74.

Krikun, J. and Orion, D. (1977) Studies of the interaction of *Verticillium dahliae* and *Pratylenchus thornei* on potato. *Phytoparasitica,* **5**, 67 (Abstract).

LaMondia, J.A. and Taylor, G.S. (1987) Influence of the tobacco cyst nematode (*Globodera tabacum*) on Fusarium wilt of Connecticut broadleaf tobacco. *Plant Disease,* **71**, 1129–32.

Loos, C.A. (1959) Symptom expression of Fusarium wilt disease of the Gros Michel banana in the presence of *Radopholus similis* (Cobb, 1893) Thorne, 1949 and *Meloidogyne incognita acrita* Chitwood, 1949. *Proceedings of the Helminthological Society of Washington,* **26**, 103–11.

Louvet, J. (1989) Microbial populations and mechanisms determining soil-suppressiveness to Fusarium wilts, in *Vascular Wilt Diseases of Plants* (eds E.C. Tjamos and C.H. Beckman), Springer-Verlag, Berlin, pp. 367–84.

Mace, M.E., Bell, A.A. and Beckman, C.H. (1976) Histochemistry and identification of disease-induced terpenoid aldehydes in Verticillium-wilt-resistant and -susceptible cottons. *Canadian Journal of Botany*, 54, 2095–9.

MacGuidwin, A.E. and Rouse, D.I. (1990a) Effect of *Meloidogyne hapla*, alone and in combination with subthreshold populations of *Verticillium dahliae*, on disease symptomology and yield of potato. *Phytopathology*, 80, 482–6.

MacGuidwin, A.E. and Rouse, D.I. (1990b) Role of *Pratylenchus penetrans* in the potato early dying disease of Russet Burbank potato. *Phytopathology*, 80, 1077–82.

MacHardy, W.E. and Beckman, C.H. (1981) Vascular wilt Fusaria: Infection and pathogenesis, in *Fusarium: Diseases, Biology and Taxonomy* (eds P.E. Nelson, T.A. Toussoun and R.J. Cook), Pennsylvania State University Press, University Park, pp. 365–90.

Mai, W.F. and Abawi, G.S. (1987) Interactions among root-knot nematodes and Fusarium wilt fungi on host plants. *Annual Review of Phytopathology*, 25, 317–38.

Mani, A. and Sethi, C.L. (1987) Interaction of root-knot nematode, *Meloidogyne incognita* with *Fusarium oxysporum* f. sp. *ciceri* and *F. solani* on chickpea. *Indian Journal of Nematology*, 17, 1–6.

Martin, M.J., Riedel, R.M. and Rowe, R.C. (1982) *Verticillium dahliae* and *Pratylenchus penetrans*: Interactions in the early dying complex of potato in Ohio. *Phytopathology*, 72, 640–4.

Martin, W.J., Newsom, L.D. and Jones, J.E. (1956) Relationship of nematodes to the development of Fusarium wilt in cotton. *Phytopathology*, 46, 285–9.

McGuire, J.M., Walters, H.J. and Slack, D.A. (1958) The relationship of root-knot nematodes to the development of Fusarium wilt in alfalfa. *Phytopathology*, 48, 344 (Abstract).

McKeen, C.D. and Mountain, W.B. (1960) Synergism between *Pratylenchus penetrans* (Cobb) and *Verticillium albo-atrum* R. and B. in eggplant wilt. *Canadian Journal of Botany*, 38, 789–94.

McKinley, R.T. and Talboys, P.W (1979) Effects of *Pratylenchus penetrans* on development of strawberry wilt caused by *Verticillium dahliae*. *Annals of Applied Biology*, 92, 347–57.

Melendéz, P.L. and Powell, N.T. (1967) Histological aspects of Fusarium wilt–root knot complex in flue-cured tobacco. *Phytopathology*, 57, 286–92.

Michell, R.E. and Powell, W.M. (1972) Influence of *Pratylenchus brachyurus* on the incidence of Fusarium wilt of cotton. *Phytopathology*, 62, 336–8.

Miller, P.M. (1975) Effect of the tobacco cyst nematode, *Heterodera tabacum*, on severity of Verticillium and Fusarium wilts of tomato. *Phytopathology*, 65, 81–2.

Minton, N.A. and Minton, E.B. (1963) Infection relationship between *Meloidogyne incognita acrita* and *Fusarium oxysporum* f. *vasinfectum* in cotton. *Phytopathology*, 53, 624 (Abstract).

Minton, N.A. and Minton, E.B. (1966) Effect of root-knot and sting nematodes on expression of Fusarium wilt of cotton in three soils. *Phytopathology*, 56, 319–22.

Morsink, F. and Rich, A.E. (1968) Interactions between *Verticillium albo-atrum* and *Pratylenchus penetrans* in the Verticillium wilt of potatoes. *Phytopathology*, 58, 401 (Abstract).

Müller, J. (1973) Zum Einfluss von *Pratylenchus penetrans* auf die Verticillium-Welke von *Impatiens balsamina*. *Zeitschrift für Pflanzenkrankheiten und Pflanzenschutz*, 80, 295–311.

Neal, D.C. (1954) The reniform nematode and its relationship to the incidence of Fusarium wilt of cotton at Baton Rouge, Louisiana. *Phytopathology*, 44, 447–50.

Negron, J.A. and Acosta, N. (1989) The *Fusarium oxysporum* f. sp. *coffeae-Meloidogyne incognita* complex in 'Bourbon' coffee. *Nematropica*, 19, 161–8.

Newhall, A.G. (1958) The incidence of Panama disease of banana in the presence of the root knot and the burrowing nematodes (*Meloidogyne* and *Radopholus*). *Plant Disease Reporter*, **42**, 853–6.

Oteifa, B.A. and Elgindi, A.Y. (1976) Potassium nutrition of cotton, *Gossypium barbadense* in relation to nematode infection by *Meloidogyne incognita* and *Rotylenchulus reniformis*, in *Fertilizer Use and Plant Health*, International Potash Institute, Bern, Switzerland, pp. 301–6.

Overman, A.J. and Jones, J.P. (1970) Effect of stunt and root knot nematodes on Verticillium wilt of tomato. *Phytopathology*, **60**, 1306 (Abstract).

Oyekan, P.O. and Mitchell, J.E. (1971) Effect of *Pratylenchus penetrans* on the resistance of a pea variety to Fusarium wilt. *Plant Disease Reporter*, **55**, 1032–5.

Pegg, G.F. (1989) Pathogenesis in vascular diseases of plants, in *Vascular Wilt Disease of Plants* (eds E.C. Tjamos and C.H. Beckman), Springer-Verlag, Berlin, pp. 51–94.

Perry, D.A. (1963) Interaction of root-knot and Fusarium wilt of cotton. *Empire Cotton Grower Review*, **40**, 41–7.

Pitcher, R.S. (1967) The host–parasite relations and ecology of *Trichodorus viruliferus* on apple roots, as observed from an underground laboratory. *Nematologica*, **13**, 547–57.

Porter, D.M. and Powell, N.T. (1967) Influence of certain *Meloidogyne* species on Fusarium wilt development in flue-cured tobacco. *Phytopathology*, **57**, 282–5.

Powell, N.T. (1971) Interactions between nematodes and fungi in disease complexes. *Annual Review of Phytopathology*, **9**, 253–74.

Price, T.V., McLeod, R.W. and Sumeghy, J.B. (1980) Studies of the interactions between *Fusarium oxysporum* f. sp. *lycopersici*, *Verticillium dahliae* and *Meloidogyne* spp. in resistant and susceptible tomatoes. *Australian Journal of Agricultural Research*, **31**, 1119–28.

Puhalla, J.E. and Howell, C.R. (1975) Significance of endo-polygalacturonase activity to symptom expression of verticillium wilt in cotton, assessed by the use of mutants of *Verticillium dahliaekleb. Physiological Plant Pathology*, **7**, 147–52.

Ranney, C.D. (1973) Cultural control, in *Verticillium Wilt of Cotton, Proceedings of Work Conference, National Cotton Pathology Research Laboratory, College Station, TX* (ed. C.D. Ranney), USDA Publication No. ARS-S-19, New Orleans, Louisiana, pp. 98–104.

Regev, U., Gutierrez, A.P., DeVay, J.E. and Ellis, C.K. (1990) Optimal strategies for management of Verticillium wilt. *Agricultural Systems*, **33**, 139–52.

Ribeiro, C.A.G. and Ferraz, S. (1983) Studies on the interaction between *Meloidogyne javanica* and *Fusarium oxysporum* f. sp. *phaseoli* on bean. *Fitopatologia Brasileira*, **8**, 439–46.

Riedel, R.M., Rowe, R.C. and Martin, M.J. (1985) Differential interactions of *Pratylenchus crenatus*, *P. penetrans*, and *P. scribneri* with *Verticillium dahliae* in potato early dying disease. *Phytopathology*, **75**, 419–22.

Ross, J.P. (1965) Predisposition of soybeans to Fusarium wilt by *Heterodera glycines* and *Meloidogyne incognita. Phytopathology*, **55**, 361–4.

Rowe, R.C. (1989) Alteration in potato cultivar response to *Verticillium dahliae* by co-infection with *Pratylenchus penetrans. American Potato Journal*, **66**, 542 (Abstract).

Rowe, R.C., Riedel, R.M. and Martin, M.J. (1985) Synergistic interactions between *Verticillium dahliae* and *Pratylenchus penetrans* in potato early dying disease. *Phytopathology*, **75**, 412–18.

Rowe, R.C., Davis, J.R., Powelson, M.L. and Rouse, D.I. (1987) Potato early dying: causal agents and management strategies. *Plant Disease*, **71**, 482–9.

Schindler, A.F., Stewart, R.N. and Semeniuk, P. (1961) A synergistic *Fusarium*-nematode interaction in carnations. *Phytopathology*, **51**, 143–6.

Scholte, K. (1990) Causes of differences in growth pattern, yield and quality of potatoes (*Solanum tuberosum* L.) in short rotations on sandy soil as affected by crop rotation, cultivar and application of granular nematicides. *Potato Research*, **33**, 181–90.

Scholte, K. and s'Jacob, J.J. (1989) Synergistic interactions between *Rhizoctonia solani* Kühn, *Verticillium dahliae* Kleb., *Meloidogyne* spp. and *Pratylenchus neglectus* (Rensch) Chitwood & Oteifa, in potato. *Potato Research*, **32**, 387–95.

Schreiber, L.R. and Green, R.J., Jr (1963) Effect of root exudates on germination of conidia and microsclerotia of *Verticillium albo-atrum* inhibited by the soil fungistatic principle. *Phytopathology*, **53**, 260–4 (Abstract).

Seti, E., Krikun, J., Orion, D. and Cohn, E. (1979) *Verticillium–Pratylenchus* interactions in potato. *Phytoparasitica*, **7**, 42.

Shanmugam, N., Vinayagamurthy, A., Rajendran, G., Muthukrishnan, T.S. and Kanadaswamy, T.K. (1977) Occurrence of nematode fungus interaction on cotton around Coimbatore. *Science and Culture*, **43**, 846–8.

Shepherd, R.L. and Kappelman, A.J. (1986) Cotton resistance to the root knot–Fusarium wilt complex. I. Relation to Fusarium wilt resistance and its implications on breeding for resistance. *Crop Science*, **26**, 228–32.

Sidhu, G. and Webster, J.M. (1977a) Predisposition of tomato to the wilt fungus (*Fusarium oxysporum lycopersici*) by the root-knot nematode (*Meloidogyne incognita*). *Nematologica*, **23**, 436–42.

Sidhu, G.S. and Webster, J.M. (1977b) The use of amino acid fungal auxotrophs to study the predisposition phenomena in the root-knot wilt fungus disease complex of tomato. *Physiological Plant Pathology*, **11**, 117–27.

Sikora, R.A. and Carter, W.W. (1987) Nematode interactions with fungal and bacterial plant pathogens: fact or fantasy, in *Vistas On Nematology* (eds J.A. Veech and D.W. Dickson), Society of Nematologists, Hyattsville, pp. 307–12.

Singh, D.B., Reddy, P.P. and Sharma, S.R. (1981) Effect of root-knot nematode *Meloidogyne incognita* on Fusarium wilt of French beans. *Indian Journal of Nematology*, **11**, 84–5.

Smits, B.G. and Noguera, R. (1982) Effect of *Meloidogyne incognita* on the pathogenicity of different isolates of *Fusarium oxysporum* on brinjal (*Solanum melongena* L.). *Agronomia Tropical*, **32**, 284–90.

Southey, J.F. (1965) Potato root eelworm, in *Plant Nematology* 2nd edn (ed. J.F. Southey), Her Majesty's Stationery Office, London, pp. 171–88.

Spiegel, Y. and Netzer, D. (1984) Effect of nitrogen form at various levels of potassium on the *Meloidogyne–Fusarium* wilt complex in muskmelon. *Plant and Soil*, **81**, 85–92.

Starr, J.L. and Veech, J.A. (1986) Susceptibility to root-knot nematodes in cotton lines resistant to the Fusarium wilt/root-knot complex. *Crop Science*, **26**, 543–6.

Storey, G.W. and Evans, K. (1987) Interactions between *Globodera pallida* juveniles, *Verticillium dahliae* and three potato cultivars, with descriptions of associated histopathologies. *Plant Pathology*, **36**, 192–200.

Sumner, D.R. and Johnson, A.W. (1973) Effect of root-knot nematodes on Fusarium wilt of watermelon. *Phytopathology*, **63**, 857–61.

Talboys, P.W. (1964) A concept of the host–parasite relationship in Verticillium wilt diseases. *Nature*, **202**, 361–4.

Tchatchova, J. and Sikora, R.A. (1983) Alteration in susceptibility of wilt resistant cotton varieties to *Verticillium dahliae* induced by *Rotylenchulus reniformis*. *Zeitschrift für Pflanzenkrankheiten und Pflanzenschutz*, **90**, 232–7.

Thomason, I.J., Erwin, D.C. and Garber, M.J. (1959) The relationship of the root-knot nematode, *Meloidogyne javanica*, to Fusarium wilt of cowpea. *Phytopathology*, **49**, 602–6.

Tisdale, H.B. and Dick, J.B. (1939) The development of wilt in a wilt-resistant and in a wilt-susceptible variety of cotton as affected by N–P–K ratio in fertilizer. *Proceedings of the Soil Science Society of America*, **4**, 333–4 (Abstract).

Townshend, J.L. and Webber, L.R. (1971) Movement of *Pratylenchus penetrans* and the moisture characteristics of three Ontario soils. *Nematologica*, **17**, 47–57.

Van Alfen, N.K. (1989) Reassessment of plant wilt toxins. *Annual Review of Phytopathology*, **27**, 533–50.

Van Gundy, S.D. (1985) Ecology of *Meloidogyne* spp.: Emphasis on environmental factors affecting survival and pathogenicity, in *An Advanced Treatise on Meloidogyne, Vol. I: Biology and Control* (eds J.N. Sasser and C.C. Carter), North Carolina State University Graphics, Raleigh, pp. 177–82.

Wallace, H.R. (1958) Movement of eelworms, I. The influence of pore size and moisture content of the soil on the migration of larvae of the beet eelworm *Heterodera schachtii* Schmidt. *Annals of Applied Biology*, **46**, 74–85.

Wang, E.L.H. and Bergeson, G.B. (1974) Biochemical changes in root exudate and xylem sap of tomato plants infected with *Meloidogyne incognita*. *Journal of Nematology*, **6**, 192–4.

Webster, J.M. (1985) Interactions of *Meloidogyne* with fungi on crop plants, in *Advanced Treatise on Meloidogyne Vol. I: Biology and Control* (eds J.N. Sasser and C.C. Carter), North Carolina State University Graphics, Raleigh, pp. 183–92.

Wheeler, T.A. (1990) The influence of soil moisture on the population dynamics of *Meloidogyne incognita* and the growth of *Nicotiana tabacum*. PhD dissertation, North Carolina State University, Raleigh, NC.

Wheeler, T.A., Rowe, R.C. and Madden, L.V. (1991) Effect of soil moisture on incidence of potato early dying and tuber yield. *Phytopathology*, **81**, 1156 (Abstract).

Wheeler, T.A., Madden, L.V., Rowe, R.C. and Riedel, R.M. (1992) Modeling of yield-loss in potato early dying caused by *Pratylenchus penetrans* and *Verticillium dahliae*. *Journal of Nematology*, **24**, 99–102.

Wilhelm, S. (1955) Longevity of the Verticillium wilt fungus in the laboratory and field. *Phytopathology*, **45**, 180–1.

Wilhelm, S. (1981) Sources and genetics of host resistance in field and fruit crops, in *Fungal Wilt Diseases of Plants* (eds M.E. Mace, A.A. Bell and C.H. Beckman), Academic Press, New York, pp. 299–376.

Yassin, A.M. (1974) Role of *Pratylenchus sudanensis* in the syndrome of cotton wilt with reference to its vertical distribution. *Sudan Agricultural Journal*, **9**, 48–52.

Zunke, U. (1990) Observations of the invasion and endoparasitic behaviour of the root lesion nematode *Pratylenchus penetrans*. *Journal of Nematology*, **22**, 309–20.

6

Interactions of nematodes with root-rot fungi

K. Evans and P.P.J. Haydock

Much of the literature dealing with the interactions of nematodes with root-rot fungi is contradictory, as noted by Sikora and Carter (1987), or deals with only selected aspects of the relationship. Where all combinations of circumstances have been reproduced and reported on, the quality and degree of interaction are often seen to change as conditions are changed. Such results are almost invariably obtained from the artificial environment of a pot experiment or a laboratory-based test system. The real test of whether interactions occur, and how agriculturally significant they are, lies in field experimentation. This can be very difficult, although Scholte and s'Jacob (1989) tried to reproduce natural conditions by using very large containers kept outside and into which nematodes and fungi were introduced and allowed to build up naturally on several potato crops. Such a procedure is very time consuming so most authors use pot tests, or laboratory experiments, despite their failure to reproduce field conditions. However, they have the advantage that conditions can be precisely controlled, and varied, and the development of disease complexes can be carefully monitored.

Such monitoring yields a wide variety of observations, including those of plant performance, nematode reproduction and fungal growth. Since it is the effect of the nematode–fungus association on plant performance that is of prime importance to a grower, the most useful terms in describing the interactions are *synergism*, which is a positive interaction (i.e. increased plant damage), and *antagonism*, which is a negative interaction (i.e. less plant damage). The baseline from which the positive and negative interactions are measured is that amount of plant damage which corresponds to the sum of the individual effects of the nematode and fungus. When the amount of damage is simply additive, it is often said that there is no interaction. Whilst this may be a statistically valid statement, there may well be considerable interaction on the roots, but no net effect of the combination of synergistic and antagonistic effects of the two organisms.

Implicit in this definition of synergism and antagonism is that they depend on the interaction of two (or more) plant pathogens (Agrios, 1969). This is not necessarily the case: Wallace (1983) referred to a discussion of synergism by Dickinson (1979), who defined the phenomenon as 'an association of two or more organisms acting at one time and effecting a change that one only is not able to make'. Wallace points out that this definition is essentially qualitative and is more relevant when one of the two interacting factors is non-pathogenic and in some way increases the disruptive effects of the

Table 6.1 Crops on which interactions between nematodes and root-rot fungi have been reported

Alfalfa	Okra
Arecanut	Pea
Banana	Peach
Betelvine	Peanut
Black pepper	Pineapple
Cardamom	Potato
Carrot	Radish
Chickpea	Rice
Chilli	Sorghum
Chrysanthemum	Soya bean
Coconut	Strawberry
Cotton	Sugar-beet
Cowpea	Sugar-cane
Eggplant	Sweet orange
French bean	Sweet potato
Ginger	Taro
Grapefruit	Tea
Kenaf	Tobacco
Kidney bean	Tomato
Kiwi fruit	Wheat
Lentil	White clover
	White jute
Maize	Yam

pathogen on the plant. Examples include predisposition of plants by nematodes to infection by pathogenic fungi without themselves causing damage
to the plants (Wallace, 1983).

The observation of damage suffered by plants under combined attack by
nematodes and root-rot fungi, and the characterization of the resultant
interaction as synergistic, antagonistic or simply additive, can be done
properly only in a factorial experiment which includes a range of inoculum
densities of both organisms, a range of inoculation times (one organism
before the other and vice versa) and a range of plant ages at the time of
treatment. Also, as Wallace (1983) points out, the environment in which the
plants are kept (and the cultivar used) can affect the results. Few reports in
the literature cover all of these factors adequately and many are based on
subjective observations. Despite this, it is clear that nematodes and root-rot
fungi do interact to a significant extent on many crops. During the literature
survey for this review, at least 45 crops were noted on which such interactions have been recorded (Table 6.1). The genera of nematodes and fungi
involved in the interactions reported on those 45 crops are listed in Table 6.2.
These interactions are more common on tropical crops, with approximately
twice as many tropical as temperate crops in the list (Table 6.1). This is

Table 6.2   Genera of nematodes and fungi involved in disease
complexes on the crops listed in Table 6.1

Nematodes	Fungi
Aphelenchoides	*Acremonium*
Criconemella	*Aspergillus*
Criconemoides	*Bipolaris*
Globodera	*Calonectria*
Heterodera	*Cephalosporium*
Hirschmanniella	*Colletotrichum*
Hoplolaimus	*Curvularia*
Macroposthonia	*Cylindrocarpon*
Meloidogyne	*Cylindrocladium*
Pratylenchus	*Fusarium*
Radopholus	*Macrophomina*
Rotylenchulus	*Nectria*
Rotylenchus	*Penicillium*
Scutellonema	*Phytophthora*
Tylenchulus	*Polyscytalum*
	Pyrenochaeta
	Pythium
	Rhizoctonia
	Sclerotium
	Thielaviopsis
	Trichoderma
	Verticillium
	Xanthomonas

mostly due to the greater range of crops which can be grown in tropical conditions but is also a reflection of the greater amount of damage that soil-borne pathogens can cause on crops grown in warmer soils.

Whatever the basis for the interaction of nematodes and fungi in inducing plant disease, the visible effects can be classified as synergistic, antagonistic or additive. We shall consider each of these reactions in detail.

6.1 ASSOCIATIONS WITH NO INTERACTION ON PLANT DAMAGE

Many papers discuss actual or suspected interactions between nematodes and root-rot fungi. However, not all reports demonstrate interaction (Table 6.3).

Jackson and Minton (1968) investigated the invasion of peanut pods by *Aspergillus flavus* in the presence of *Pratylenchus brachyurus* to establish if there was a relationship between the two organisms in field microplots. In trials from 1965 to 1967, while numbers of *A. flavus* and *A. niger* in the kernels were unaffected by the presence of lesion nematodes, total numbers of all fungi in kernels at pod maturity were increased. The authors concluded that the results showed no interaction between *P. brachyurus* and *A. flavus*.

Abawi and Barker (1984) researched the effects of tomato cultivar, soil temperature and levels of *Meloidogyne incognita* on root necrosis and *Fusarium* wilt. Factorial analysis of data from glasshouse pot trials showed no interaction between low levels of the nematode and *F. oxysporum* irrespective of the resistance status of the tomato plant to the nematode.

Investigating the epidemiology of maize root-rot in South Africa, Chambers (1987) isolated nematodes and fungi from naturally infected field

Table 6.3 Nematode–fungus associations with either no effect from at least one of the organisms or, at the most, an additive (A) effect on plant damage

Crop	Nematode	Fungus	Additive?	Reference
Kidney bean	*Heterodera glycines*	*Fusarium oxysporum*		Abawi and Jacobsen (1984)
Maize	*Pratylenchus* spp.	*Helminthosporium pedicellatum*	A	Chambers (1987)
		Fusarium moniliforme	A	
Peanut	*Pratylenchus brachyurus*	*Aspergillus flavus*		Jackson and Minton (1968)
Soya bean	*H. glycines*	(Stem canker)		Pacumbaba and Tadesse (1990)
Sugar-beet	*Heterodera schachtii*	*Pythium aphanidermatum*	A	Whitney (1974)
Tomato	*Meloidogyne incognita*	*F. oxysporum*	A	Abawi and Barker (1984)
Wheat	*Pratylenchus thornei*	*Bipolaris sorokiniana*	A	Doyle *et al.* (1987)
		Fusarium graminearum	A	

plots. *Helminthosporium pedicellatum*, *Fusarium moniliforme* and *Praty-lenchus* spp. were the most commonly isolated organisms. The numbers of root lesion nematodes were not significantly correlated with fungus frequency or root-rot development. It was suspected that an interaction had occurred but due to the small nematode infestation the effect could not be quantified.

Doyle *et al.* (1987) described the results of six field experiments started in 1978 to investigate the cause of wheat yield decline in New South Wales, Australia. Early experiments showed the presence of *Pratylenchus thornei* in soils where wheat yield decline was a problem. Application of aldicarb improved yield (32–78%) by reducing nematode populations but had no effect on root-rot. Fumigation with methyl bromide had greater effects on yield than aldicarb because it controlled both nematodes and common root-rot, whereas aldicarb only controlled nematodes. Nematodes and root-rot fungi were both implicated in wheat yield decline but interactions were not demonstrated.

In glasshouse trials, Abawi and Jacobsen (1984) showed that there were no interactions between *Heterodera glycines*, *Fusarium* and *Pythium* spp. on kidney bean at a range of *H. glycines* population densities from 1 to 100 eggs cm^{-3} of soil.

It is possible that the number of cases in which no interactions are reported may underestimate the frequency with which they are observed, as such results may be considered less exciting and less worthy of publication.

6.2 ASSOCIATIONS WITH AN ADDITIVE EFFECT ON PLANT DAMAGE

Additive effects occur when the plant damage observed corresponds to the sum of the effects of the nematode and fungus separately (Table 6.3). Additive effects have been reported in glasshouse experiments but in practice they are difficult to distinguish from synergistic interactions, except in carefully controlled factorial trials. Many papers assume the reaction to be synergistic without proof that it is not simply additive.

Doyle *et al.* (1987), in field experiments in Australia, investigated the cause of wheat yield decline. *Pratylenchus thornei* was thought to be involved with *Bipolaris sorokiniana* in causing root-rot but, where observed, the effects were at most additive. Abawi and Barker (1984) showed in glasshouse trials that when the initial population density of *Meloidogyne incognita* on tomato roots was increased there was a striking increase in root necrosis. They suggested that soil-borne organisms such as *Fusarium* spp. found in association with the roots could be exacerbating damage caused by the nematodes, but reported no more than additive effects.

Whitney (1974) used factorial glasshouse experiments to investigate the effects of *Pythium ultimum* and *P. aphanidermatum* with and without *Heterodera schachtii* on root-rot of sugar-beet. One in four yield tests revealed a significant synergistic interaction between *P. ultimum* and *H. schachtii* but the effects of *P. aphanidermatum* and *H. schachtii* on yield in all tests were only additive.

6.3 ASSOCIATIONS WITH A SYNERGISTIC EFFECT ON PLANT DAMAGE

Identification of synergistic effects is not always easy but Table 6.4 lists those associations where synergism has been reasonably proved or, at least, strongly suggested. Table 6.5 lists other associations where synergistic effects usually occur, but sometimes effects are antagonistic or nematode or fungus reproduction is impaired. From a practical point of view, it is a synergistic effect on yield that is important. Frequently, however, yields are not recorded or would be impractical to take or even meaningless because of the system in which the plant has been grown. In these cases, synergistic effects are sometimes reported on another measurement of plant growth and sometimes on disease symptom (usually root-rotting) development. Indeed, the synergistic effects are sometimes restricted to symptom development rather than effects on growth.

6.3.1 Examples

Tables 6.4 and 6.5 show that certain species of nematodes and fungi are more frequently involved in disease interactions than are others. The tables contain no fewer than 17 separate situations in which *Meloidogyne incognita* is involved in disease complexes of the synergistic type, with *M. javanica* involved in nine and *Radopholus similis* in six. Of the fungi, *Rhizoctonia solani* is involved in 18 separate complexes, with *Fusarium oxysporum* and *F. solani* involved in eight and six respectively.

Meloidogyne incognita

The frequency of involvement of nematodes and fungi in disease complexes is reflected in the number of crops on which such complexes are recorded and, as the single most destructive nematode species in the world, it is not surprising that *Meloidogyne incognita* has been so frequently reported in disease complexes.

On chickpea, *Meloidogyne incognita* interacted with *Fusarium oxysporum* to reduce plant weight (Kumar *et al.*, 1988), with greater damage occurring if the nematode infection was established before inoculation of the fungus (Mani and Sethi, 1987), an effect also noted with *F. solani*. *Pythium aphanidermatum* and *Rhizoctonia solani* were both found to interact with *M. incognita* on chilli, causing some loss of nematode resistance in the two cultivars tested (Hasan, 1985), whereas the interaction of *P. debaryanum* with this nematode appeared to be due to a physiological response of the plants to nematode infection, making the roots more susceptible to invasion

Table 6.4 Nematode–fungus associations with a synergistic effect on plant damage

Crop	Nematode	Fungus	Reference
Alfalfa	Various	*Pythium ultimum*	Townshend (1984)
Arecanut	*Radopholus similis*	*Cylindrocarpon obtusisporum*	Sundararaju and Koshy (1987)
Banana	*R. similis*	*Cylindrocarpon musae*	Booth and Stover (1974)
	R. similis	*Fusarium solani*	Stover (1966)
		Rhizoctonia sp.	
Chickpea	*Meloidogyne incognita*	*Fusarium oxysporum*	Kumar *et al.* (1988)
	Meloidogyne javanica	*F. oxysporum*	Upadhyay and Dwivedi (1987)
Chilli	*M. incognita*	*Rhizoctonia solani*	Hasan (1985)
		Pythium aphanidermatum	
Chrysanthemum	*Pratylenchus coffeae*	*R. solani*	Hasan (1988)
Clover	*Heterodera daverti*	*Fusarium avenaceum*	Nordmeyer and Sikora (1983a,b)
Cotton	Various	*R. solani*	Brodie and Cooper (1964)
Cowpea	*M. incognita*	*R. solani*	Varshney *et al.* (1987)
	M. incognita	*R. solani*	Khan and Husain (1989)
	Rotylenchulus reniformis		
Ginger	*M. incognita*	*F. solani*	Doshi and Mathur (1987)
Grapefruit	*R. similis*	*Fusarium* spp.	Feder and Feldmesser (1961)
Maize	*Hoplolaimus aegypti*	*Cephalosporium maydis*	Koura and Satour (1987)
		Fusarium moniliforme	
	Meloidogyne spp.	*Pythium* spp.	Sumner *et al.* (1985)
	Paratrichodorus minor		
	Pratylenchus brachyurus	*F. moniliforme*	Jordaan *et al.* (1987)
Okra	*M. incognita*	*R. solani*	Golden and Van Gundy (1975)

Pea	M. incognita	R. solani	Husain et al. (1985)
	Pratylenchus crenatus	Thielaviopsis basicola	Green et al. (1983)
Peach	Criconemella xenoplax	Various	Nyczepir and Lewis (1984)
Peanut	Macroposthonia ornata	Cylindrocladium spp.	Diomande and Beute (1981a, 1981b)
	Meloidogyne hapla	Pythium myriotylum	Garcia and Mitchell (1975a, 1975b)
	Meloidogyne arenaria	Sclerotium rolfsii	Good et al. (1958)
	P. brachyurus	F. solani	Hamada et al. (1985)
Pepper	M. incognita	Various	Godfrey (1936)
Pineapple	Meloidogyne sp.	R. solani	Grainger and Clark (1963)
Potato	Globodera sp.	R. solani	Dunn and Hughes (1967)
	Globodera sp.	Polyscytalum pustulans	Kotcon et al. (1985)
	Pratylenchus penetrans	R. solani	Khan and Müller (1982)
Radish	M. hapla	Collectotrichum coccodes	
Rice	Hirschmanniella oryzae	R. solani	Gokulapalan and Nair (1983, 1986)
Sorghum	Various	R. solani	
Soya bean	Heterodera glycines	Various	Cuarezma-Teran (1985)
	Hoplolaimus galeatus	F. solani	Roy et al. (1989); Hershman et al. (1990)
	M. incognita	Various	McGawley et al. (1984)
	M. incognita	R. solani	Abu-El-Amayem et al. (1985)
	M. javanica	Various	Dianese et al. (1986)
Strawberry	P. penetrans	Rhizoctonia fragariae	LaMondia and Martin (1989)
Sugar-beet	Heterodera schachtii	P. ultimum	Whitney (1974)
	H. schachtii	R. solani	Polychronopoulos et al. (1969)
Sugar-cane	M. javanica	Curvularia lunata	Khurana and Singh (1971)
Sweet orange	Tylenchulus semipenetrans	F. solani	Van Gundy and Tsao (1963)

Table 6.4 *(cont.)*

Crop	Nematode	Fungus	Reference
Taro	*Hirschmanniella miticausa*	*Corticium solani*	Bridge *et al.* (1983)
Tea	*Pratylenchus loosi*	'Soft-root-rot'	Arulpragasam and Addaickan (1983)
Tobacco	*Meloidogyne* spp.	*Phytophthora parasitica*	Sasser *et al.* (1955)
	M. incognita	*Trichoderma harzianum*	Melendez and Powell (1969)
	P. brachyurus	*P. parasitica*	Inagaki and Powell (1969)
Tomato	*M. incognita*	*F. oxysporum*	Noguera and Smits (1982); Smits and Noguera (1982); Gonzalez (1982); Noguera (1983); Kleineke-Borchers and Wyss (1982); El-Sherif and Elwakil (1991)
	M. incognita	*R. solani*	Golden and Van Gundy (1975); Chahal and Chhabra (1984)
	M. javanica	*Aspergillus niger*	Nath and Kamalwanshi (1989)
		Curvularia trifolii	
		Rhizoctonia bataticola	
		Rhizopus nigricans	
White jute	*M. incognita*	*R. bataticola*	Mishra *et al.* (1988)
Yam	*Scutellonema bradys*	*F. oxysporum*	Goodey (1935)
		R. solani	

Table 6.5 Nematode–fungus associations with a synergistic effect on plant damage but sometimes with a reduced effect on plant damage or an antagonistic effect on nematode or fungus reproduction

Crop	Nematode	Fungus	Reference
Alfalfa	Meloidogyne hapla	Phytophthora megasperma	Griffin et al. (1988)
	M. hapla	Fusarium oxysporum	Griffin and Thyr (1988)
Banana	Radopholus similis	Fusarium moniliforme	Pinochet and Stover (1980)
		Cylindrocarpon musae Acremonium stromaticum	
Chickpea	Meloidogyne incognita	F. oxysporum Fusarium solani	Mani and Sethi (1987)
	Meloidogyne javanica	F. oxysporum	Goel and Gupta (1986a)
	M. javanica	Rhizoctonia bataticola	Goel and Gupta (1986b)
Chrysanthemum	Pratylenchus coffeae	Pythium aphanidermatum	Hasan (1988)
Coconut	R. similis	Cylindrocarpon effusum	Koshy and Sosamma (1987)
Ginger	M. incognita	P. aphanidermatum	Doshi and Mathur (1987)
	M. incognita	Pythium myriotylum	Lanjewar and Shukla (1985)
Lentil	M. javanica	Macrophomina phaseolina	Tiyagi et al. (1988)
Peanut	Meloidogyne arenaria	Aspergillus flavus	Patel et al. (1986)
Rice	Heterodera oryzicola	Sclerotium rolfsii	Jayaprakash and Rao (1984)
Soya bean	Heterodera glycines	Calonectria crotalariae	Overstreet et al. (1990)
Sugar-cane	Hoplolaimus indicus	F. moniliforme	Nath et al. (1976)
	Pratylenchus zeae	Pythium graminicola	Santo and Holtzmann (1970)
Tomato	Globodera sp.	Pyrenochaeta lycopersici	Roy (1968)
	M. javanica	Fusarium sp. Rhizoctonia sp.	Nath et al. (1984)

by the fungus (Brodie and Cooper, 1964). Inoculation of cowpea with *M. incognita* and *R. solani* led to breakdown of resistance to both organisms (Khan and Husain, 1989), and the greatest decrease in plant dry weight occurred if the nematode was inoculated two weeks before the fungus (Varshney *et al.*, 1987). Eggplant was not susceptible to *F. oxysporum* unless *M. incognita* was also present (Smits and Noguera, 1982). Although ginger was susceptible to root-rotting caused by *P. aphanidermatum*, or by *F. solani*, prior infection with *M. incognita* appeared to prevent rotting caused by *P. aphanidermatum* whereas the rotting caused by *F. solani* became more severe (Doshi and Mathur, 1987). Rotting of ginger roots by *P. myriotylum* was equally severe in the presence or absence of *M. incognita* but the presence of the fungus decreased nematode reproduction (Lanjewar and Shukla, 1985). Several species of root-rot fungi seemed to interact with *M. incognita* (and *M. hapla*) on maize, and soil fumigation usually decreased disease severity (Sumner *et al.*, 1985). This was also reported by Golden and Van Gundy (1975) when they fumigated soil containing *M. incognita* and *R. solani* and in which okra or tomato plants were grown. Peas were also damaged by *M. incognita* or *R. solani* but plant growth was depressed even further when plants were inoculated with both organisms, with the maximum effect occurring when the two were inoculated simultaneously (Husain *et al.*, 1985). Abu-El-Amayem *et al.* (1985) showed an interaction of the same two organisms on soya beans, whilst Dianese *et al.* (1986) suggested that *M. incognita* may interact with *Cylindrocladium clavatum* to cause root disease of soya bean. On tobacco, *M. incognita* infection allowed *Phytophthora parasitica* to colonize roots earlier and more extensively than in nematode-free or mechanically wounded plants (Sasser *et al.*, 1955). Tomatoes have frequently been reported to suffer the effects of disease complexes, with *M. incognita* interacting with *F. oxysporum* (Gonzalez, 1982; Kleineke-Borchers and Wyss, 1982; Noguera and Smits, 1982; El-Sherif and Elwakil, 1991) and with *R. solani* (Chahal and Chhabra, 1984; Hasan and Khan, 1985). On white jute plants, combined inoculation of *M. incognita* and *R. bataticola* markedly increased the incidence of root-rot over that found with either organism alone (Mishra *et al.*, 1988).

Other species of Meloidogyne

Meloidogyne hapla appeared to exacerbate the loss of alfalfa seedlings in soil contaminated with *Pythium ultimum* (Townshend, 1984), and to be essential for the infection of alfalfa by *Fusarium oxysporum*, although this fungus also appeared to suppress nematode reproduction (Griffin and Thyr, 1988). Root-rot of alfalfa caused by *Phytophthora megasperma* was also made more severe by *M. hapla*, especially when inoculation of the nematode preceded that of the fungus (Griffin *et al.*, 1988). Inoculation of chickpea plants with *Meloidogyne javanica* before adding *F. oxysporum* caused maximum root damage by decreasing host plant resistance to the fungus (Upadhyay and Dwivedi, 1987), a result also found by Goel and Gupta (1986a) and for *M. javanica* and *Rhizoctonia bataticola* on chickpea (Goel and Gupta, 1986b). *Macrophomina phaseolina* interacted with *Meloidogyne javanica* on lentil to cause most damage when the two organisms were inoculated simultaneously but the fungus inhibited nematode reproduction, especially when it was inoculated first (Tiyagi *et al.*, 1988). *Meloidogyne arenaria* interacted with *Aspergillus flavus* to cause increased root disease of peanut and this was accompanied by reduced nematode multiplication (Patel *et al.*, 1986). This species of nematode also interacted synergistically with *Pythium myriotylum* on peanut (Garcia and Mitchell, 1975a). *Cylindrocladium crotalariae* causes black rot of peanut roots and severity of this disease was much enhanced by the presence of M. *hapla*, even on Cylindro-cladium black rot-resistant roots (Diomande and Beute, 1981a), but *Macroposthonia ornata* was less effective at predisposing peanut roots to infection by this fungus (Diomande and Beute, 1981b). Khan and Müller (1982) used gnotobiotic culture to investigate interactions between *R. solani* and *Meloidogyne hapla* on radish and showed that prior infection of roots by the nematode helped the fungus colonize the roots. On sugar-cane, rotting of the root system was increased when both *Curvularia lunata* and M. *javanica* were present, with greatest damage occurring when they were inoculated together but with significant damage even when the inoculations were ten days apart (Khurana and Singh, 1971). M. *javanica* inoculated at sowing time along with three test fungi increased the percentage of damping-off of tomato seedlings, especially with *R. solani*; *P. debaryanum* was the most serious pathogen in its own right (Nath *et al.*, 1984). *Meloidogyne javanica* also interacted with fungi not usually thought of as pathogenic, such that *Rhizopus nigricans* and *Penicillium digitatum*, in the presence of the nematode, caused significant root necrosis (Nath and Kamalwanshi, 1989).

Root lesion nematodes (Pratylenchus *spp.*)

The amounts of rotting caused in chrysanthemum roots by *Pythium aphanidermatum* and *Rhizoctonia solani* were increased in the presence of *Pratylenchus coffeae*, and were further increased when plants were attacked by all three organisms (Hasan, 1988). Nematode reproduction, however, was decreased in the presence of *Pythium aphanidermatum* and increased in the presence of *R. solani*, such that when both fungi were present nematode reproduction was essentially unaffected. A combination of *Pratylenchus brachyurus* and *Pratylenchus zeae* interacted with the root-rot fungus, *Fusarium moniliforme*, on maize to cause more severe effects on plant growth than from nematodes or fungus alone (Jordaan *et al.*, 1987). The minor pathogen, *Thielaviopsis basicola*, assisted, and was assisted by, *Pratylenchus crenatus* in penetrating pea roots (Green *et al.*, 1983). *Pratylenchus penetrans* increased the infection levels of *Colletotrichum coccodes* and *R. solani* on Russet Burbank potato roots but the interactions had no effects on yields, even in the presence of the wilt fungus *Verticillium dahliae* (Kotcon *et al.*, 1985). Black root-rot of strawberries, caused by *Rhizoctonia* spp., was also exacerbated by *Pratylenchus penetrans*, more so at higher nematode inoculum levels (LaMondia and Martin, 1989). Unknown disease agents, believed to be a fungus or fungi, cause a soft root-rot in tea bushes under certain conditions and the disease seems to depend on *Pratylenchus loosi* as a predisposing factor (Arulpragasam and Addaickan, 1983). The development of black shank disease in tobacco, following invasion of the roots by *Phytophthora parasitica*, was enhanced by inoculation with *Pratylenchus brachyurus* up to one week before fungus inoculation. Addition of nematodes three weeks before fungal inoculation delayed symptom appearance whereas simple mechanical wounding of roots enhanced symptom development (Inagaki and Powell, 1969).

Radopholus similis

Cylindrocarpon obtusisporum caused no appreciable damage to arecanut seedlings on its own but, when introduced three weeks after *Radopholus similis*, root lesions were more extensive than those caused by the nematode alone (Sundararaju and Koshy, 1987). However, the fungus inhibited multiplication of the nematode so root lesions and effects on plant growth were actually less except when the nematode was introduced three weeks prior to the fungus. Stover (1966) suggested that *Fusarium solani* and *Rhizoctonia* sp. may contribute to the extension of lesions in banana roots caused by *R. similis* and even to the death of roots, whilst Pinochet and Stover (1980) found *F. solani*, *F. moniliforme* and *C. musae* to be commonly associated with banana root lesions. Experimentally, they also found that *Acremonium stromaticum* caused little damage by itself but was destructive to roots in the presence of *R. similis*. All of these fungi were pathogenic in wounded banana rhizomes, with *C. musae* the most pathogenic. However, it seemed not to be pathogenic unless the rhizomes were wounded or first invaded by *R. similis* (Booth and Stover, 1974). Coconut seedlings were also severely damaged by *R. similis* and, although the nematode seemed to be necessary for substantial invasion by *C. effusum* to occur, the damage to the plant from the two organisms was not significantly greater than from the nematode alone. Nematode reproduction, however, was decreased in the presence of the fungus (Koshy and Sosamma, 1987). In contrast, combined infection of grapefruit seedlings by *R. similis* and *Fusarium* spp. resulted in more severe damage than from the nematode or fungi alone (Feder and Feldmesser, 1961).

Cyst nematodes

Despite the similarity of life cycles of cyst and root-knot nematodes, there are many fewer records of interactions of cyst nematodes with fungal pathogens. Dunn and Hughes (1964) showed interactions between *Globodera rostochiensis* and *Rhizoctonia solani* or *Colletotrichum atramentarium* on tomatoes. The severity of the interaction between *Heterodera daverti* and *Fusarium oxysporum* on subterranean clover (Nordmeyer and Sikora, 1983a) depended greatly on the inoculation sequence of the two organisms. If the nematode was added before the fungus, the yield loss was less than the additive effects of the individual pathogens; simultaneous inoculation and prior inoculation of the fungus gave a greater than additive (i.e. synergistic) effect. *H. glycines* has been reported to interact with *F. solani* (Roy et al., 1989; Hershman et al., 1990) and *Calonectria crotalariae* (Overstreet et al., 1990) on soya beans. Seedling blight of rice is caused by *Sclerotium rolfsii* but greater seedling mortality occurs if the plants are first inoculated with *H. oryzicola*. Addition of the fungus followed by the nematode also causes more plant mortality than the fungus alone, but considerably fewer nematodes invade the roots in this case (Jayaprakash and Rao, 1984).

Other nematodes

Criconemella xenoplax was associated with *Fusarium* spp. and *Pythium* spp. on roots of peach trees in orchards with a history of peach tree short life, and may have been involved in an interaction (Nyczepir and Lewis, 1984). A possible interaction occurred with *Hirschmanniella miticausa* and *Corticum solani* causing 'mitimiti' disease of taro, reported by Bridge *et al.* (1983) Another species, *Hirschmanniella oryzae*, interacted with *Rhizoctonia solani* to intensify the amount of sheath blight of rice (Gokulapalan and Nair, 1983, 1986). Several species of *Hoplolaimus* have been reported in disease complexes: *Hoplolaimus aegypti* with *Cephalosporium maydis* and *F. moniliforme* on maize (Koura and Satour, 1987), *Hoplolaimus galeatus* with *Rhizoctonia*, *Fusarium* and *Macrophomina* spp. on soya bean (McGawley *et al.*, 1984) and *Hoplolaimus indicus* and *F. moniliforme* on sugar-cane (Nath *et al.*, 1976).

6.3.2 Mechanisms of synergistic interactions

The spectrum of interactions extends from those where fungi that are normally non-pathogenic become pathogenic to plants in the presence of nematodes (e.g. *Aspergillus niger/Meloidogyne javanica* on tomatoes; Nath and Kamalwanshi, 1989), to those where an already severe pathogen becomes even more damaging in the presence of the nematode (e.g. *Fusarium solani/Heterodera glycines* on soya beans; Roy *et al.*, 1989). Split-root techniques have been used to demonstrate systemic, physiological effects in complex diseases involving fungi and nematodes, but whilst some of these effects have involved fungi capable of causing root-rotting, the fungi have usually also been vascular parasites causing wilt diseases and therefore are beyond the remit of this chapter.

When attacked by root-rot fungi, plant root systems become less extensive and efficient and the crop is placed under stresses which can be further exacerbated by environmental conditions. For example, the soft root-rot of tea, apparently dependent on predisposition by feeding of *Pratylenchus loosi* on roots, was first noticed in plantations which were recovering from pruning and which were subjected to a long dry spell of weather (Arulpragasam and Addaickan, 1983). In contrast, Hershman *et al.* (1990) reported that soya bean sudden death syndrome, involving *Heterodera glycines* and *Fusarium solani*, was more likely to develop in conditions of high soil moisture, presumably because such conditions aided the infection process. Leguminous crops are also likely to suffer additionally when the degree of colonization by nitrogen-fixing *Rhizobium* is reduced, for example on peas (Husain *et al.*, 1985).

The roles of nematodes in disease complexes listed by Bergeson (1972) cover the mechanisms of synergistic interactions in general terms and these are considered below, with examples to illustrate each mechanism. However, many authors simply speculate on the mechanism that may have operated in their work and rarely present informative data on the details of the association.

Nematodes facilitate access to the root for the fungus

It has occasionally proved possible to demonstrate that nematodes interact with fungi in complex diseases by providing access to the root tissues through the wounds that they cause. Proof of such a simple relationship has been achieved by imitating the nematode damage by mechanical wounding of the roots. Inagaki and Powell (1969) wounded tobacco roots mechanically to show that such treatment allowed symptoms of black shank infection to develop as quickly as when plants were inoculated with *Pratylenchus brachyurus*.

The mechanical access that nematodes provide to root tissue may be more than is provided by a simple wound, and there are many reports of fungal disease organisms growing in and extending the lesions caused by burrowing nematode or root lesion nematodes, and growing along the invasion tracks left by invading juveniles of cyst and root-knot nematodes. For instance, *Cylindrocarpon musae* is only common in the lesions formed in banana roots by *Radopholus similis* (Booth and Stover, 1974) and Polychronopoulos *et al.* (1969) showed not only that *Rhizoctonia solani* would grow along *Heterodera schachtii* juvenile invasion tracks in sugar-beet, but also that increased lateral root production following nematode invasion also provided entry sites for fungi. Lastly, nematodes with a sedentary life style, such as cyst and root-knot nematodes, have enlarged females which usually rupture the root cortex and thereby provide entry sites for fungi. Since it takes about three to four weeks from invasion for this to occur, this may at least partly explain why inoculation of nematodes some time before the fungus leads to maximum synergistic effect. In the case of root-knot nematodes, penetration through ruptures formed in the root by females may be facilitated by the fungus growing more extensively on galled rather than healthy root tissue, as occurred with *Meloidogyne incognita* and *R. solani* on tomato roots (Golden and Van Gundy, 1975). Of course, as galled roots degenerate they will also develop cracks around the knots which will provide invasion access for fungi.

*Changes in the rhizosphere attract or stimulate nematode
or fungal pathogens*

The first evidence that modification of the rhizosphere, probably brought about by changes in root exudation, is influencing the development of a nematode or fungal pathogen is usually an increase in density of one or the other on or in the roots. For instance, Gokulapalan and Nair (1983) recorded many more *Hirschmanniella oryzae* on roots of rice infected by *Rhizoctonia solani* than on healthy roots, and Jordaan *et al.* (1987) found more root lesion nematodes in maize roots infected by *Fusarium moniliforme*, which they attributed to increased attraction and/or penetration of the nematodes into the roots. This attraction etc. might operate in a number of ways. *F. oxysporum*, when present in large amounts, made alfalfa roots more attractive to *Pratylenchus penetrans* and this effect was associated with an increased output of carbon dioxide (Edmunds and Mai, 1966, 1967). There may also be more nutrients in root exudates from plants attacked by nematodes. Exudates from tomato roots infected by *Meloidogyne incognita* increased the in vitro germination rate of *F. oxysporum* microconidia and contained more carbohydrates and reducing sugars, of which the latter had a positive effect on fungal growth (Kleineke-Borchers and Wyss, 1982). Noguera and Smits (1982) found a similar stimulation of the fungus but also found fewer colonies of actinomycetes antagonistic to *F. oxysporum* in the tomato root rhizosphere when the roots were infected by *M. incognita*. Gonzalez (1982) also showed that extracts of *M. incognita*-infected tomato roots would stimulate growth and germination of *F. oxysporum*, and found that these extracts contained more carbohydrates and amino acids than those from healthy roots.

Nematode or fungus infection causes physiological changes in the host

Nordmeyer and Sikora (1983a) worked with *Heterodera daverti* and clover and showed that *Fusarium avenaceum* infection of the roots allowed not only more nematode penetration but also more nematode development than in control plants, an effect which could be reproduced by pretreating plants with *F. avenaceum* culture filtrates. However, the changes wrought in the host may be more direct, a suggestion made by Sasser *et al.* (1955), who concluded that the role of *Meloidogyne incognita* in promoting *Phytophthora parasitica* infection of tobacco was more than just the wounding of root tissues, since mechanical wounding of roots would not reproduce the effect of nematodes. Brodie and Cooper (1964) found the same with *Rhizoctonia solani* and *Pythium debaryanum* when mechanical wounding of cotton seedlings failed to increase their susceptibility to either pathogen. They also found that sporangial production of *P. debaryanum* was almost ten times greater in the presence of juice expressed from root-knot galls than in the presence of juice from healthy roots. These observations indicated that the nematodes made the roots a better environment for fungal development, perhaps most simply by increasing the nutrient supply available. Such a mechanism seemed to operate in tomato roots attacked by *M. incognita*, where *R. solani* colonized the nutrient-rich nematode feeding cells very heavily (Golden and Van Gundy, 1975), probably the same mechanism as with *R. solani* and *M. hapla* on radish (Khan and Müller, 1982). Further evidence for physiological predisposition of plants to fungus attack by nematodes comes from Nath *et al.* (1984), who were unable to reproduce the effects of *M. javanica* on invasion of tomatoes by a range of fungi by mechanical wounding, and Noguera (1983) suggested that *M. incognita*-infested tomato roots become generally susceptible to invasion by *F. oxysporum*. The existence of a general, systemic effect on susceptibility to this fungus was confirmed by El-Sherif and Elwakil (1991), who used split tomato root systems and inoculated one half with *M. incognita* and the other with the fungus. An increase in susceptibility to fungal invasion was evident in the half of the root system free of nematodes. Thus, there is much evidence to indicate that physiological changes caused by nematodes increase plant susceptibility to fungi, but little evidence to suggest the actual mechanisms involved.

In the same way that nematodes can cause changes in the host to predispose it to fungal attack, fungi can predispose plants to nematode attack. Again, there is much circumstantial evidence of increased nematode population densities compared with fungus-free controls (e.g. Varshney *et al.*, 1987) but little to suggest a mechanism by which nematode reproduction could be increased. Possible exceptions to this are the reports by Edmunds and Mai (1966), who suggested that fungal enzymes may help *Pratylenchus penetrans* to penetrate alfalfa roots by loosening the root cells, and Nordmeyer and Sikora (1983b) who showed that *Fusarium avenaceum* produced pectin methylesterase. Clover roots exposed to a culture filtrate showed an increased ion efflux and *Heterodera daverti* juveniles were attracted to root diffusates from roots treated in this way.

Nematode or fungus infection causes breakdown of resistance in the host

If a cultivar has been specifically produced to be resistant to a particular pathogen it is important that the resistance be stable and not broken by another pathogen. Unfortunately, there are many examples in the literature of a fungus or nematode breaking crop cultivar resistance to the other. There are probably more examples of nematodes breaking resistance to fungal diseases but this may be an accident of history in that cultivars with resistance to fungal disease were probably produced before nematode-resistant cultivars became available and there has therefore been more opportunity to observe breakdown of resistance to fungi. Examples of this are given by Kleineke-Borchers and Wyss (1982), who showed that *Fusarium oxysporum*-resistant tomatoes are less resistant in the presence of *Meloidogyne incognita*, and Upadhyay and Dwivedi (1987), who reported an *F. oxysporum*-resistant chickpea cultivar losing its resistance when infected by *M. javanica*.

Hasan (1985) reported loss of resistance to *Meloidogyne incognita* in two chilli cultivars infected with *Rhizoctonia solani* or *Pythium aphanidermatum*, and Hasan and Khan (1985) found that tomato cultivars could lose resistance to *M. incognita* in the presence of *R. solani* and *Sclerotium rolfsii*. Both of these examples used cultivars with resistance which was identified in glasshouse screening systems but lost the resistance in field or microplot situations. This underlines the importance of examining cultivar performance in a natural situation rather than the artificial environment of a laboratory-based assay. Khan and Husain (1989) reported experiments with cultivars of cowpeas in which resistance to both nematodes and fungi was broken by the other pathogen.

6.4 ASSOCIATIONS WITH AN ANTAGONISTIC EFFECT ON PLANT DAMAGE

Antagonistic interactions may be more common than the number of reports (Table 6.6) would suggest but synergistic or additive interactions are more obvious and more important to growers, and are therefore more likely to be reported. In antagonistic interactions there is a dynamic balance between the effects of the organisms involved, and this balance is disturbed in synergistic interactions.

6.4.1 Examples

Perez Sendin *et al.* (1986) inoculated soya beans with *Rhizoctonia solani* and *Meloidogyne incognita* and found that the fungal infection index for the roots was significantly less than when the fungus alone was inoculated. When sugar-cane plants were inoculated with *M. incognita* or *Pratylenchus*

Table 6.6 Nematode–fungus associations in which plant damage is less than would be expected from the additive effect of the two organisms (i.e. antagonistic associations)

Crop	Nematode	Fungus	Reference
Cowpea	*Heterodera cajani*	*Rhizoctonia bataticola*	Walia and Gupta (1986a)
Kidney bean	*Meloidogyne incognita*	*Rhizoctonia solani*	Costa Manso and Huang (1986)
Soya bean	*Heterodera glycines*	*Fusarium solani*	Killebrew *et al.* (1988)
	M. incognita	*R. solani*	Perez Sendin *et al.* (1986)
Sugar-beet	*Heterodera schachtii*	*F. oxysporum*	Jorgenson (1970)
Sugar-cane	*M. incognita* *Pratylenchus zeae*	*Pythium graminicola*	Valle-Lamboy and Ayala (1980)
Tomato	*M. javanica*	*F. oxysporum* *F. solani*	Qadri and Saleh (1990)
Wheat	*Aphelenchoides hamatus*	*Fusarium culmorum*	Rössner and Urland (1983)

zeae, growth was much reduced. *Pythium graminicola* also reduced growth (somewhat less than the nematodes) but when either species of nematode was inoculated with the fungus their combined effect on growth was considerably less than additive. When both species of nematode were inoculated with the fungus, plants grew as well as uninoculated controls even though there was considerable nematode reproduction (Valle-Lamboy and Ayala, 1980). In contrast, Santo and Holtzmann (1970) found much reduced reproduction of *Pratylenchus zeae* in the presence of *Pythium graminicola* on sugar-cane. There was little effect on cowpea growth when either M. *incognita* or R. *solani* was inoculated separately or together, but nematode reproduction was less when the two organisms were added together (Kanwar *et al.*, 1987). A similar effect on reproduction of M. *javanica* on peanuts was found when R. *bataticola* or *Fusarium solani* was added with the nematodes, R. *bataticola* being the more antagonistic to the nematodes (Sakhuja and Sethi, 1986). Several species of fungus reduced the effects of *Heterodera schachtii* on sugar-beet growth and M. *javanica* on tomato growth and also reduced nematode reproduction (Qadri and Saleh, 1990). Similar results were reported by Jorgenson (1970), who found that, although H. *schachtii* and F. *oxysporum* individually reduced sugar-beet growth, the presence of the fungus reduced nematode invasion and reproduction and improved plant growth. If R. *solani* was added to cowpea before H. *cajani*, nematode reproduction was severely depressed (Walia and Gupta, 1986a, 1986b) and only when the fungus was added two weeks after the nematode was there an effect on plant growth (Walia and Gupta, 1986a). Disease severity on soya beans was not substantially changed when F. *solani* and H. *glycines* were inoculated together, as compared to their inoculation individually, but seedling stand was improved (Killebrew *et al.*, 1988). A different type of interaction was reported by Rössner and Urland (1983), who found that cereal plants were regularly colonized by mycophagous *Aphelenchoides* spp. and that A. *hamatus* would cause a quick collapse of cultures of cereal foot-rot organisms. Indeed, damage to wheat by F. *culmorum* was significantly reduced by the nematodes in pot tests. Choi *et al.* (1988) also exploited this capability of mycophagous nematodes when they applied large numbers of *Aphelenchus avenae* to soil in which cucumber seedlings were growing and effectively suppressed pre-emergence damping-off due to R. *solani*.

6.4.2 Mechanisms of antagonistic interactions

Since the fungi and nematodes under consideration are all root parasites, an obvious (but rarely observed and reported) mechanism by which they would antagonize each other is competition for root space. This and other possible mechanisms can perhaps be classified as follows.

1. Competition for root space: the timing of inoculation or the opportunity for root invasion could be critical in giving one organism an advantage

over the other. Also, prior inoculation of one may result in a reduction of root mass available for colonization by the other.

2. Prior inoculation of a fungus may lead to the production of nematicidal or nematistatic metabolites which would lessen competition from the nematodes.
3. Fungi may infect nematodes and kill them or at least impair their reproduction.
4. Nematodes may impair fungal growth and reproduction by feeding on the fungi.

Competition for root space

Whilst it might seem logical for two separate organisms which parasitize roots to be in competition for space within the roots, there are no reports which clearly identify competition between nematodes and fungi for root space as a mechanism influencing reproductive success. However, there are many reports in which the presence of one influences reproductive success of the other and competition for space may be important in these inter-actions. Such reports are dominated by examples in which the presence of a fungus limits the reproduction of a nematode. They can only be broadly classified, perhaps into those where nematode feeding leads to root lesions (Santo and Holtzmann, 1970; Nath *et al.*, 1976; Koura and Satour, 1987; Sundararaju and Koshy, 1987) and those involving sedentary endoparasitic nematodes (Husain *et al.*, 1985; Lanjewar and Shukla, 1985; Goel and Gupta, 1986a; Sakhuja and Sethi, 1986; Perez Sendin *et al.*, 1986; Walia and Gupta, 1986a, 1986b; Kanwar *et al.*, 1987; Mani and Sethi, 1987; Griffin and Thyr, 1988; Tiyagi *et al.*, 1988; Qadri and Saleh, 1990).

There are far fewer reports of the presence of a nematode influencing reproductive success of a fungus but Valle-Lamboy and Ayala (1980) found that *Meloidogyne incognita* curtailed development of *Pythium graminicola*, perhaps by 'decreasing the available area of fungal development'. Also, Costa Manso and Huang (1986) found that infection by *M. incognita* seemed to give plants of *Phaseolus vulgaris* protection against *Rhizoctonia solani*.

Fungus produces metabolites toxic to nematodes

Because it is such a clearly defined event and an essential phase of the nematode life cycle, ability to hatch normally is frequently used as an assay for potentially nematicidal compounds. Culture filtrates taken from cultures of candidate fungi have been used by many workers to show that fungi produce metabolites which prevent hatching of and even kill nematode juveniles. For instance, Dahiya and Singh (1985) showed that *Aspergillus niger* culture filtrate killed juveniles of *Meloidogyne* spp. and also interfered with hatching. Khan *et al.* (1984b) used culture filtrates of eight species of *Aspergillus* and showed that they all decreased hatching and killed hatched juveniles of *M. incognita*. Similar effects with culture filtrates of *Sclerotium rolfsii* and *Rhizoctonia solani* were found by Ali (1989).

In order to try to identify the metabolites responsible for reduced hatching of nematodes, Ciancio *et al.* (1988) assayed the effects of a range of *Fusarium* toxins on hatching of *Meloidogyne incognita*. Of the nine that were tested, five decreased hatching significantly and the most toxic was T2 toxin. Mani *et al.* (1986) adopted a more direct approach to identifying the toxins from *F. solani* that immobilized *M. incognita* and were able to identify a series of long-chain paraffins as being involved. Although other metabolites were also toxic, they were not present in sufficient concentration in the extracts to allow their characterization. Vaishnav *et al.* (1985) also used an activity assay to show that culture filtrates of *Aspergillus* spp. would affect the activity of *M. arenaria*.

There are other aspects of the nematode life cycle which can be used as a basis for assaying toxicity of fungal metabolites. Walia and Swarup (1985) showed that culture filtrates of three species of nematophagous fungi decreased hatching of *Meloidogyne incognita* and *Heterodera zeae* and decreased penetration of tomato roots by *M. incognita*. Also, *Aspergillus niger* and *Rhizoctonia solani* culture filtrates decreased the numbers of *Meloidogyne* juveniles invading tomato roots and partially suppressed their reproduction (Khan *et al.*, 1984a). Haseeb and Alam (1984) measured the reproduction of six different species of plant-parasitic nematodes on tomato plants grown in soil treated with *R. solani* culture filtrate and found that the reproduction of all of them was decreased.

All of the effects reported above are directly on the nematodes but Fattah and Webster (1989) showed that culture filtrates of two *Fusarium* spp. and *Colletotrichium coccodes* would affect (and cause deterioration of) the structure of *Meloidogyne javanica* feeding cells in tomato roots. These effects were noted from filtrates which did not affect the nematodes directly, and represent yet another route by which fungal metabolites can interfere with nematode development.

Fungi infect nematodes

Several species of fungi thought of as plant parasites have recently been shown to be capable of infecting and killing plant parasitic nematodes. This includes species involved in disease complexes with nematodes. In addition, species of fungus known best for their ability to kill nematodes have been shown capable of colonizing plant roots, although they may not cause much damage to the plant.

Lal *et al.* (1982) showed that *Fusarium solani* reduced the number of cysts of *Heterodera zeae* on maize and the number of eggs that the cysts contained, whilst Khan and Husain (1986) reported *F. solani* infecting *Meloidogyne incognita* from eggplants. Moussa and Hague (1988) showed that *F. oxysporum* impaired the reproduction of *M. incognita* on soya bean by reducing the number of females and increasing the proportion of males that formed, attributing these effects to invasion and eventual destruction of the nematode feeding cells by the fungus. *Paecilomyces lilacinus* will infect many species of nematode and kill them, and Khan and Husain (1988) found that it would infect and kill *M. incognita* and *Rotylenchulus reniformis* on cowpea. In addition, it reduced the damage that these nematodes cause to cowpea in conjunction with *Rhizoctonia solani*, both through a direct effect on the nematodes and an antagonistic effect towards *R. solani*.

A survey of fungal species associated with *Meloidogyne incognita* eggs in a soya bean field showed that *Fusarium oxysporum* and *Paecilomyces lilacinus* were the commonest species infecting the eggs (Morgan-Jones *et al.*, 1984), whilst a test of ability of 14 fungi isolated from *Heterodera glycines* cysts to colonize soya bean roots showed that isolates of both of these species (and 11 others) were capable of infecting the roots (Stiles and Glawe, 1989). *Acremonium strictum* isolated from field populations of *H. schachtii* was also capable of infecting seven out of nine plant species tested (Nigh *et al.*, 1980). The identification of such fungi is very difficult and races of each, which may vary in their preference for plant roots or nematodes, are common. This variation in host preference may, in fact, occur in many of the fungal species that colonize plant roots, such that many species thought of primarily as plant parasites are actually capable of parasitizing nematodes to a greater or lesser extent (D.H. Crump, personal communication).

Nematodes feed on fungi

Zunke *et al.* (1986) confirmed, by detailed observation, that *Aphelenchoides hamatus* would feed on species of plant-pathogenic fungi, and Bird *et al.* (1989) made similar observations on *A. hylurgi* growing on fungal pathogens of wheat. However, in achieving control of damping off of cucumber caused by *Rhizoctonia solani* (referred to earlier), Choi *et al.* (1988) had to apply more than 5×10^4 individuals of *Aphelenchus avenae* per 500 ml of soil. Thus, suppression of this type may occur naturally only rarely and is unlikely to become an important control strategy.

6.5 INFECTIONS OF PLANT PARTS OTHER THAN ROOTS

Many of the fungi that cause root-rots are, to at least some extent, opportunistic in that they will use as a substrate any suitable material that becomes available to them. Because they are soil borne, the plant parts which they contact are usually roots but can also be stem bases, where they can cause foot-rots, and storage organs such as tubers, corms, etc. Several of these have been mentioned already but the spectrum of nematode and fungal species and types of plant organs that may suffer attack are particularly well illustrated by examples from the peanut crop.

McDonald *et al.* (1979) showed that infestation of peanut seeds with the test nematode, *Aphelenchoides arachidis*, probably facilitated entry of soil fungi which then became established in the tissues of the seeds. Application of fungicidal seed-dressing improved germination of nematode-free seed but not nematode-infected seed, suggesting that nematodes not only allowed more seeds to become infected but also permitted the fungal infection to become more deep seated and difficult to eradicate.

Peanut pod rot caused by *Pythium myriotylum* was made more severe when pods were exposed to a combination of the fungus with *Meloidogyne arenaria* (Garcia and Mitchell, 1975b), and peanut peg rot caused by *Sclerotium rolfsii* was associated with *Pratylenchus brachyurus* in the field (Good *et al.*, 1958).

6.6 DISCUSSION

A great problem with much of the literature on nematode–fungus interactions is that experimental conditions which do not resemble the real situation have frequently been used. Some authors have been so concerned to demonstrate an interaction that they have adjusted the amounts and timings of inoculation of two organisms and the observations that they made until an interaction appears. In fact, many natural situations involve simultaneous exposure of a crop to the two organisms (Sikora and Carter, 1987).

Once an interaction has been identified as a field problem, a decision may have to be taken on whether to attempt control of the nematode, the fungus or both. When the role of the nematode is clearly to predispose the crop to fungal attack it may be more appropriate to control the nematode, as nematodes are usually capable of causing damage in their own right. The use of nematode resistance may, however, not be an appropriate short-term measure for nematode control as wounding of the roots may occur before the plant's resistance mechanism operates against the nematode. A good example of this is found with potato cyst nematodes and wilt disease of potatoes. When potato cultivars resistant to *Globodera pallida* were exposed to this nematode, *Verticillium dahliae* or both organisms, one clone in particular was very susceptible to *V. dahliae*; although this clone was resistant and tolerant to the nematode alone, it was also very susceptible to the interaction of nematode and fungus (Lacey *et al.*, 1985). Thus, growing this clone on land contaminated with both the nematode and the fungus could give an impression of extreme intolerance of the nematode. This would not be desirable in a cultivar produced specifically for its nematode resistance. Such an example underlines the fact that breeders may have to produce cultivars with resistance to two separate disease organisms when the two are involved in a disease complex.

Integrated systems of pest and disease management may have to be adopted, taking into account both the nematodes and the fungi. Eventually, it may be possible to exploit some of the antagonistic effects of fungi towards nematodes described above. From the spectrum of fungal parasites of roots it may be possible to select species, races or isolates that have little or no plant pathogenicity but sufficient 'rhizosphere competence' and pathogenicity towards target nematode species to be able to exert good nematode control. However, a great deal of work is still required before we fully understand the complex relationships between nematodes and root-rot fungi.

REFERENCES

Abawi, G.S. and Barker, K.R. (1984) Effects of cultivar, soil temperature, and population levels of *Meloidogyne incognita* on root necrosis and Fusarium wilt of tomatoes. *Phytopathology*, **74**, 433–8.

Abawi, G.S. and Jacobsen, B.J. (1984) Effect of initial inoculum densities of *Heterodera glycines* on growth of soybean and kidney bean and their efficiency as hosts under greenhouse conditions. *Phytopathology*, **74**, 1470–4.

Abu-El-Amayem, M.M., El-Shoura, M.Y., Radwan, M.A., Ahmed, A.H. and Abd-El-All, A. (1985) Joint action effects of some nematicides and benomyl against *Meloidogyne incognita* and *Rhizoctonia solani* on soybean. *Mededelingen van de Faculteit Landbouwwetenschappen Rijksuniversiteit Gent*, **50**, 839–50.

Agrios, G.N. (1969) *Plant Pathology*, Academic Press, New York.

Ali, A.H.H. (1989) The effect of culture filtrates of *Rhizoctonia solani* and *Sclerotium rolfsii* on hatching and juvenile mortality of *Meloidogyne javanica*. *Japanese Journal of Nematology*, **18**(7), 36–8.

Arulpragasam, P.V. and Addaickan, S. (1983) Soft root rot: A new root disease of tea? *Tea Quarterly*, **52**(2), 52–5.

Bergeson, G.B. (1972) Concepts of nematode–fungus associations in plant disease complexes: A review. *Experimental Parasitology*, **32**, 301–14.

Bird, A.F., Bird, J., Fortuner, R. and Moen, R. (1989) Observations on *Aphenenchoides hylurgi* Massey, 1974 feeding on fungal pathogens of wheat in Australia. *Revue de Nématologie*, **12**, 27–34.

Booth, C. and Stover, R.H. (1974) *Cylindrocarpon musae* sp. nov., commonly associated with burrowing nematode (*Radopholus similis*) lesions on bananas. *Transactions of the British Mycological Society*, **63**, 503–7.

Bridge, J., Mortimer, J.J. and Jackson, G.V.H. (1983) *Hirschmanniella miticausa* n. sp. (Nematoda: Pratylenchidae) and its pathogenicity on taro (*Colocasia esculenta*). *Revue de Nématologie*, **6**, 285–90.

Brodie, B.B. and Cooper, W.E. (1964) Relation of parasitic nematodes to post emergence damping-off of cotton. *Phytopathology*, **54**, 1023–7.

Chahal, P.P.K. and Chhabra, H.K. (1984) Interaction of *Meloidogyne incognita* with *Rhizoctonia solani* on tomato. *Indian Journal of Nematology*, **14**, 56–7.

Chambers, K.R. (1987) Epidemiology of maize root rot in South Africa. *Journal of Phytopathology*, **118**, 84–93.

Choi, D.R., Ishibashi, N. and Tanaka, K. (1988) Possible integrated control of soil insect pests, soil-borne diseases and plant nematodes by mixed application of fungivorous and entomogenous nematodes. *Bulletin of the Faculty of Agriculture, Saga University*, **65**, 27–35.

Ciancio, A., Logrieco, A., Lamberti, F. and Bottalico, A. (1988) Nematicidal effects of some *Fusarium* toxins. *Nematologia Mediterranea*, **16**, 137–8.

Costa Manso, E.S.B.G. and Huang, C.S. (1986) Interacao entre *Meloidogyne incognita* (Raca 4) e isolados de *Rhizoctonia solani*, em *Phaseolus vulgaris* cv. Rico 23. *Fitopatologia Brasileira*, **11**, 857–64.

Cuarezma-Teran, J.A. (1985) Nematodes and fungi associated with a sorghum root disease complex. *Dissertation Abstracts International, B. Sciences and Engineering*, 45(9), 2759.

Dahiya, J.S. and Singh, D.P. (1985) Inhibitory effects of *Aspergillus niger* culture filtrate on mortality and hatching of larvae of *Meloidogyne* spp. *Plant and Soil*, 86, 145–6.

Dianese, J.C., Ribeiro, W.R.C. and Urben, A.F. (1986) Root-rot of soybean caused by *Cylindrocladium clavatum* in central Brazil. *Plant Disease*, 70(10), 977–80.

Dickinson, C.H. (1979) External synergisms among organisms inducing disease, in *Plant Disease*, Vol. IV (eds J.G. Horsfall and E.B. Cowling), Academic Press, New York, pp. 97–111.

Diomande, M. and Beute, M.K. (1981a) Effects of *Meloidogyne hapla* and *Macroposthonia ornata* on Cylindrocladium black rot of peanut. *Phytopathology*, 71, 491–6.

Diomande, M. and Beute, M.K. (1981b) Relation of *Meloidogyne hapla* and *Macrospothonia ornata* populations to Cylindrocladium black rot in peanuts. *Plant Disease*, 65, 339–42.

Doshi, A. and Mathur, S. (1987) Symptomatology, interaction and management of rhizome rot of ginger by xenobiotics. *Korean Journal of Plant Pathology*, 26, 261–5.

Doyle, A.D., McLeod, R.W., Wong, P.T.W., Hetherington, S.E. and Southwell, R.J. (1987) Evidence for the involvement of the root lesion nematode *Pratylenchus thornei* in wheat yield decline in northern New South Wales. *Australian Journal of Experimental Agriculture*, 27, 563–70.

Dunn, E. and Hughes, W.A. (1964) Interrelationship of the potato root eelworm, *Heterodera rostochiensis* Woll., *Rhizoctonia solani* Kühn and *Colletotrichum atramentarium* (B. & Br.) Toub., on the growth of the tomato plant. *Nature, London*, 201, 413–14.

Dunn, E. and Hughes, W.A. (1967) Interactions of *Oospora pustulans, Rhizoctonia solani* and *Heterodera rostochiensis* on the potato. *European Potato Journal*, 10, 327–8.

Edmunds, J.E. and Mai, W.F. (1966) Effect of *Trichoderma viride, Fusarium oxysporum* and fungal enzymes upon the penetration of alfalfa roots by *Pratylenchus penetrans*. *Phytopathology*, 56, 1132–5.

Edmunds, J.E. and Mai, W.F. (1967) Effect of *Fusarium oxysporum* on movement of *Pratylenchus penetrans* toward alfalfa roots. *Phytopathology*, 57, 468–71.

El-Sherif, A.G. and Elwakil, M.A. (1991) Interaction between *Meloidogyne incognita* and *Agrobacterium tumefaciens* or *Fusarium oxysporum* f. sp. *lycopersici* on tomato. *Journal of Nematology*, 23, 239–42.

Fattah, F.A. and Webster, J.M. (1989) Ultrastructural modifications of *Meloidogyne javanica* induced giant cells caused by fungal culture filtrates. *Revue de Nématologie*, 12, 197–210.

Feder, W.A. and Feldmesser, J. (1961) The spreading decline complex: The separate and combined effects of *Fusarium* spp. and *Radopholus similis* on the growth of Duncan grapefruit seedlings in the greenhouse. *Phytopathogy*, 51, 724–6.

Garcia, R. and Mitchell, D.J. (1975a) Interactions of *Pythium myriotylum* with *Fusarium solani, Rhizoctonia solani* and *Meloidogyne arenaria* in pre-emergence damping-off of peanut. *Plant Disease Reporter*, 59, 665–9.

Garcia, R. and Mitchell, D.J. (1975b) Synergistic interactions of *Pythium myriotylum* with *Fusarium solani* and *Meloidogyne arenaria* in pod rot of peanut. *Phytopathology*, 65, 832–3.

Godfrey, G.H. (1936) The pineapple root system as affected by the root-knot nematode. *Phytopathology*, 26, 408–28.

Goel, S.R. and Gupta, D.C. (1986a) Interaction of *Meloidogyne javanica* and *Fusarium oxysporum* f. sp. *ciceri* on chickpea. *Indian Phytopathology*, 39, 112–14.

Goel, S.R. and Gupta, D.C. (1986b) Interaction of *Meloidogyne javanica* and *Rhizoctonia bataticola* on chickpea (*Cicer arietinum* L.). *Indian Journal of Nematology*, **16**, 133–4.

Gokulapalan, C. and Nair, M.C. (1983) Field screening for sheath blight and rice root nematode resistance. *International Rice Research Newsletter*, **8**(6), 4.

Gokulapalan, C. and Nair, M.C. (1986) Role of rice root nematode in the severity of sheath blight disease of rice in Kerala. *Indian Phytopathology*, **39**, 436–8.

Golden, J.K. and Van Gundy, S.D. (1975) A disease complex of okra and tomato involving the nematode, *Meloidogyne incognita*, and the soil-inhabiting fungus, *Rhizoctonia solani*. *Phytopathology*, **65**, 265–73.

Gonzalez, R. (1982) Crecimento y germinacion de clamidosporas de *Fusarium oxysporum* f. sp. *lycopersici* en extractos radiculares de plantas de tomate infectadas con *Meloidogyne incognita*. *Agronomia Tropical*, **30**, 305–13.

Good, J.M., Boyle, L.W. and Hammons, R.D. (1958) Studies of *Pratylenchus brachyurus* on peanuts. *Phytopathology*, **48**, 530–5.

Goodey, T. (1935) Observations on a nematode disease of yams. *Journal of Helminthology*, **13**(3), 173–90.

Grainger, J. and Clark, M.R.M. (1963) Interactions of *Rhizoctonia* and potato root eelworm. *European Potato Journal*, **6**, 131–2.

Green, C.D., Ghumra, M.F. and Salt, G.A. (1983) Interaction of root-lesion nematodes, *Pratylenchus thornei* and *P. crenatus*, with the fungus *Thielaviopsis basicola* and a grey sterile fungus on roots of pea (*Pisum sativum*). *Plant Pathology*, **32**, 281–8.

Griffin, G.D. and Thyr, B.D. (1988) Interaction of *Meloidogyne hapla* and *Fusarium oxysporum* f. sp. *medicaginis* on alfalfa. *Phytopathology*, **78**, 421–5.

Griffin, G.D., Gray, F.A. and Johnson, D.A. (1988) Effect of *Meloidogyne hapla* on resistance and susceptibility of alfalfa to *Phytophthora megasperma* f. sp. *medicaginis*. *Report of the 31st North American Alfalfa Improvement Conference, held at Beltsville, Maryland, USA, 19–23 June 1988* (eds J.B. Montray and J.H. Elgin), p. 23.

Hamada, M., Hirakata, K. and Uchida, T. (1985) Influence of Southern root-knot nematode, *Meloidogyne incognita*, on the occurrence of root rot of pepper (*Piper nigrum* L.) caused by *Fusarium solani* f. sp. *piperi*. *Proceedings of the Kanto-Tosan Plant Protection Society*, **32**, 236–7.

Hasan, A. (1985) Breaking resistance in chilli to root-knot nematode by fungal pathogens. *Nematologica*, **31**, 210–17.

Hasan, A. (1988) Interaction between *Pratylenchus coffeae* and *Pythium aphanidermatum* and/or *Rhizoctonia solani* on chrysanthemum. *Journal of Phytopathology*, **123**, 227–32.

Hasan, A. and Khan, M.N. (1985) The effect of *Rhizoctonia solani*, *Sclerotium rolfsii*, and *Verticillium dahliae* on the resistance of tomato to *Meloidogyne incognita*. *Nematologia Mediterranea*, **13**, 133–6.

Haseeb, A. and Alam, N.M. (1984) Soil population of plant parasitic nematodes infesting tomato in relation to metabolites of *Rhizoctonia solani*. *Indian Journal of Plant Pathology*, **2**, 189–90.

Hershman, D.E., Hendrix, J.W., Stuckey, R.E. and Bachi, P.R. (1990) Influence of planting date and cultivar on soybean sudden death syndrome in Kentucky. *Plant Disease*, **74**, 761–6.

Husain, S.I., Khan, T.A. and Jabri, M.R.A. (1985) Studies on root-knot, root rot and pea mosaic virus complex of *Pisum sativum*. *Nematologia Mediterranea*, **13**, 103–9.

Inagaki, H. and Powell, N.T. (1969) Influence of the root-lesion nematode on black shank symptom development in flue-cured tobacco. *Phytopathology*, **59**, 1350–5.

Jackson, C.R. and Minton, N.A. (1968) Pod invasion by fungi in the presence of lesion nematodes in Georgia. *Oléagineux*, **23**, 531–4.

Jayaprakash, A. and Rao, Y.S. (1984) Cyst nematode, *Heterodera oryzicola* and seedling blight fungus *Sclerotium rolfsii* disease complex in rice. *Indian Journal of Nematology*, **14**, 58–9.

Jordaan, E.M., Loots, G.C., Jooste, W.J. and de Waele, D. (1987) Effects of root-lesion nematodes (*Pratylenchus brachyurus* Godfrey and *P. zeae* Graham) and *Fusarium moniliforme* Sheldon alone or in combination, on maize. *Nematologica*, **33**, 213–19.

Jorgenson, E.C. (1970) Antagonistic interaction of *Heterodera schachtii* Schmidt and *Fusarium oxysporum* (Woll.) on sugarbeets. *Journal of Nematology*, **2**, 393–8.

Kanwar, R.S., Gupta, D.C. and Walia, K.K. (1987) Interaction of *Meloidogyne javanica* and *Rhizoctonia solani* on cowpea. *Nematologia Mediterranea*, **15**, 385–6.

Khan, M.W. and Müller, J. (1982) Interaction between *Rhizoctonia solani* and *Meloidogyne hapla* on radish in gnotobiotic culture. *Libyan Journal of Agriculture*, **11**, 133–40.

Khan, T.A. and Husain, S.I. (1986) Parasitism of *Meloidogyne incognita* by *Fusarium solani*. *International Nematology Network Newsletter*, **3(2)**, 11–13.

Khan, T.A. and Husain, S.I. (1988) Studies on the efficacy of *Paecilomyces lilacinus* as biocontrol agent against a disease complex caused by the interaction of *Rotylenchulus reniformis*, *Meloidogyne incognita* and *Rhizoctonia solani* on cowpea. *Nematologia Mediterranea*, **16**, 229–31.

Khan, T.A. and Husain, S.I. (1989) Relative resistance of six cowpea cultivars as affected by the concomitance of two nematodes and a fungus. *Nematologia Mediterranea*, **17**, 39–41.

Khan, T.A., Azam, M. F. and Husain, S.I. (1984a) Effect of fungal filtrates of *Aspergillus niger* and *Rhizoctonia solani* on penetration and development of root-knot nematodes and the plant growth of tomato var. Marglobe. *Indian Journal of Nematology*, **14**, 106–9.

Khan, T.A., Husain, S.I. and Azam, M.F. (1984b) Effect of culture filtrates of eight species of *Aspergillus* on the hatching and mortality of *Meloidogyne incognita*. *Indian Journal of Nematology*, **14**, 51–4.

Khurana, S.M.P. and Singh, S. (1971) Interrelationship of a fungus, *Curvularia lunata* with root-knot nematode, *Meloidogyne javanica* in sugarcane seedling blight. *Annals of the Phytopathological Society of Japan*, **37**, 313–15.

Killebrew, J.F., Roy, K.W., Lawrence, G.W., McLean, K.S. and Hodges, H.H. (1988) Greenhouse and field evaluation of *Fusarium solani* pathogenicity to soybean seedlings. *Plant Disease*, **72**, 1067–70.

Kleineke-Borchers, A. and Wyss, U. (1982) Investigations on changes in susceptibility of tomato plants to *Fusarium oxysporum* f. sp. *lycopersici* after infection by *Meloidogyne incognita*. *Zeitschrift für Pflanzenkrankheiten und Pflanzenschutz*, **89**, 67–78.

Koshy, P.K. and Sosamma, V.K. (1987) Pathogenicity of *Radopholus similis* on coconut (*Cocos nucifera* L.) seedlings under greenhouse and field conditions. *Indian Journal of Nematology*, **17**, 108–18.

Kotcon, J.B., Rouse, D.I. and Mitchell, J.E. (1985) Interactions of *Verticillium dahliae*, *Colletotrichum coccodes*, *Rhizoctonia solani* and *Pratylenchus penetrans* in the early dying syndrome of Russet Burbank potatoes. *Phytopathology*, **75**, 68–73.

Koura, F.H. and Satour, M.M. (1987) Interaction of *Hoplolaimus aegypti* with soil pathogenic fungi *Cephalosporium maydis* and *Fusarium moniliforme* in the root rot complex of maize. *Annals of Agricultural Science, Ain-Shams University*, **32**, 1849–55.

Kumar, R., Ahmad, S. and Saxena, S.K. (1988) Disease complex in chickpea involving *Meloidogyne incognita* and *Fusarium oxysporum*. *International Nematology Network Newsletter*, 5(3), 12–14.

Lacey, C.N.D., Jellis, G.J., Currell, S.B. and Thomson, A.J. (1985) Changes in the apparent tolerance of potato cultivars to *Globodera* spp. in the presence of *Verticillium dahliae*, in *EAPR Abstracts of Conference Papers. 9th Triennial Conference of the European Association for Potato Research* (eds A. Winiger and A. Stöckli), pp. 359–60.

Lal, A., Mathur, V.K. and Agarwal, P.C. (1982) Studies on the effect of *Fusarium solani* parasitism on *Heterodera zeae*. *Nematologica*, 28, 447–50.

LaMondia, J.A. and Martin, S.B. (1989) The influence of *Pratylenchus penetrans* and temperature on black root rot of strawberry by binucleate *Rhizoctonia* spp. *Plant Disease*, 73, 107–10.

Lanjewar, R.D. and Shukla, V.N. (1985) Parasitism and interaction between *Pythium myriotylum* and *Meloidogyne incognita* in soft rot complex of ginger. *Indian Journal of Nematology*, 15, 170–3.

Mani, A. and Sethi, C.L. (1987) Interaction of root-knot nematode, *Meloidogyne incognita* with *Fusarium oxysporum* f. sp. *ciceri* and *F. solani* on chickpea. *Indian Journal of Nematology*, 17, 1–6.

Mani, A., Sethi, C.L. and Devkumar, (1986) Isolation and identification of nematoxins produced by *Fusarium solani* (Mart) Sacc. *Indian Journal of Nematology*, 16, 247–51.

McDonald, D., Bos, W.S. and Gumel, M.H. (1979) Effects of infections of peanut (groundnut) seed by the testa nematode *Aphelenchoides arachidis*, on seed infection by fungi and on seedling emergence. *Plant Disease Reporter*, 63, 464–7.

McGawley, E.C., Winchell, K.L. and Berggren, G.T. (1984) Possible involvement of *Hoplolaimus galeatus* in a disease complex of 'Centennial' soybean. *Phytopathology*, 74, 831.

Melendez, P.L. and Powell, N.T. (1969) The influence of *Meloidogyne* on root decay in tobacco caused by *Pythium* and *Trichoderma*. *Phytopathology*, 59, 1348 (Abstract).

Mishra, C., Som, D. and Singh, B. (1988) Interaction of *Meloidogyne incognita* and *Rhizoctonia bataticola* in white jute (*Corchorus capsularis*). *Indian Journal of Agricultural Sciences*, 58, 234–5.

Morgan-Jones, G., White, J.F. and Rodriguez-Kabana, R. (1984) Fungal parasites of *Meloidogyne incognita* in an Alabama soybean field soil. *Nematropica*, 14, 93–6.

Moussa, E.M. and Hague, N.G.M. (1988) Influence of *Fusarium oxysporum* f. sp. *glycines* on the invasion and development of *Meloidogyne incognita* on soybean. *Revue de Nématologie*, 11, 437–9.

Nath, R. and Kamalwanshi, R.S. (1989) Role of non-pathogenic fungi in presence of root-knot nematode in damping-off of tomato seedlings. *Indian Journal of Nematology*, 19, 84–5.

Nath, R.P., Singh, R.K. and Haider, M.G. (1976) Combined effect of *Fusarium moniliforme* and *Hoplolaimus indicus* on sugarcane plants. *Sugarcane Pathologists' Newsletter*, 17, 24–5.

Nath, R., Khan, M.N., Kamalwanshi, R.S. and Dwivedi, R.P. (1984) Influence of root-knot nematode, *Meloidogyne javanica* on pre- and post-emergence damping-off of tomato. *Indian Journal of Nematology*, 14, 135–40.

Nigh, E.A., Thomason, I.J. and Van Gundy, S.D. (1980) Identification and distribution of fungal parasites of *Heterodera schachtii* eggs in California. *Phytopathology*, 70, 884–91.

Noguera, R. (1983) Influence of *Meloidogyne incognita* on the colonization of *Fusarium oxysporum* f. sp. *lycopersici* in tomatoes. *Agronomia Tropical*, 33, 103–9.

Noguera, R. and Smits, B.G. (1982) Variations in the rhizosphere microflora of tomatoes infected with *Meloidogyne incognita*. *Agronomia Tropical*, **32**, 147–54.

Nordmeyer, D. and Sikora, R.A. (1983a) Studies on the interaction between *Heterodera daverti*, *Fusarium avenaceum* and *F. oxysporum* on *Trifolium subterraneum*. *Revue de Nématologie*, **6**, 193–8.

Nordmeyer, D. and Sikora, R.A. (1983b) Effect of a culture filtrate from *Fusarium avenaceum* on the penetration of *Heterodera daverti* into roots of *Trifolium subterraneum*. *Nematologica*, **29**, 88–94.

Nyczepir, A.P. and Lewis, S.A. (1984) Incidence of *Fusarium* and *Pythium* spp. in peach feeder roots as related to dibromochloropropane application for control of *Criconemella xenoplax*. *Plant Disease*, **68**, 497–9.

Overstreet, C., McGawley, E.C. and Russin, J.S. (1990) Interactions between *Calonectria crotalariae* and *Heterodera glycines* on soybean. *Journal of Nematology*, **22**, 496–505.

Pacumbaba, R.P. and Tadesse, W. (1990) Effect of multiple disease infestation on the agronomic performance of a soybean cultivar. *Phytopathology*, **80**, 1039.

Patel, H.R., Vaishnav, M.U. and Dhruj, I.U. (1986) Effect of interaction between *Meloidogyne arenaria* and *Aspergillus flavus* on groundnut. *Indian Journal of Mycology and Plant Pathology*, **16**, 103–7.

Perez Sendin, M. de L.A., Fernandez, M., Ortega, J. and Gonzalez, J. (1986) Relacion de los nematodos *Meloidogyne incognita* y *Rotylenchulus reniformis* con el hongo *Rhizoctonia solani* en la soya. *Ciencias de la Agricultura*, **29**, 34–8.

Pinochet, J. and Stover, R.H. (1980) Fungi lesions caused by burrowing nematodes on bananas and their root and rhizome rotting potential. *Tropical Agriculture, Trinidad*, **57**, 227–32.

Polychronopoulos, A.G., Houston, B.R. and Lownsbery, B.F. (1969) Penetration and development of *Rhizoctonia solani* in sugar beet seedlings infected with *Heterodera schachtii*. *Phytopathology*, **59**, 482–5.

Qadri, A.N. and Saleh, H.M. (1990) Fungi associated with *Heterodera schachtii* (Nematoda) in Jordan. II. Effect on *H. schachtii* and *Meloidogyne javanica*. *Nematologica*, **36**, 104–13.

Rössner, J. and Urland, K. (1983) Mycophagous nematodes of the genus *Aphelenchoides* from the stem base of cereal plants and their action against foot rot diseases of cereals. *Nematologica*, **29**, 454–62.

Roy, A.K. (1968) The formation of giant cells in tomato roots infected by the potato cyst nematode, *Heterodera rostochiensis*, and the grey sterile fungus, G.S.F. *Nematologica*, **14**, 313–14.

Roy, K.W., Lawrence, G.W., Hodges, H.H., McLean, K.S. and Killebrew, J.F. (1989) Sudden death syndrome of soybean: *Fusarium solani* as incitant and relation of *Heterodera glycines* to disease severity. *Phytopathology*, **79**, 191–7.

Sakhuja, P.K. and Sethi, C.L. (1986) Multiplication of *Meloidogyne javanica* as affected by *Fusarium solani* and *Rhizoctonia bataticola* on groundnut. *Indian Journal of Nematology*, **16**, 1–3.

Santo, G.S. and Holtzmann, O.V. (1970) Interrelationships of *Pratylenchus zeae* and *Pythium graminicola* on sugarcane. *Phytopathology*, **50**, 1537.

Sasser, J.N., Lucas, G.B. and Powers, H.R. (1955) The relationship of root-knot nematodes to black-shank resistance in tobacco. *Phytopathology*, **45**, 459–61.

Scholte, K. and s'Jacob, J.J. (1989) Synergistic interactions between *Rhizoctonia solani*, *Verticillium dahliae*, *Meloidogyne* spp. and *Pratylenchus neglectus* in potato. *Potato Research*, **32**, 387–95.

Sikora, R.A. and Carter, W.W. (1987) Nematode interactions with fungal and bacterial plant pathogens: Fact or fantasy, in *Vistas on Nematology* (eds J.A. Veech and D.W. Dickson), Society of Nematologists, Hyattsville, pp. 307–12.

Smits, B.G. and Noguera, R. (1982) Effect of *Meloidogyne incognita* on the pathogenicity of different isolates of *Fusarium oxysporum* on brinjal (*Solanum melongena* L.). *Agronomia Tropical*, **32**, 285–90.

Stiles, C.M. and Glawe, D.A. (1989) Colonization of soybean roots by fungi isolated from cysts of *Heterodera glycines*. *Mycologia*, **81**, 797–9.

Stover, R.H. (1966) Fungi associated with nematode and non-nematode lesions on banana roots. *Canadian Journal of Botany*, **44**, 1703–10.

Sumner, D.R., Dowler, C.C., Johnson, A.W., Chalfant, R.B., Glaze, N.C., Phatak, S.C. and Epperson, J.E. (1985) Effect of root diseases and nematodes on yield of corn in an irrigated multiple-cropping system with pest management. *Plant Disease*, **69**, 382–7.

Sundararaju, P. and Koshy, P.K. (1987) Separate and combined effects of *Radopholus similis* and *Cylindrocarpon obtusisporum* on arecanut seedlings. *Indian Journal of Nematology*, **17**, 301–5.

Tiyagi, S.A., Zaidi, S.B.I. and Alam, M.M. (1988) Interaction between *Meloidogyne javanica* and *Macrophomina phaseolina* on lentil. *Nematologia Mediterranea*, **16**, 221–2.

Townshend, J.L. (1984) Inoculum densities of five plant parasitic nematodes in relation to alfalfa seedling growth. *Canadian Journal of Plant Pathology*, **6**, 309–12.

Upadhyay, K.D. and Dwivedi, K. (1987) Root-knot nematode, *Meloidogyne javanica* breaks wilt resistance in chickpea variety 'Avrodhi'. *Current Science*, **56(17)**, 915–16.

Vaishnav, M.U., Patel, H.R. and Dhruj, I.U. (1985) Effect of culture filtrates of *Aspergillus* spp. on *Meloidogyne arenaria*. *Indian Journal of Nematology*, **15**, 116–17.

Valle-Lamboy, S. and Ayala, A. (1980) Pathogenicity of *Meloidogyne incognita* and *Pratylenchus zeae*, and their association with *Pythium graminicola* on roots of sugarcane in Puerto Rico. *Journal of Agriculture of University of Puerto Rico*, **64**, 338–47.

Van Gundy, S.D. and Tsao, P.H. (1963) Growth reduction of citrus seedlings by *Fusarium solani* as influenced by the citrus nematode and other soil factors. *Phytopathology*, **53**, 488–9.

Varshney, V.P., Khan, A.M. and Saxena, S.K. (1987) Interaction between *Meloidogyne incognita*, *Rhizoctonia solani* and *Rhizobium* species on cowpea. *International Nematology Network Newsletter*, **4(3)**, 11–14.

Walia, K.K. and Gupta, D.C. (1986a) Antagonism between *Heterodera cajani* and *Rhizoctonia solani* on cowpea (*Vigna unguiculata* (L.) Walp). *Indian Journal of Nematology*, **16**, 41–3.

Walia, K.K. and Gupta, D.C. (1986b) Effect of the fungus, *Rhizoctonia bataticola* on the population development of pigeon-pea cyst nematode, *Heterodera cajani* on cowpea (*Vigna unguiculata*). *Indian Journal of Nematology*, **16**, 131–2.

Walia, K.K. and Swarup, G. (1985) Effect of some fungi on nematode hatching and larval root penetration. *Indian Journal of Nematology*, **15**, 174–6.

Wallace, H.R. (1983) Interactions between nematodes and other factors on plants. *Journal of Nematology*, **15**, 221–7.

Whitney, E.D. (1974) Synergistic effect of *Pythium ultimum* and the additive effect of *P. aphanidermatum* with *Heterodera schachtii* on sugar beet. *Phytopathology*, **64**, 380–3.

Zunke, U., Rössner, J. and Wyss, U. (1986) Parasitierungsverhalten von *Aphelenchoides hamatus* an fünf verschiedenen phytopathogenen Pilzen und an *Agaricus campestris*. *Nematologica*, **32**, 194–201.

7

Interactions between nematodes in cohabitance

Jonathan D. Eisenback

Plant-parasitic nematodes frequently parasitize plants in mixed populations of two or more genera, species or races. They occur in polyspecific communities because of their persistence, polyphagous feeding habits, wide distribution and weak interspecific competition (Oostenbrink, 1966). Although competition is weak, the nematode community is very dynamic, and its members are constantly interacting with each other. Interaction, the effect that one population has on another, may enhance or inhibit nematode reproduction, which is ultimately the driving force in interspecific competition.

Feeding habitats, survival mechanisms and ecological requirements of the different parasitic forms of nematodes vary considerably. The presence of a particular species in a location is related to its method of dissemination, host suitability, host range, interactions with other organisms, cropping history and other factors (Khan and Khan, 1990; Niblack, 1989; Norton, 1978; Norton and Niblack, 1991; Oostenbrink, 1966). Some of the nematodes that occur in polyspecific communities may be relics from previous host crop species, but these species may also be competing for the same host (Jones and Perry, 1978). Competition is usually strongest between organisms that are most alike with respect to their physiological demands on the host. A single species may dominate if it is better adapted to the host or to the ecological conditions. Species usually coexist if they utilize different ecological niches or have dissimilar life cycles.

The coexistence of two species of the same genus of microbivorous nematodes was explained by differences in the pattern of colonization, competitive ability and survival mechanisms in the various phases of population development (Sohlenius, 1988). They survived successive subculturing concomitantly for five years on agar plates. Although competition was apparently weak, pure cultures developed greater populations than mixed cultures.

Species of plant-parasitic nematodes interact ecologically as competitors and aetiologically in their effect on plant growth and crop production. These two aspects of nematode interactivity may be related since the amount of disease is often correlated with population density. Concomitant populations of nematodes may increase the difficulty of predicting the relationship between the numbers of nematodes and crop performance (Green and Dennis, 1981). Also, plant susceptibility to damage is not necessarily related to nematode reproduction (Oostenbrink, 1966). Therefore, ecological and aetiological aspects of nematode–nematode interactions are separate phenomena (Eisenback, 1985; Eisenback and Griffin, 1987).

7.1 ECOLOGICAL INTERACTIONS

Interactions between nematodes most commonly interfere with the well-being of one or both species, but they may also be beneficial to one or both. Competition occurs between species when the reproductive capacity of one species is greater individually than when combined with another species (Brewer, 1978). Adverse interactions may result from the physical destruction or spatial occupancy of feeding sites, or by changes in host physiology which alter its suitability (Norton, 1978).

According to the theory of competitive exclusion, two species cannot occupy the same ecological niche (Gause, 1934). Two species attempting to do so compete for limited resources until the species with the competitive advantage predominates. Competition is related to host suitability, pathogenicity and persistence, and may be both density and time dependent (Brewer, 1978).

The dynamic nature of the soil environment makes it difficult to define and delineate niches; therefore, application of the competitive exclusion principle is difficult to apply to communities of plant-parasitic nematodes. Interactions may occur between species that occupy different niches if they are in close proximity or simply feeding on the same root system (Brewer, 1978; Norton, 1978). When competition between species is less than competition among individuals within a species, the two species can cohabit (Brewer, 1978). The interaction between two species is not always antagonistic. The relationship may be beneficial for one or both species as the result of a mechanical or physiological alteration of the host. The plant may be changed into a more suitable host by enhanced nutritional value or reduced natural defences (Norton, 1978).

The nature of parasitism greatly influences the competition between species. Competition is more severe between species with similar feeding habits, and competitive advantage seems to increase as the host–parasite relationship becomes more complex or as persistence increases.

Sometimes species may commonly occur together (Boag and Topham, 1985). Analysis of data from a large survey of common species of plant-parasitic nematodes revealed several positive correlations between virus-vector species and non-vector species. A few negative relationships were also revealed. Thus, the presence of one species may be useful to predict the absence or presence of another species.

7.1.1 Ectoparasites

Ectoparasites have been placed into ecological groups according to their feeding habits. Surface-feeders parasitize only epidermal cells and root hairs. They have a relatively short stylet, and their association with the host is not very specialized or long-lived. A separate group of ectoparasites with short stylets may penetrate the host with the anterior portion of their bodies and feed on cortical tissues. Deep-feeding ectoparasites have relatively long stylets and feed several cell layers deep. The host–parasite relationship is somewhat complex because hypertrophy and hyperplasia are incited, but the feeding period is relatively short.

The mode of parasitism affects the competition between species. Surface-feeders that feed on epidermal cells and root hairs are more antagonistic to each other than they are to the deep-feeding ectoparasites. Likewise, ectoparasites that establish a more complex host–parasite relationship are more competitive than the primitive surface-feeders (Boag and Alphey, 1988; Johnson, 1970; Johnson and Nusbaum, 1968; Misra and Das, 1979). Interactions among ectoparasites are summarized in Table 7.1.

In greenhouse experiments *Tylenchorhynchus claytoni* was antagonistic to *Paratrichodorus minor* on one cultivar of corn, and *P. minor* was antagonistic to *T. claytoni* on three cultivars (Johnson and Nusbaum, 1968). More competition occurred between *T. martini* and *Criconemella ornata* than between *T. martini* and *Belonolaimus longicaudatus* or *C. ornata* and *B. longicaudatus* on six varieties of Bermudagrass (Johnson, 1970). Reactions varied according to cultivar, but competition was more intense between the shallow-feeding species.

Interaction between species may be affected by host suitability. *Helicotylenchus pseudorobustus* and *Criconemella similis* were antagonistic to *Paratylenchus projectus* on soya bean. Soya bean was a good host for *H. pseudorobustus* and *C. similis*, but a poor host for *P. projectus* (McGawley and Chapman, 1983).

Pathogenicity and competition for feeding sites may affect interactions between concomitant populations of deep-feeding ectoparasites. The highly pathogenic *Belonolaimus longicaudatus* suppressed reproduction of the less virulent *Dolichodorus heterocephalus* on corn, but *D. heterocephalus* had no effect on the reproduction of *B. longicaudatus* (Rhoades, 1985).

Population density and environmental factors may alter the interaction between ectoparasites. *Paratylenchus neoamblycephalus* suppressed low numbers of *Criconemella xenoplax* at 22°C, and high numbers of *C. xenoplax* suppressed low numbers of *P. neoamblycephalus*. High populations of *P. neoamblycephalus* were antagonistic to low populations of *C. xenoplax* at 26°C, and high populations of *C. xenoplax* greatly suppressed low populations of *P. neoamblycephalus*. Species were more competitive at higher densities even though the optimum temperature for

C. xenoplax reproduction was 26°C and 20°C for *P. neoamblycephalus* (Braun *et al.*, 1975).

Interactions between ectoparasites can stimulate the reproduction of one or both species. Reproduction of both *Hoplolaimus columbus* and *Scutellonema brachyurum* was mutually stimulated in concomitant populations on cotton (Kraus-Schmidt and Lewis, 1981). Likewise, the presence of *Hoplolaimus galeatus* stimulated the reproduction of *Belonolaimus longicaudatus* on cotton (Yang *et al.*, 1976). The mechanisms involved in this stimulation of reproduction are unknown.

7.1.2 Ecto- and migratory endoparasites

Migratory endoparasitic nematodes alter root morphology and physiology as they move through plant tissue, and are therefore often antagonistic to ectoparasitic nematodes (Amosu and Taylor, 1975; Chapman, 1959; Cuarezma-Teran and Trevathan, 1985; Pinochet *et al.*, 1976; Upadhyay and Swarup, 1981). Interactions between ectoparasitic and migratory endoparasitic nematodes are summarized in Table 7.2.

The suitability of the host often plays an important role in the interaction. Reproduction of *Tylenchorhynchus martini* was suppressed 75–90% by *Pratylenchus penetrans* on red clover. Red clover was a good host for both species (Chapman, 1959). No interaction was observed between *T. agri* and *P. penetrans* on creeping bentgrass; bentgrass was an excellent host for *T. agri* and a poor host for *P. penetrans* (Sikora *et al.*, 1979). *P. minor* or *Helicotylenchus dihystera* suppressed reproduction by *P. brachyurus* on most varieties of soya bean, but *P. brachyurus* suppressed *H. dihystera* on two varieties (Johnson and Nusbaum, 1968). *Paratrichodorus minor* and *Pratylenchus zeae* were mutually stimulatory on several corn varieties (Johnson and Nusbaum, 1968). Although exact mechanisms involved are unknown, small changes in host suitability may increase the rate of reproduction of certain nematodes.

Interactions between nematodes may be time dependent. *Tylenchorhynchus agri* suppressed reproduction by *Pratylenchus penetrans* on red clover initially for three months, but had no effect after five months (Amosu and Taylor, 1975).

Feeding by one species may alter the attraction of the roots to other nematode species, or their feeding may interfere with another nematode's ability to penetrate the root. *Tylenchorhynchus claytoni* inhibited penetration by 25–90% of *Pratylenchus penetrans* on tobacco cultivar 'WS117'. Split-root experiments demonstrated that translocatable factors altered the attractiveness of the roots to *P. penetrans*. Long-term effects were not determined by these short-duration tests (Miller and McIntyre, 1975).

Table 7.1 Summary of interactions between ectoparasitic nematodes

Nematode combination	Test host	Dominant species	General response	Comments	Reference
Paratrichodorus minor + *Tylenchorhynchus claytoni*	Corn	*T. claytoni* on 1 cv. *P. minor* on 3 cvs Neither on 1 cv.	Antagonistic Antagonistic	50 days, greenhouse tests	Johnson and Nusbaum (1968)
T. martini + *Criconemella ornata*	Bermuda grass	*T. martini*	Antagonistic		Johnson (1970)
T. martini + *Belonolaimus longicaudatus*	Bermuda grass	None	None		
C. ornata + *B. longicaudatus*	Bermuda grass	None	None		
Paratylenchus projectus + *Helicotylenchus pseudorobustus*	Soya bean	*H. pseudorobustus*	Antagonistic	50 days, greenhouse tests	McGawley and Chapman (1983)
P. projectus + *C. similis* *C. similis* + *H. pseudorobustus*	Soya bean Soya bean	*C. similis* None	Antagonistic	Soya bean is a good host for *C. similis*, a moderate host for *H. pseudorobustus*, and poor host for *P. projectus*	

Species combination	Crop	Associated species/condition	Interaction	Comments	Reference
Paratylenchus nanus + *Rotylenchus robustus*	Grass	*P. nanus*	Antagonistic	*P. nanus* was also less pathogenic	Boag and Alphey (1988)
High numbers of *P. neoamblycephalus* + low number of *C. xenoplax*	Plum	*P. neoamblycephalus*	Antagonistic	Optimum temp. for *P. neoamblycephalus* is 20°C, and 26°C for *C. xenoplax*	Braun *et al.* (1975)
Low numbers of *P. neoamblycephalus* + high number of *C. xenoplax*	Plum	*P. neoamblycephalus* at 26°C / *C. xenoplax* at 20°C	Antagonistic / Antagonistic		
Hoplolaimus columbus + *Scutellonema brachyurum*	Cotton	None	Stimulatory	Both species were stimulated after 90 days	Kraus-Schmidt and Lewis (1981)
B. longicaudatus + *Dolichodorus heterocephalus*	Corn	*B. longicaudatus*	Antagonistic	30 weeks, greenhouse tests	Rhoades (1985)
B. longicaudatus + *H. galeatus*	Cotton	None	Stimulatory	Only *B. longicaudatus* was stimulated; 63 days, greenhouse	Yang *et al.* (1976)

Table 7.2 Summary of interactions between ectoparasitic and migratory endoparasitic nematodes

Nematode combination	Test host	Dominant species	General response	Comments	Reference
Tylenchorhynchus martini + Pratylenchus penetrans	Red clover	P. penetrans	Antagonistic	267 days, greenhouse tests	Chapman (1959)
T. agri + P. penetrans	Creeping bentgrass	None	Neutral	10 months, greenhouse tests; excellent host for T. agri, but poor host for P. penetrans	Sikora et al. (1972)
P. zeae + Quinisulcius acutus	Sorghum	None	Antagonistic	42 days, greenhouse tests	Cuarezma-Teran and Trevathan (1985)
P. brachyurus + Paratrichodorus minor	Soya bean	P. minor	Antagonistic	50 days, greenhouse tests	Johnson and Nusbaum (1968)
P. brachyurus + Helicotylenchus dihystera	Soya bean	H. dihystera on most; P. brachyurus on 2 cvs			
T. agri + P. penetrans	Red clover	T. agri	Antagonistic	Suppression after 3 months but not after 5 or 7 months	Amosu and Taylor (1975)

T. claytoni + P. penetrans	Tobacco	T. claytoni	Antagonistic	Penetration of P. penetrans inhibited by a translocatable factor; tests were only several days in duration	Miller and McIntyre (1975)
P. minor + Pratylenchus zeae	Corn	None	Stimulatory	P. zeae populations were higher with P. minor on 2 cultivars of corn. P. minor populations higher with P. zeae on all 5 cultivars of corn. P. brachyurus stimulated P. minor on 3 cultivars of soya bean. 50 days, greenhouse tests	
T. vulgaris + P. zeae	Corn	None	Antagonistic		Upadhyay and Swarup (1981)
Xiphinema index + P. vulnus	Grape	P. vulnus	Antagonistic		Pinochet et al. (1976)

7.1.3 Ecto- and sedentary endoparasites

Ectoparasitic nematodes feed at very different sites than do sedentary endo-parasites, and the two often coexist on the same plant without affecting each other (Haque and Mukhopadhyaya, 1979; Misra and Das, 1979). Sometimes, however, their relationship may be suppressive or beneficial for one or both species. Interactions between ectoparasitic and sedentary endoparasitic nematodes are summarized in Table 7.3.

Ectoparasites feeding on the preferred penetration sites of the infective stages of sedentary endoparasites may be a limiting factor in the success of the sedentary endoparasite. Ectoparasites may also damage the root system and thus indirectly reduce the number of feeding sites available for the endoparasites. *Meloidogyne naasi* was inhibited on creeping bentgrass by *Paratrichodorus minor*, which preferred to feed near the root tips and thus limited the number of infection sites for the root-knot nematode (Sikora et al., 1972; Sikora et al., 1979). *Tylenchorhynchus agri* was also antago-nistic to *M. naasi*. *M. incognita* was inhibited by *T. vulgaris* on bajra (Vaishnav and Sethi, 1978), by *T. brassicae* on vegetables (Khan and Khan, 1986), by *T. brassicae* on cauliflower (Khan et al., 1987), and by *Criconemella ornata, Hoplolaimus indicus* and *T. nudus* on brinjal (Misra and Das, 1979). Likewise, *Criconemoides xenoplax* inhibited *M. hapla* on Concord grape (Santo and Bolander, 1977) and *P. minor* suppressed *M. javanica* on tomato (Van Gundy and Kirkpatrick, 1975). *H. columbus* antagonized *M. incognita* on cotton, probably by altering the host physio-logy rather than by competing for feeding sites (Kraus-Schmidt and Lewis, 1981). The interaction between *H. columbus* and *M. incognita* was density dependent (Guy and Lewis, 1987a). *H. columbus* penetrated soya bean roots infected with *M. incognita* more easily than non-infected roots (Guy and Lewis, 1987b). Root-knot-infected roots were also more attractive to *H. columbus* than non-infected roots. *H. galeatus* increased from non-detectable levels and gradually replaced *M. incognita* as the dominant plant-parasitic nematode species during six growing seasons of cotton (Bird et al., 1974). At a different location, *H. galeatus* dominated a portion of the field, but *M. incognita* dominated the remaining portion; seemingly the two species were mutually exclusive. Cucumbers inoculated with *Aphelenchus avenae* and *M. incognita* had fewer galls than when inoculated with root-knot nematodes alone (Choi et al., 1988; Ishibashi and Choi, 1991).

Sedentary endoparasites can suppress ectoparasites even though they are separated by plant tissue, probably by physiological mechanisms. *Xiphinema americanum* was dominated by *Meloidogyne hapla* for four years in an alfalfa field. *M. hapla* may have limited the number of feeding sites for *Xiphinema*, or the root-knot nematode may have altered the physiology of the alfalfa so that it was no longer a suitable host for the dagger nematode (Norton, 1969). Likewise *M. incognita* suppressed *Scutel-lonema brachyurum* on cotton (Kraus-Schmidt and Lewis, 1981), limited

Table 7.3 Summary of interactions between ectoparasitic and sedentary endoparasitic nematodes

Nematode combination	Test host	Dominant species	General response	Comments	Reference
Paratrichodorus minor + Meloidogyne naasi	Creeping bentgrass	P. minor	Antagonistic	6 weeks, greenhouse tests. Competition for feeding sites	Sikora et al. (1972, 1979)
Tylenchorhynchus agri + M. naasi	Creeping bentgrass	T. agri	Antagonistic	Inhibition of root growth	
T. vulgaris + M. incognita	Bajra	T. vulgaris	Antagonistic		Vaishnav and Sethi (1978)
T. brassicae + M. incognita	Eggplant	None	Antagonistic		Khan et al. (1986b)
Criconemella ornata + M. incognita	Brinjal	C. ornata	Antagonistic		Misra and Das (1979)
Hoplolaimus indicus + M. incognita	Brinjal	H. indicus	Antagonistic		
T. nudus + M. incognita	Brinjal	T. nudus	Antagonistic		
C. xenoplax + M. hapla	Grape	C. xenoplax	Antagonistic		Santo and Bolander (1977)
P. minor + M. javanica	Tomato	P. minor	Antagonistic		Van Gundy and Kirkpatrick (1975)
H. columbus + M. incognita	Cotton	H. columbus	Antagonistic	Physiological alteration of the host. 60 days, greenhouse tests	Kraus-Schmidt and Lewis (1981)
M. incognita + Scutellonema brachyurum	Cotton		Antagonistic		
M. incognita + T. brassicae	Vegetables	T. brassicae	Antagonistic	Sex ratio of M. incognita increased toward males	Khan and Khan (1986)

Table 7.3 (Cont)

Nematode combination	Test host	Dominant species	General response	Comments	Reference
M. incognita + T. agri	Red clover	M. incognita	Antagonistic	3, 5, 7 months greenhouse tests	Amosu and Taylor (1975)
M. hapla + Xiphinema americanum	Alfalfa	M. hapla	Antagonistic	4 years field tests. Limited feeding sites and physiological changes	Norton (1969)
Hoplolaimus galeatus + M. incognita	Cotton	H. galeatus or M. incognita	Antagonistic	Edaphic factors are important	Bird et al. (1974)
Globodera tabacum + T. claytoni	Tobacco	None	Mutually antagonistic	Mutually suppressive in greenhouse tests, 4 weeks.	Miller and Wihrheim (1968)
	Tobacco	G. tabacum	Antagonistic	Field tests, cyst nematode is more persistent	
M. incognita + Hoplolaimus galeatus	Cotton	None	Mutually antagonistic		Yang et al. (1976)
M. incognita + Belonolaimus longicaudatus	Cotton	None	Mutually antagonistic		
M. incognita + C. ornata	Brinjal	None	Mutually antagonistic		Misra and Das (1979)
M. javanica + Hemicycliophora arenaria	Tomato	None	Mutually antagonistic		Van Gundy and Kirkpatrick (1975)

Combination	Crop		Interaction	Comments	Reference
M. incognita + H. columbus	Cotton	H. columbus	Stimulatory	H. columbus reproduction was stimulated	Kraus-Schmidt and Lewis (1981)
M. incognita + H. columbus	Soya bean	H. columbus	Antagonistic	H. columbus inhibited M. incognita; M. incognita had no effect on H. columbus. H. columbus is attracted to and penetrates more easily roots infected with M. incognita	Guy and Lewis (1987a, 1987b)
M. hapla + C. xenoplax	Grape	C. xenoplax	Stimulatory	C. xenoplax reproduction was stimulated	Santo and Bolander (1977)
M. incognita + T. vulgaris	Corn	(See comments)	Stimulatory	Penetration by M. incognita was enhanced	Kaul and Sethi (1982)
M. incognita + T. brassicae	Tomato	None	Mutually antagonistic	Greenhouse tests	Khan et al. (1986a)
M. incognita + S. brachyurum	Cotton	M. incognita	Stimulatory	M. incognita reproduction was stimulated	Kraus-Schmidt and Lewis (1981)
Rotylenchulus reniformis + T. brassicae	Tomato	None	Mutually antagonistic	Sex ratio increased toward males	Khan and Khan (1986)
Heterodera zeae + T. vulgaris	Corn	None	Mutually antagonistic	Prior inoculation inhibited penetration of other species	Kaul and Sethi (1982)
Aphelenchus avenae + M. incognita	Cucumber	A. avenae	Antagonistic	A. avenae inhibited reproduction by M. incognita	Choi et al. (1988); Ishibashi and Choi (1991)

Tylenchorhynchus brassicae on tomato (Khan and Haq, 1979) and antagonized *T. agri* on red clover (Amosu and Taylor, 1975). In the greenhouse, *T. claytoni* and *Globodera tabacum* were mutually antagonistic, but in the field the persistent cyst nematode *G. tabacum* dominated the ectoparasite (Miller and Wihrheim, 1968). *Meloidogyne incognita* and *Hoplolaimus galeatus* or *Belonolaimus longicaudatus* on cotton (Yang *et al.*, 1976), *M. incognita* and *Criconemella ornata* on brinjal (Misra and Das, 1979) and *M. javanica* and *Hemicycliophora arenaria* on tomato (Van Gundy and Kirkpatrick, 1975) were mutually antagonistic.

Interactions between ectoparasites and sedentary endoparasites may be beneficial to one or both species. *M. incognita* stimulated reproduction of *Hoplolaimus columbus* on cotton (Kraus-Schmidt and Lewis, 1981). *Tylenchorhynchus vulgaris* on bajra (Vaishnav and Sethi, 1978), and *M. hapla* enhanced the reproduction of *Criconemoides xenoplax* on Concord grape (Santo and Bolander, 1977). The mechanisms involved are unknown but may be physiological and/or physical.

7.1.4 Migratory endoparasites

Migratory endoparasites generally utilize the same feeding sites and are probably very competitive with each other, although it is not uncommon to find concomitant populations. *Pratylenchus* species often occur as mixed populations, but the difficulty of identifying individuals to species makes studies of interactions between species very difficult. Host suitability may be an important factor in their interaction. Interactions between concomitant populations of migratory endoparasitic nematodes are summarized in Table 7.4.

Pratylenchus penetrans inhibited reproduction by *P. alleni* on soya bean, and *P. alleni* altered the sex ratio in *P. penetrans* towards females (Ferris *et al.*, 1967). Field observations that *P. coffeae* suppressed *Scutellonema bradys* on guinea yam were supported in greenhouse tests (Acosta and Ayala, 1976). Climatic or edaphic factors may be involved in the interaction of concomitant populations. *P. coffeae* and *Radopholus similis* were mutually suppressive on citrus, but fine-textured soil favoured *P. coffeae* and coarse-textured soil favoured *R. similis* (O'Bannon *et al.*, 1976).

7.1.5 Migratory endo- and sedentary endoparasites

Migratory endoparasites disrupt plant tissue and often disturb feeding by sedentary endoparasites. *Meloidogyne incognita* was greatly inhibited by *Pratylenchus major* on pineapple (Guerount, 1968), by *P. brachyurus* on cotton (Gay and Bird, 1973), and by *Pratylenchus* spp. on tobacco (Graham *et al.*, 1964). *Heterodera glycines* was antagonized by *P. scribneri* on soya bean (Lawn and Noel, 1990). The interactions between migratory endoparasitic and sedentary endoparasitic nematodes are summarized in Table 7.5.

Table 7.4 Summary of interactions between migratory endoparasitic nematodes

Nematode combination	Test host	Dominant species	General response	Comments	Reference
Pratylenchus penetrans + *P. alleni*	Soya bean	*P. penetrans*	Antagonistic	Female-to-male sex ratio was higher in *P. penetrans*	Ferris *et al.* (1967)
Scutellonema bradys + *P. coffeae*	Guinea yam	*P. coffeae*	Antagonistic	Greenhouse tests supported field observations	Acosta and Ayala (1976)
Radopholus similis + *P. coffeae*	Citrus	None	Mutually antagonistic	10–15 months, greenhouse tests. *P. coffeae* dominated in fine-textured soils; *R. similis* in coarse soils	O'Bannon *et al.* (1976)

Table 7.5 Summary of interactions between migratory and sedentary endoparasitic nematodes

Nematode combination	Test host	Dominant species	General response	Comments	Reference
Pratylenchus major + *Meloidogyne* spp.	Pineapple	*P. major*	Antagonistic		Guerout (1968)
P. brachyurus + *M. incognita*	Cotton	*P. brachyurus*	Antagonistic		Gay and Bird (1973)
Pratylenchus spp. + *M. incognita*	Tobacco	*Pratylenchus* spp.	Antagonistic	Field observations	Graham *et al.* (1964)
M. incognita + *P. brachyurus*	Tomato	*M. incognita*	Antagonistic	*M. incognita* suppresses penetration by *P. penetrans*	Gay and Bird (1973)
M. incognita + *P. brachyurus*	Cotton	*P. brachyurus*	Stimulatory	*M. incognita* stimulates penetration by *P. penetrans*	
P. penetrans + *M. naasi*	Creeping bentgrass	*P. penetrans*	Antagonistic	Initially antagonistic but no effect after 10 months	Sikora *et al.* (1972)
M. incognita + *P. penetrans*	Red clover	*M. incognita*	Antagonistic	Initially neutral but antagonistic after 3 months	Amosu and Taylor (1975)
M. incognita + *P. penetrans*	Tomato	None or *M. incognita*	Mutually antagonistic	Split-root experiments; translocatable factor antagonistic to *P. penetrans*	Estores and Chen (1970, 1972)
			Antagonistic		
M. incognita + *P. brachyurus*	Tobacco	None	Mutually antagonistic	Susceptible cultivar 'NC2326'	Johnson and Nusbaum (1970)

Combination	Crop		Interaction	Susceptible cultivar 'Hicks'	Reference
M. incognita + *P. brachyurus*	Tobacco	*M. incognita*	Antagonistic	Resistant cultivars 'NC95' and 'NC2512'	
M. incognita + *P. brachyurus*	Tobacco	*P. brachyurus*	Stimulatory	Resistant cultivars 'Hicks' and 'NC95'	
M. hapla + *P. brachyurus*	Tobacco	*P. brachyurus*	Antagonistic		
M. hapla + *P. brachyurus*	Tobacco	None	Neutral	Cultivars 'NC2326' and 'NC2512'	
M. incognita + *P. penetrans*	Red clover	*M. incognita*	Antagonistic	*P. penetrans* deposited 37% fewer eggs in the presence of low numbers of *M. incognita* and 57% with high numbers	Chapman and Turner (1972, 1975)
Globodera tabacum + low numbers of *P. penetrans*	Tobacco	*G. tabacum*	Antagonistic	*G. tabacum* generally occurs monospecifically	Miller (1970); Miller and Wihrheim (1968)
Globodera tabacum + high numbers of *P. penetrans*	Tobacco	*P. penetrans*	Antagonistic		
M. incognita + *P. vulnus*	Grape	*M. incognita*	Antagonistic	*M. incognita* inoculated 1 month before *P. vulnus*	Chitamber and Raski (1984)
M. incognita + *P. brachyurus*	Soya bean	None	Mutually antagonistic	Density dependent	Herman *et al.* (1988)
M. incognita + *P. brachyurus*	Soya bean	*M. incognita*	Antagonistic	Split-root systems; translocatable factor inhibited reproduction of *P. brachyurus*	Herman *et al.* (1988)
M. javanica + *P. sefaensis*	Corn cowpea	*M. javanica*	Antagonistic	Penetration by *P. safaensis* inhibited	Egunjobi *et al.* (1986)

Table 7.5 (Cont)

Nematode combination	Test host	Dominant species	General response	Comments	Reference
Heterodera glycines + *P. scribneri*	Soya bean	*P. scribneri*	Antagonistic	Interaction not affected by temperature	Lawn and Noel (1990)
P. coffeae + *Tylenchulus semipenetrans*	Citrus	None	Mutually antagonistic	Field observations	Kaplan and Timmer (1982)
Radopholus similis + *M. incognita*	Black pepper	None	Mutually antagonistic	Greenhouse experiments	Sheela and Venkitesan (1981)
M. incognita + *P. brachyurus*	Tobacco	*P. brachyurus*	Stimulatory	*M. incognita* increased the reproduction of *P. brachyurus*	Johnson and Nusbaum (1970)
M. naasi + *P. penetrans*	Creeping bentgrass	*P. penetrans*	Stimulatory	*M. naasi* increased the reproduction of *P. penetrans*	Sikora *et al.* (1972)
Ditylenchus dipsaci + *M. hapla*	Alfalfa	None	Stimulatory	Resistance was reduced by infection with *D. dipsaci*	Griffin (1972, 1980)
Rotylenchulus reniformis + *P. sefaensis*	Corn, cowpea	*R. reniformis*	Stimulatory	Penetration by *R. reniformis* enhanced	Egunjiobi *et al.* (1986)
Ditylenchus dipsaci + *M. hapla*	Alfalfa, crop bean, tomato, sugar-beet, sweet clover, wheat	None	None	No effect on nematode reproduction; synergistic effect on suppression of plant growth	Griffin (1987)

Migratory endoparasites often penetrate the host faster or they inhibit the penetration by the sedentary species (Gay and Bird, 1973; Turner and Chapman, 1971, 1972). Sometimes, however, the sedentary species may penetrate faster (Freckman and Chapman, 1972). The role that sedentary endoparasites play in penetration of the migratory species may be affected by host suitability. Prior inoculation by *Meloidogyne incognita* inhibited penetration by *Pratylenchus brachyurus* in tomato, a better host for *M. incognita* than *P. brachyurus*; but in cotton, a better host for *P. brachyurus* than *M. incognita*, prior inoculation by *M. incognita* slightly stimulated penetration by *P. brachyurus* (Gay and Bird, 1973).

Interactions between migratory and sedentary endoparasitic nematodes may be time dependent. *Meloidogyne incognita* was initially inhibited by *Pratylenchus penetrans* on creeping bentgrass, but had no effect after ten months (Sikora *et al.*, 1972). *P. penetrans* on red clover was not affected by *M. incognita* after three months, but was suppressed after five months (Amosu and Taylor, 1975).

Differences in host tissue preference and nematode feeding mechanisms may limit interaction between migratory and sedentary endoparasites. *Ditylenchus dipsaci* and *Heterodera schachtii* were not affected by each other, probably because they occupied different sites in the host (Griffin, 1987).

Nematode interactions can be affected by the timing of inoculations. Reproduction of *Pratylenchus vulnus* was greatly inhibited after 125 days when *Meloidogyne incognita* was inoculated one month prior to *P. vulnus*, but in simultaneous inoculations the inhibition was delayed until 250 days (Chitamber and Raski, 1984).

Migratory endoparasites are less advanced parasites than sedentary endoparasites which establish a complex relationship with the host and alter plant physiology. This change in physiology often affects the suitability of the host for the migratory endoparasite. *Meloidogyne incognita* induced a translocatable factor in split-root experiments that greatly inhibited the reproduction of *Pratylenchus penetrans*. *P. penetrans* also suppressed *M. incognita* but a translocatable factor was not involved. The inhibition may have been caused by disruption of feeding sites created by the migratory endoparasite (Estores and Chen, 1970, 1972). On corn, *M. javanica* limited penetration by *P. sefaensis* (Egunjobi *et al.*, 1986).

Host suitability affects competition between species. *Pratylenchus brachyurus* and *Meloidogyne incognita* were mutually suppressive on *M. incognita*-susceptible tobacco cultivar 'NC2326' (Johnson and Nusbaum, 1970). On susceptible cultivar 'Hicks', *P. brachyurus* had no effect on *M. incognita* but *M. incognita* inhibited *P. brachyurus*. On root-knot-resistant cultivars 'NC95' and 'NC2512', *M. incognita* stimulated the reproduction of *P. brachyurus*. In comparison *M. hapla* was suppressed by *P. brachyurus* on Hicks and NC95, but not on NC2326 or NC2515. *P. brachyurus* was inhibited by *M. hapla* on Hicks and NC2326, but not on NC95 and NC2512. Prior infection by *Ditylenchus dipsaci* on alfalfa reduced the resistance to *M. hapla* (Griffin, 1980).

The effect of host suitability on nematode interactions is density dependent. Low and high numbers of *Meloidogyne incognita* caused 37% and 57%, respectively, fewer eggs to be deposited by *Pratylenchus penetrans* (Chapman and Turner, 1972, 1975). Population density may also affect interactions that are mutually suppressive. *Globodera tabacum* was inhibited by high numbers of *P. penetrans* on tobacco in the greenhouse, and low numbers of *P. penetrans* inhibited *G. tabacum*; however, in the field *G. tabacum* usually occurs in monospecific communities (Miller, 1970; Miller and Wihrheim, 1968).

The species that predominates in antagonistic interactions between sedentary and migratory endoparasites is usually determined by climatic or edaphic factors. *Pratylenchus coffeae* and *Tylenchulus semipenetrans* on citrus (Kaplan and Timmer, 1982), *Radopholus similis* and *Meloidogyne incognita* on black pepper (Sheela and Venkitesan, 1981), and *P. brachyurus* and *M. incognita* on soya bean (Herman *et al.*, 1988) were mutually suppressive.

Migratory and sedentary endoparasites may also not compete with each other (Griffin, 1983) or their interaction may be stimulatory to one or both species. Reproduction by *Pratylenchus brachyurus* on tobacco was higher in the presence of *Meloidogyne incognita* (Johnson and Nusbaum, 1970). Likewise, *M. naasi* stimulated reproduction of *P. brachyurus* on creeping bentgrass (Sikora *et al.*, 1972). Prior infection of a root-knot-resistant cultivar of alfalfa by *Ditylenchus dipsaci* resulted in the reduction of resistance to *M. hapla* (Griffin, 1972, 1980).

7.1.6 Sedentary endoparasites

Sendentary endoparasites are highly specialized parasites and have a long-lasting relationship with their host. Competition between species is generally mutually suppressive because they utilize the same sites for feeding (Norton, 1978), and often cause drastic physiological changes in the host tissues. Interactions between sedentary endoparasitic nematodes are summarized in Table 7.6. Interactions that occur between species of the same genus are discussed in a separate section.

Meloidogyne species may inhibit, have no effect, or stimulate *Heterodera* species. *H. oryzicola* was inhibited by *M. graminicola* on rice in simultaneous and sequential inoculations (Rao and Prasad, 1981). *M. incognita* and *H. cajani* on cowpea (Sharma and Sethi, 1978) and *M. hapla* and *H. schachtii* on sugar-beet (Jatala and Jensen, 1976a,b, 1983) did not interact; however, prior inoculation of *M. hapla* on sugar-beet stimulated *H. schachtii*.

Heterodera species may inhibit or have no effect on *Meloidogyne* species. *H. zeae* inoculated three days prior to *M. incognita* suppressed reproduction

Table 7.6 Summary of interactions between sedentary endoparasitic nematodes

Nematode combination	Test host	Dominant species	General response	Comments	Reference
Meloidogyne graminicola + Heterodera oryzicola	Rice	M. graminicola	Antagonistic	52 days, greenhouse tests	Rao and Prasad (1981)
M. incognita + H. cajani	Cowpea	None	Neutral		Sharma and Sethi (1978)
M. hapla + H. schachtii	Sugar-beet	None	None or stimulatory	Prior infection by M. hapla stimulated infection by H. schachtii	Jatala and Jensen (1976a, 1976b)
M. hapla + H. schachtii	Tomato	H. schachtii	Antagonistic		Griffin and Waite (1982)
M. hapla + H. schachtii	Tomato	M. hapla	Antagonistic	M. hapla inoculated 20 days prior to H. schachtii	Griffin (1985)
M. hapla + H. schachtii	Tomato	H. schachtii	Antagonistic	Simultaneous inoculations; inhibition more pronounced as temperature increased	
Low numbers of M. incognita + H. glycines	Soya bean	None (see comments)	Neutral		Ross (1964)
High numbers of M. incognita + H. glycines	Soya bean		Antagonistic or stimulatory	Antagonistic early in the season; stimulatory late in the season	
M. incognita + H. glycines	Soya bean	(See comments)	Antagonistic	Low populations of cyst suppressed by M. incognita	Niblack et al. (1986)
M. incognita + H. zeae	Corn	(See comments)	Antagonistic	Prior inoculation of H. zeae suppressed M. incognita	Kaul and Sethi (1982)

Table 7.6 (Cont)

Nematode combination	Test host	Dominant species	General response	Comments	Reference
M. graminicola + H. oryzicola	Rice	M. graminicola	Antagonistic	30 days, greenhouse tests	Rao and Prasad (1981)
M. incognita + H. cajani	Cowpea	M. incognita	Antagonistic or mutually antagonistic	Density dependent	Sharma and Sethi (1978)
H. schachtii + M. hapla	Sugar-beet	H. schachtii	Antagonistic or stimulatory	Prior inoculation of H. schachtii suppressed M. hapla; whereas prior inoculation with M. hapla stimulated H. schachtii	Jatala and Jensen (1976b, 1983)
M. incognita + Rotylenchulus reniformis	Soya bean	M. incognita	Antagonistic		Singh (1976)
M. incognita + R. reniformis	Grape	None	Mutually antagonistic	M. incognita is more competitive over time	Rao and Seshadri (1981)
M. incognita + R. reniformis	Black gram	M. incognita	Antagonistic		Mishra and Gaur (1981)
Low numbers of M. incognita + R. reniformis	Tomato	R. reniformis	Antagonistic	Density dependent	Kheir and Osman (1977); Winoto and Lim (1972)
High numbers of M. incognita + R. reniformis	Tomato	M. incognita	Antagonistic		
M. javanica + R. reniformis	Cowpea	None	Neutral	R. reniformis more competitive initially, but M. javanica is not affected over time	Rao and Prasad (1971); Taha and Kassab (1979, 1980)

Nematode combination	Crop	Predominant species	Interaction	Comments	Reference
Low number of R. reniformis + M. incognita	Sweet potato	R. reniformis	Antagonistic	Density dependent	Thomas and Clark (1980, 1981, 1983a, 1983b)
High numbers of M. incognita + R. reniformis	Sweet potato	M. incognita	Antagonistic	Density dependent	Taha and Sultan (1977)
Tylenchulus semipenetrans + R. reniformis	Grape	None	Mutually antagonistic		Pathak et al. (1985)
M. incognita + R. reniformis	Pigeon pea	R. reniformis	Antagonistic	Density dependent	Khan and Khan (1986)
M. incognita + R. reniformis	Vegetables	None	Antagonistic	Sex ratios increased toward males	Khan et al. (1985, 1986a)
M. incognita + R. reniformis	Tomato	None	Antagonistic	Inoculum density dependent	Khan et al. (1986b)
	Eggplant		Mutually antagonistic		Khan et al. (1987)
	Cauliflower		Mutually antagonistic		
M. incognita + R. reniformis	Cowpea	(See comments)	Stimulatory	Cultivar S-488 lost resistance to R. reniformis in the presence of M. incognita	Khan and Husain (1989)
M. incognita + R. reniformis	Cowpea	(See comments)	Stimulatory	M. incognita increased the reproduction of R. reniformis	
Nacobbus aberrans + H. schachtii	Sugar-beet	N. aberrans	Antagonistic	Time dependent	Inserra et al. (1984)
N. aberrans + M. hapla	Sugar-beet	N. aberrans	Antagonistic	Time dependent	

of *M. incognita* (Kaul and Sethi, 1982). *M. hapla* was inhibited by *H. schachtii* on tomato (Griffin and Waite, 1982) and sugar-beet (Jatala and Jensen, 1976a, 1976b, 1983).

The relationship between the root-knot and cyst nematodes may be density and time dependent. Low populations of *Meloidogyne incognita* had no effect on *Heterodera glycines* on soya bean, but high populations were antagonistic early and stimulating late in the season (Ross, 1964). *M. incognita* was detrimental to *H. cajani* on cowpea when established first, but prior infection by *H. cajani* was detrimental to *M. incognita* (Sharma and Sethi, 1978). In comparison, simultaneous inoculations were density dependent, but generally mutually antagonistic. Likewise, prior inoculation of *H. schachtii* on sugar-beet suppressed *M. hapla* reproduction, but previous infection by *M. hapla* stimulated the cyst nematode (Jatala and Jensen, 1976a,b, 1983).

Temperature may affect the interaction between cyst and root-knot nematode species. The inhibition of *Meloidogyne hapla* by *Heterodera schachtii* on tomato increased as the plant growth temperatures were raised (Griffin, 1985). Interactions between *Meloidogyne* and *Rotylenchulus* species can be mutually antagonistic (Khan *et al.*, 1985) or suppressive to just one species (Pathak *et al.*, 1985). *R. reniformis* was inhibited by *M. incognita* on soya bean, but the root-knot nematode was not affected by the reniform nematode (Singh, 1976). Similar interactions occurred on black gram (Mishra and Gaur, 1981), but on grape the two species were mutually antagonistic, although *M. incognita* was more competitive over time (Rao and Seshadri, 1981). *R. reniformis* generally limited reproduction of *M. incognita* on tomato, except high populations of root-knot nematode inhibited reproduction of the reniform nematode (Kheir and Osman, 1977; Winoto and Lim, 1972). *R. reniformis* initially inhibited *M. javanica* on tomato, but the root-knot nematode was more competitive over time (Taha and Kassab, 1979, 1980). *R. reniformis* has a short life cycle which allowed it to dominate *M. javanica* after 60 days, but not after 90 days (Rao and Prasad, 1971).

Interactions between reniform and root-knot nematodes may be time and density dependent. *Meloidogyne incognita* was inhibited by low levels of *Rotylenchulus reniformis* on sweet potato, and low levels of *M. incognita* had no effect on the reniform nematode. High levels of *M. incognita* suppressed *R. reniformis* and the root-knot nematode was not affected. Each species was capable of dominating, depending upon the initial density (Thomas and Clark, 1980, 1981, 1983a, 1983b). The suitability of the host also plays an important role in the interaction of species (Khan *et al.*, 1986a, 1986b, 1987).

Interactions between *Tylenchulus semipenetrans* and *Rotylenchulus reniformis* on grape were mutually antagonistic. Populations resulting from concomitant inoculations were significantly lower than populations from monospecific inoculations (Taha and Sultan, 1977). Likewise, *Nacobbus aberrans* was antagonistic to both *Heterodera schachtii* and *Meloidogyne hapla*, especially the root-knot nematode (Inserra *et al.*, 1984). The root-knot and cyst nematodes did not affect the reproduction of *N. aberrans*, which was less aggressive than the other two species.

7.1.7 Interactions between species of the same genus

Reports about interactions between two species of the same genus are most common in *Meloidogyne* and *Heterodera*. A few cases of interactions between species of *Globodera* and *Pratylenchus* have been reported. Interactions between species of the same genus are summarized in Table 7.7.

Competition between mixtures of *Pratylenchus* species were antagonistic to *P. alleni* on soya bean, and the female-to-male sex ratio was altered in *P. penetrans* towards males (Ferris *et al.*, 1967). Although mixtures of *Pratylenchus* are common in the field, studies about interactions of species in this genus are difficult because many times it is nearly impossible to identify an individual specimen to species.

Two or more root-knot nematode species are commonly found in the same field, root system or gall (Minz and Strich-Harari, 1959). Factors other than competition may be important in the domination of a particular species. Temperature and other climatic factors may be important because certain species may be better adapted to cooler temperatures, whereas other species are more common in warmer climates (Taylor *et al.*, 1982).

Meloidogyne javanica and *M. hapla* were dominated by *M. incognita* at high temperatures, but *M. javanica* dominated *M. incognita* and *M. hapla* at low temperatures (Minz and Strich-Harari, 1959). At high temperatures, simultaneous inoculations of *M. incognita* and *M. hapla* resulted in a population of 90% *M. incognita* and 10% *M. hapla*. At low temperatures, the population was only 57% *M. incognita* (Chapman, 1965). Competition between species can be affected by factors other than temperature. *M. javanica* dominated *M. hapla* at 20°C, even though low temperatures generally favour *M. hapla* (Minz and Strich-Harari, 1959). *M. hapla* suppressed *M. incognita* on root-knot-resistant tobacco cultivars 'NC95' and 'NC2512' at moderate temperatures (Johnson and Nusbaum, 1970). Competition may be density dependent, particularly for *M. hapla* (Kinloch and Allen, 1972). In simultaneous inoculations of *M. incognita* and *M. hapla*, *M. incognita* penetrated the roots faster than *M. hapla*. All of the feeding sites were occupied and destroyed by the hypersensitive reaction (Johnson and Nusbaum, 1970). Likewise, prior inoculation of *M. incognita* on *M. arenaria*-susceptible soya bean resulted in a decrease in the numbers of galls and eggs produced by *M. arenaria* (Ibrahim, 1987; Ibrahim and Lewis, 1986). However, prior inoculation of *M. arenaria* or *M. hapla* on root-knot-resistant tobacco masked the resistance to *M. incognita* (Eisenback, 1983; Ibrahim and Lewis, 1986). This effect on resistance may be a factor in the persistence of *M. incognita* in tobacco fields where resistance

Table 7.7 Summary of interactions between species of the same genus

Nematode combination	Test host	Dominant species	General response	Comments	Reference
Pratylenchus penetrans + *P. alleni*	Soya bean	*P. penetrans*	Antagonistic	Female-to-male sex ratio was higher in *P. penetrans*	Ferris *et al.* (1967)
Meloidogyne javanica + *M. incognita*	Tomato	*M. javanica* or *M. incognita*	Antagonistic	Greenhouse tests; *M. javanica* dominated at low temperatures, *M. incognita* at high temperatures	Minz and Strich-Harari (1959)
M. javanica + *M. hapla*	Tomato	*M. javanica*	Antagonistic	*M. javanica* dominated at high and low temperatures	Chapman (1965); Kinloch and Allen (1972)
Meloidogyne javanica + *M. incognita*	Tomato	*M. incognita* race 2	Antagonistic	Competition was not intense	Khan and Haider (1991)
M. incognita + *M. hapla*	Tomato	*M. incognita*	Antagonistic	Temperature affected level of dominance	Chapman (1965)
M. incognita + *M. hapla*	Tobacco	*M. incognita*	Antagonistic	*M. incognita* penetrated faster and 'occupied' all of the feeding sites	Johnson and Nusbaum (1970)
M. incognita + *M. arenaria*	Tobacco	(See comments)	Stimulatory	Field resistance to *M. incognita* overcome by infection with *M. arenaria*	Tedford *et al.* (1986)
M. incognita + *M. arenaria*	Soya bean	(See comments)	Antagonistic	*M. arenaria*-susceptible cultivar; reproduction on *M. arenaria* reduced by infection with *M. incognita*	Ibrahim and Lewis (1986)

Species combination	Crop/Host		Interaction	Comments	Reference
M. incognita + M. arenaria	Soya bean	(See comments)	Antagonistic	M. arenaria-resistant cultivar; eggs and galls reduced by prior inoculation with M. incognita	Ibrahim and Lewis (1986)
Globodera rostochiensis + G. pallida	Potato	G. rostochiensis	Antagonistic	G. pallida becomes more competitive when small; populations from New Zealand	Marshall (1989)
G. rostochiensis + G. pallida	Potato	G. pallida	Antagonistic	Populations from England	Parrot et al. (1976)
G. rostochiensis + G. pallida	Potato	(See comments)		More competition between individuals of G. pallida than with G. rostochiensis	Seinhorst and Oostrom (1989)
G. rostochiensis + G. pallida	Potato	(See comments)	Hybridization	Fertile hybrids produced with intermediate morphology	Miller (1983)
G. rostochiensis + G. pallida	Potato	(See comments)	Hybridization	Naturally occurring hybrids	Olsson (1985, 1988)
Heterodera schachtii + H. glycines	Soya bean + sugar-beet	(See comments)	Hybridization	Fertile hybrids produced	Potter and Fox (1965)
Heterodera schachtii + H. glycines	Soya bean + sugar-beet	(See comments)	Hybridization	Fertile hybrids produced with intermediate morphology and host response	Miller (1988)
H. glycines + H. cruciferae	Soya bean + cabbage + lespedeza	(See comments)	Hybridization	Fertile hybrids produced with intermediate morphology and host response	Miller (1989)

Table 7.7 (Cont)

Nematode combination	Test host	Dominant species	General response	Comments	Reference
M. hapla + M. hapla	Sugar-beet	(See comments)	Neutral	Double inoculations 10 days apart increased	Jatala and Jensen (1976b)
M. hapla + M. hapla	Tomato	(See comments)	Inhibitory	Diploid isolate more competitive than polyploid	Triantaphyllou (1991)
H. schachtii + H. schachtii	Sugar-beet	(See comments)	Stimulatory	Infections 3–5 times	

has been utilized for several years (Tedford *et al.*, 1986). Double inoculations of *M. hapla* on sugar-beet had little effect on each other, whereas double inoculations of *Heterodera schachtii* on sugar-beet that were ten days apart caused a three to fivefold increase in the number of infections as compared to that of single inoculations (Jatala and Jensen, 1972). Some races of root-knot nematodes appear to be more competitive than others. *M. javanica* and races of *M. incognita* were usually mutually inhibitory, although race 2 of *M. incognita* was more competitive than the other three host races (Khan and Haider, 1991).

Globodera rostochiensis and *G. pallida* virtually occupy the same ecological niche and cannot coexist (Parrot *et al.*, 1976). They may occur together in varying proportions for different periods of time, but eventually one species dominates (Kort and Bakker, 1980). In New Zealand *G. rostochiensis* was antagonistic to *G. pallida*, but at very low populations *G. pallida* was more competitive and was probably able to persist at very low population densities (Marshall, 1989). However, in England, *G. pallida* was more competitive than *G. rostochiensis* (Parrot *et al.*, 1976). *G. rostochiensis* may hatch more freely and survive less efficiently at lower soil temperatures. Conversely, in The Netherlands, more competition occurred between individuals of *G. pallida* than with mixtures of *G. rostochiensis* (Seinhorst, 1986; Seinhorst and Oostrom, 1989).

Competition between species within the same genus is probably related to feeding sites, relative hatching rates, and other genetic differences. Whatever the reason, concomitant populations of these two sibling species complicate the use of resistance and the predicting yield loss based on initial populations (Jones and Perry, 1978).

Competition between species in the same genus may be further complicated by interspecific hybridization. *Globodera rostochiensis* and *G. pallida* produced fertile hybrids in greenhouse studies (Miller, 1983). Interspecific crosses of *Globodera* species were apparent in field populations based on host specificity and the length of styles of second-stage juveniles (Kort and Jaspers, 1973) and spicule morphology of males (Olsson, 1985, 1988). Morphologically divergent hybrids indicated that cross-breeding between *G. rostochiensis* and *G. pallida* occurred naturally in the field. Likewise, interspecific crosses of *Heterodera* species were successful in greenhouse tests (Miller, 1983). Fertile hybrids were obtained from crosses of *H. schachtii* and *H. glycines* (Potter and Fox, 1965; Miller, 1988) and *H. glycines* and *H. cruciferae* (Miller, 1989). Hybrids of *H. schachtii* and *H. cruciferae* were intermediate in their morphology, and they were able to reproduce on both sugar-beet and soya bean, whereas their parents could only reproduce on one host or the other (Miller, 1988). Likewise, hybrids of *H. glycines* and *H. cruciferae* reproduced on soya bean, cabbage and sugar-beet; but *H. glycines* reproduced only on soya bean and *H. cruciferae* reproduced on cabbage, but not on soya bean or sugar-beet (Miller, 1989).

7.1.8 Interactions between races of the same species

Races of the same species also interact with each other as they compete for feeding sites in the host (Price *et al.*, 1976). Since they are not sexually isolated, races also hybridize with each other (Price *et al.*, 1978). Multiple matings further complicate the role that competition plays in interactions between species (Green *et al.*, 1970; Triantaphyllou and Esbenshade, 1990). In the field it seems as if nematodes' movements are restricted so that natural inbreeding produces isolates that may be quite variable in morphology and parasitic capability. Comparisons of 11 isolates of *Heterodera glycines* from one location showed that each isolate was morphologically distinct and also a separate physiological race (Miller, 1970). Crosses between the isolates were successful and the hybrids had intermediate characteristics of their parents. Likewise, crosses of two races of *H. avenae* produced viable offspring with parasitic capabilities that were genetically inherited (Andersen, 1965). Races of *Ditylenchus dipsaci* may cross and produce fertile hybrids (Ericksson, 1965, 1974; Webster, 1967). The host ranges of the hybrids may be similar to one or both parents, or they may be able to parasitize additional hosts that neither parent could (Sturhan, 1966). Thus, competition between races and their hybrids may be important in the selection and development of additional host races within the species (Sturhan, 1971).

Cytological forms of a single species can interact with each other. The frequency of a tetraploid isolate of *Meloidogyne hapla* was reduced from 50% to about 9% by competition from a diploid isolate after six generations on tomato (Triantaphyllou, 1991). Attempts to cross two cytological forms revealed that they were reproductively isolated, which may provide stability to the tetraploids and increase their chances for successful establishment in nature (Triantaphyllou, 1991).

7.2 AETIOLOGICAL INTERACTION

Studies on the causes of disease often consider a single nematode species as the entity responsible for both the symptoms and the effect on plant growth (Norton, 1978). However, because nematodes rarely occur in monospecific communities (Oostenbrink, 1966), they may interact with each other to alter the course of the disease (Duncan and Ferris, 1981, 1982, 1983). If the amount of disease caused by both nematodes is less than the combined effect of each alone, the interaction is negative (antagonistic); if it is more, the interaction is positive (synergistic); and if it is the same, there is no interaction (Burrows, 1987; Wallace, 1983). The interactions of concomitant populations of nematodes that affect disease expression are summarized in Table 7.8.

Sometimes interactions may be difficult to detect in greenhouse experiments because they may be density dependent. In microplot studies with *Meloidogyne incognita* and *Heterodera glycines* on soya bean, yield varied

Table 7.8 Summary of nematode–nematode interactions that affect disease expression

Nematode combination	Additional pathogens	Test host	Plant growth response and comments	Reference
Negative interactions				
Meloidogyne incognita + Pratylenchus penetrans	None	Tomato	M. incognita suppressed growth more by itself than when combined with P. penetrans	Estores and Chen (1970)
M. incognita + Rotylenchulus reniformis	None	Black gram	Growth was less in combined inoculations than in single inoculations	Mishra and Gaur (1981)
M. incognita + R. reniformis	None	Grape	Growth was less in combined inoculations than in single inoculations	Rao and Seshadri (1981)
Positive interactions				
M. hapla + H. schachtii	None	Tomato	Combined inoculations suppressed root growth by 65, 64 and 61% below that of controls and single inoculations	Griffin and Waite (1982)
M. javanica + M. incognita	None	Tobacco	Combined inoculations significantly suppressed growth of cultivar 'Hicks', whereas single inoculations did not	Paez et al. (1976)
M. hapla + Ditylenchus dipsaci	None	Alfalfa	Resistance of cultivar 'Vernal 298' to D. dipsaci was reduced by infection by M. hapla	Griffin (1980)
M. arenaria + M. incognita M. hapla + M. incognita	None	Tobacco	Resistance of cultivar 'NC95' to M. incognita was reduced by prior infection with M. arenaria or M. hapla	Eisenback (1983)
M. incognita + M. javanica	None	Soya bean	Galling was more severe with both species than with either alone	McGawley and Winchell (1987)

from slightly less to slightly more than additive, depending on the level of nematode inoculum (Ross, 1964).

7.2.1 Negative interactions

Less disease occurs where there is strong competition between nematode species. The less pathogenic species reduced the number of infection sites for the more pathogenic species (Duncan and Ferris, 1981, 1982, 1983). Other factors, such as physiological changes within the plant, may also be involved in the relationship. Negative interactions of nematode species usually involve very strong, pathogenic competitors. Either of the two species can predominate depending upon the inoculum level and other factors, but the two species usually cannot coinhabit. *Meloidogyne incognita* suppressed the growth of tomato more by itself than in combination with *Pratylenchus penetrans* (Estores and Chen, 1970). Similar responses occurred with *M. incognita* and *Rotylenchulus reniformis* on black gram (Mishra and Gaur, 1981) and grape (Rao and Seshadri, 1981).

7.2.2 Positive interactions

Nematodes may predispose a plant to attack by another organism (Pitcher, 1978; Powell, 1971, 1979). The mechanisms involved are not fully understood, but may involve a change in host physiology that makes it easier to penetrate and establish a parasitic relationship. Interruption of defence mechanisms may also be involved. All of the positive interactions that have been reported for concomitant populations of nematodes involve a species of *Meloidogyne* and some other endoparasite.

Heterodera schachtii and *Meloidogyne hapla* suppressed growth of tomato by 65%, 64% and 61% below that of uninoculated controls and single inoculations of either *M. hapla* or *H. schachtii*, respectively (Griffin and Waite, 1982). Likewise, growth of tobacco was suppressed more by a combination of *M. javanica* and *M. incognita* than by either species alone (Paez et al., 1976).

The interaction of concomitant populations of nematodes may be time dependent. Combined inoculations of *Heterodera schachtii*, *Meloidogyne hapla* and *Nacobbus aberrans* significantly suppressed growth of sugar-beet (Inserra et al., 1984), and *M. naasi*, *Pratylenchus penetrans* and *Tylenchorhynchus agri* were more pathogenic on creeping bentgrass in combination than in monospecific inoculations (Sikora et al., 1972). The greatest inhibition of growth occurred when *T. agri* and *Paratrichodorus minor* preceded *M. naasi* by three weeks (Sikora et al., 1979).

A combination of *Ditylenchus dipsaci* and *Meloidogyne hapla* reduced the growth of alfalfa cultivar 'Ranger' which is susceptible to both nematodes (Griffin, 1980). The two nematodes did not affect the resistance or susceptibility of alfalfa to *D. dipsaci*, but the resistance to *M. hapla* of cultivar

'Vernal 298' was reduced by infection with *D. dipsaci*. Resistance in tobacco cultivar 'NC 95' to *M. incognita* race 1 was reduced by prior infection by *M. arenaria* or *M. hapla* (Eisenback, 1983). Prior infection with *M. javanica* or *M. incognita* race 4 had no effect on the resistance to *M. incognita* race 1. Likewise, *M. incognita* and *M. javanica* interacted to produce severe galling on soya bean, whereas *M. javanica* alone excited only moderate galling and single inoculations of *M. incognita* did not induce galls (McGawley and Winchell, 1987).

7.2.3 Interaction with other pathogens

Nematodes often interact with other organisms to cause plant disease (Pitcher, 1978; Powell, 1971, 1979). When two nematodes are present with another organism they may interact to make the disease even more severe. The mechanical damage and the physiological changes made by the nematodes may be involved in the mechanism of the interaction. Interactions between concomitant populations of nematodes and plant pathogens are summarized in Table 7.9.

Meloidogyne incognita, *Belonolaimus longicaudatus* and *Pythium aphanidermatum* caused more suppression of the growth of chrysanthemum than did any pathogen alone or any single nematode and fungus combination (Johnson and Littrell, 1970). In jute, *M. incognita* and *Hoplolaimus indicus* interacted with a fungus to inhibit plant growth the most and to produce the most severe disease symptoms (Haque and Mukhopadhyaya, 1979). *M. incognita* and *Rotylenchulus reniformis* plus the fungus *Rhizoctonia solani* on cowpea caused more plant disease than either pathogen alone or any two organisms combined (Khan and Husain, 1989). *Pratylenchus penetrans* and *P. coffeae* interacted with *Fusarium* to produce more wilt symptoms and suppression of plant growth than did either nematode alone or in combination with the fungus (Hirano and Kawamura, 1972). Reproduction of *P. coffeae* was stimulated by the presence of the fungus.

7.3 CONCLUSIONS

The nematode community is dynamic and its members are constantly interacting with each other. This interaction may affect the population dynamics of the species involved and the relationship between plant growth as affected by feeding and parasitism by the nematode community. Concomitant populations of nematodes can interact with each other to affect their reproductive ability and they can alter the aetiology of disease. Although the information presented herein is useful, most previous studies have been qualitative and provide limited information. It is therefore impossible to predict how a given combination of nematode species will affect plant growth and nematode population dynamics. The nature of the host–parasite relationship, differences in pathogenicity, and environmental effects undoubtedly play an

Table 7.9 Summary of nematode–nematode interactions that affect disease expression with an additional pathogen

Nematode combination	Additional pathogens	Test host	Plant growth response and comments	Reference
Interactions with other pathogens				
Meloidogyne incognita + Rotylenchulus reniformis	Rhizoctonia solani	Cowpea	All three cause more plant disease than any combination organisms combined	Khan and Husain (1989)
M. incognita + Belonolaimus longicaudatus	Pythium aphanidermatum	Chrysanthemum	All three pathogens interacted to cause more growth suppression than did any pathogen alone or any single nematode and fungus combination	Johnson and Littrell (1970)
M. incognita + Hoploiaimus indicus	Macrophomina phaseoli	Jute	All three pathogens produced the most severe disease symptoms	Haque and Mukhopadhyaya (1979)
Pratylenchus penetrans + P. coffeae	Fusarium oxysporum	Tomato	All three pathogens caused the greatest inhibition of growth and more severe wilt symptoms	Hirano and Kawamura (1972)
Radopholus similis + M. incognita	Fusarium sp.	Pepper	Suppression of plant growth and yellowing symptoms were most severe when all 3 pathogens were present	Mustika (1984)

important role in nematode–nematode interactions. More interactions may be detected if additional levels of these factors are used. Therefore, interpretation of data is difficult and conclusions are precarious. Nevertheless, the following general conclusions are listed in order to summarize the complexity of interactions of nematodes in cohabitance.

1. Nematode–nematode interactions can be studied from both ecological and aetiological viewpoints.
2. Interactions can be stimulatory or inhibitory to one or both species.
3. Competition between nematode species is generally weak.
4. Competition between species can restrict the distribution of some species.
5. Competition is more intense between species with similar feeding habits.
6. Competitive advantage increases as the host–parasite relationship becomes more complex. Hence, endoparasites are more competitive than ectoparasites, and sedentary endoparasites are more competitive than migratory endoparasites.
7. Competition between species may be density and time dependent.
8. Environmental factors can modify the effects of competition.
9. Host suitability is a key factor in interactions and is often responsible for one species dominating another.
10. Mechanisms of competition may include mechanical destruction or physical occupation of feeding sites or induced physiological changes in the host's suitability or attractiveness.
11. Generally, the amount of disease caused by two or more nematode species is additive or antagonistic, but sometimes it can be synergistic.
12. Strong competitors cause less disease when in combination with another species than when alone.
13. Most synergistic nematode–nematode interactions involve a sedentary endoparasite.
14. Nematode–nematode interactions can become more complex when other organisms are also involved in the aetiology of the disease.
15. Interactions among nematodes are important in nature, and more precise experiments are needed if their effects on the variability of crop growth are to be quantified.

REFERENCES

Acosta, N. and Ayala, A. (1976) Effects of *Pratylenchus coffeae* and *Scutellonema bradys* alone and in combination on Guinea Yam (*Discorea rotunda*). *Journal of Nematology*, 8, 315–17.

Amosu, J.O. and Taylor, D.P. (1975) Interaction of *Meloidogyne hapla*, *Pratylenchus penetrans* and *Tylenchorhynchus agri* on Kenland clover, *Trifolium pratense*. *Indian Journal of Nematology*, 4, 124–31.

Andersen, S. (1965) Heredity of race 1 or race 2 in *Heterodera avenae*. *Nematologica*, 11, 121–4.

Bird, G.W., Brooks, O.L. and Perry, C.E. (1974) Dynamics of concomitant field populations of *Hoplolaimus columbus* and *Meloidogyne incognita. Journal of Nematology*, **6**, 190–4.

Boag, B. and Alphey, T.J.W. (1988) Can interspecific competition be utilized to control plant-parasitic nematodes? *Nematologica*, **34**, 257–8.

Boag, B. and Topham, P.B. (1985) The use of associations of nematode species to aid the detection of small numbers of virus-vector nematodes. *Plant Pathology*, **34**, 20–4.

Braun, A.L., Mojtahedi, H. and Lownsbery, B.F. (1975) Separate and combined effects of *Paratylenchus neoamblycephalus* and *Criconemoides xenoplax* on Myrobalan plum. *Phytopathology*, **6**, 328–30.

Brewer, R. (1978) *Principles of Ecology*. W.B. Saunders, Philadelphia.

Burrows, P.M. (1987) Interaction concepts for analysis of responses to mixtures of nematode populations, in *Vistas on Nematology* (eds J.A. Veech and D.W. Dickson), Society of Nematologists, Hyattsville, pp. 82–93.

Chapman, R.A. (1959) Development of *Pratylenchus penetrans* and *Tylenchorhynchus martini* on red clover and alfalfa. *Phytopathology*, **49**, 357–9.

Chapman, R.A. (1965) Infection of single root systems by larvae of two coincident species on root-knot nematodes. *Nematologica*, **12**, 89.

Chapman, R.A. and Turner, D.R. (1972) Effect of entrant *Meloidogyne incognita* on reproduction on concomitant *Pratylenchus penetrans* in red clover. *Journal of Nematology*, **4**, 221.

Chapman, R.A. and Turner, D.R. (1975) Effect of *Meloidogyne incognita* on reproduction of *Pratylenchus penetrans* in red clover and alfalfa. *Journal of Nematology*, **7**, 6–10.

Chitamber, J.J. and Raski, D.J. (1984) Reactions of grape rootstocks to *Pratylenchus vulnus* and *Meloidogyne* spp. *Journal of Nematology*, **16**, 166–70.

Choi, D.R., Ishibashi, N. and Tanaka, K. (1988) [Possible integrated control of soil insect pests, soil-borne diseases, and plant nematodes by mixed application of fungivorous and entomogenous nematodes]. *Bulletin of the Faculty of Agriculture*, no. 65, 27–35.

Cuarezma-Teran, J.A. and Trevathan, L.E. (1985) Effects of *Pratylenchus zeae* and *Quinisulcius acutus* alone and in combination on sorghum. *Journal of Nematology*, **17**, 169–74.

Duncan, L. and Ferris, H. (1981) Preliminary considerations of a model of multiple nematode species-plant growth relationships. *Journal of Nematology*, **13**, 435.

Duncan, L. and Ferris, H. (1982) Interactions between phytophagous nematodes, in *Nematodes in Soil Ecosystems* (ed. D. Freckman), University of Texas Press, Austin, pp. 29–51.

Duncan, L. and Ferris, H. (1983) Validation of a model for prediction of host damage by two nematode species. *Journal of Nematology*, **15**, 227–34.

Egunjobi, O.A., Akonde, P.T. and Caveness, F.E. (1986) Interaction between *Pratylenchus sefaensis*, *Meloidogyne javanica* and *Rotylenchulus reniformis* in sole and mixed crops of maize and cowpea. *Revue de Nematologie*, **9**, 61–70.

Eisenback, J.D. (1983) Loss of resistance in tobacco cultivar 'NC95' by infection of *Meloidogyne arenaria* or *M. hapla. Journal of Nematology*, **15**, 478.

Eisenback, J.D. (1985) Interactions among concomitant populations of nematodes, in *An Advanced Treatise on Meloidogyne, Vol. I: Biology and Control* (eds J.N. Sasser and C.C. Carter), North Carolina State University Graphics, Raleigh, pp. 193–213.

Eisenback, J.D. and Griffin, G.D. (1987) Interactions with other nematodes, in *Vistas on Nematology* (eds J.A. Veech and D.W. Dickson), Society of Nematologists, Hyattsville, pp. 313–20.

Ericksson, K.M. (1965) Crossing experiments with races of *Ditylenchus dipsaci* on callus tissue cultures. *Nematologica*, **11**, 244–8.

Ericksson, K.M. (1974) Intraspecific variation in *Ditylenchus dipsaci*. 1. Compatibility tests with races. *Nematologica*, 20, 147–62.

Estores, R.A. and Chen, T.A. (1970) Interaction of *Pratylenchus penetrans* and *Meloidogyne incognita acrita* as cohabitants on tomatoes. *Phytopathology*, 60, 1291.

Estores, R.A. and Chen, T.A. (1972) Interactions of *Pratylenchus penetrans* and *Meloidogyne incognita acrita* as cohabitants on tomato. *Journal of Nematology*, 4, 170–4.

Ferris, V.R., Ferris, J.M. and Bernard, R.L. (1967). Relative competition on two species of *Pratylenchus* in soybeans. *Nematologica*, 13, 143.

Freckman, D.W. and Chapman, R.A. (1972) Infection of red clover seedlings by *Heterodera trifolii* Goffart and *Pratylenchus penetrans* (Cobb). *Journal of Nematology*, 4, 23–8.

Gause, G. (1934) *The Struggle for Existence*. Williams and Wilkins, Baltimore.

Gay, C.M. and Bird, G.W.(1973) Influence of concomitant *Pratylenchus brachyurus* and *Meloidogyne* spp. on root penetration and population dynamics. *Journal of Nematology*, 5, 212–17.

Graham, T.W., Ford, Z.T. and Currin, R.E. (1964) Response of root-knot resistant tobacco to the nematode diesease complex caused by *Pratylenchus* spp. and *Meloidogyne incognita acrita*. *Phytopathology*, 54, 205–10.

Green, C.D. and Dennis, E.B. (1981) An analysis of the variability in yield of pea crops attacked by *Heterodera goettingiana*, *Helicotylenchus vulgaris*, and *Pratylenchus thornei*. *Plant Pathology*, 30, 65–72.

Green, C.D., Greet, D.N. and Jones, F.G.W. (1970) The influence of multiple mating on the reproduction and genetics of *Heterodera rostochiensis* and *H. schachtii*. *Nematologica*, 16, 309–26.

Griffin, G.D. (1972) Interaction of *Meloidogyne hapla* and *Ditylenchus dipsaci* on root-knot resistant alfalfa. *Phytopathology*, 62, 1103.

Griffin, G.D. (1980) Interrelationship of *Meloidogyne hapla* and *Ditylenchus dipsaci* on resistant and susceptible alfalfa. *Journal of Nematology*, 12, 287–93.

Griffin, G.D. (1983) The interrelationship of *Heterodera schachtii* and *Ditylenchus dipsaci* on sugarbeet. *Journal of Nematology*, 15, 426–32.

Griffin, G.D. (1985) Interrelationship of *Heterodera schachtii* and *Meloidogyne hapla* on tomato. *Journal of Nematology*, 17, 385–8.

Griffin, G.D. (1987) Interaction of *Ditylenchus dipsaci* and *Meloidogyne hapla* on resistant and susceptible plant species. *Journal of Nematology*, 19, 441–6.

Griffin, G.D. and Waite, W.W. (1982) Pathological interaction of a combination of *Heterodera schachtii* and *Meloidogyne hapla* on tomato. *Journal of Nematology*, 14, 182–7.

Guerount, R. (1968) Competition *Pratylenchus brachyurus–Meloidogyne* n. sp. dans les cultures d'ananas de Cote d'lvoire, *Reports 8th Nematology Symposium*, Antibes, 8–14 September 1965. E.J. Brill, pp. 64–9.

Guy, D.W., Jr and Lewis, S.A. (1987a) Interaction between *Meloidogyne incognita* and *Hoplolaimus* on Davis soybean. *Journal of Nematology*, 19, 346–51.

Guy, D.W., Jr and Lewis, S.A. (1987b). Selective migration and root penetration by *Meloidogyne incognita* and *Hoplolaimus columbus* on soybean roots in vitro. *Journal of Nematology*, 19, 390–2.

Haque, M.S. and Mukhopadhyaya, M.C. (1979) Pathogenicity of *Macrophomina phaseoli* on jute in the presence of *Meloidogyne incognita* and *Hoplolaimus indicus*. *Journal of Nematology*, 11, 318–21.

Herman, M., Hussey, R.S. and Boerma, H.R. (1988) Interactions between *Meloidogyne incognita* and *Pratylenchus brachyurus* on soybean. *Journal of Nematology*, 20, 79–84.

Hirano, K. and Kawamura, T. (1972) Incidence of complex diseases caused by

Pratylenchus penetrans or P. coffeae and Fusarium wilt fungus in tomato seedlings. Technical Bulletin, Faculty of Horticulture, Chiba University, 20, 37–43.

Ibrahim, I.K.A. (1987) Interaction between Meloidogyne arenaria and M. incognita on tobacco. Nematologia Mediterranea, 15, 287–91.

Ibrahim, I.K.A. and Lewis, S.A. (1986) Interrelationships of Meloidogyne arenaria and Meloidogyne incognita on tolerant soybean. Journal of Nematology, 18, 106–11.

Inserra, R.V., Griffin, G.D., Vovlas, N., Anderson, J.L. and Kerr, E.D. (1984) Relationships between Heterodera schachtii, Meloidogyne hapla, and Nacobbus aberrans on sugarbeet. Journal of Nematology, 16, 135–40.

Ishibashi, N. and Choi, D.R. (1991) Biological control of soil pests by mixed application of entomopathogenic and fungivorous nematodes. Journal of Nematology, 23, 175–81.

Jatala, P. and Jensen, H.J. (1972) Self-interactions of Meloidogyne hapla and Heterodera schachtii on Beta vulgaris. Journal of Nematology, 4, 226.

Jatala, P. and Jensen, H.J. (1976a) Parasitism of Beta vulgaris by Meloidogyne hapla and Heterodera schachtii alone and in combination. Journal of Nematology, 8, 200–5.

Jatala, P. and Jensen, H.J. (1976b) Self-interactions of Meloidogyne hapla and of Heterodera schachtii on Beta vulgaris. Journal of Nematology, 8, 43–8.

Jatala, P. and Jensen, H.J. (1983) Influence of Meloidogyne hapla Chitwood, 1949 on development and establishment of Heterodera schachtii Schmidt, 1871 on Beta vulgaris L. Journal of Nematology, 15, 564–6.

Johnson, A.W. (1970) Pathogenicity and interaction of three nematode species on six Bermudagrasses. Journal of Nematology, 2, 36–41.

Johnson, A.W. and Littrell, R.H. (1970) Pathogenicity of Pythium aphanidermatum to chrysanthemum in combined inoculations with Belonolaimus longicaudatus or Meloidogyne incognita. Journal of Nematology, 2, 255–9.

Johnson, A.W. and Nusbaum, C.J. (1968) The activity of Tylenchorhynchus claytoni, Trichodorus christiei, Pratylenchus brachyurus, P. zeae, and Helicotylenchus dihystera in single and multiple inoculations on corn and soybean. Nematologica, 14, 9.

Johnson, A.W. and Nusbaum, C.J. (1970) Interactions between Meloidogyne incognita, M. hapla, and Pratylenchus brachyurus in tobacco. Journal of Nematology, 2, 334–40.

Jones, F.G.W. and Perry, J.N. (1978) Modeling populations of cyst-nematodes. Journal of Applied Ecology, 15, 349–71.

Kaplan, D.T. and Timmer, L.W. (1982) Effects of Pratylenchus coffeae–Tylenchulus semipenetrans interactions on nematode population dynamics in citrus. Journal of Nematology, 14, 368–73.

Kaul, R.K. and Sethi, C.L. (1982) Effect of simultaneous inoculations and prior establishment of Heterodera zeae and Meloidogyne incognita, and Tylenchorhynchus vulgaris singly and in combination on penetration of H. zeae and M. incognita into maize roots. Indian Journal of Nematology, 11, 100.

Khan, M.W. and Haider, S.R. (1991) Interaction of Meloidogyne javanica with different races of Meloidogyne incognita. Journal of Nematology, 23, 298–305.

Khan, M.W. and Haq, S. (1979) Interaction between Meloidogyne incognita and Tylenchorhynchus brassicae on tomato. Libyan Journal of Agriculture, 8, 181–6.

Khan, R.M. and Khan, M.W. (1986) Antagonistic behaviour of root-knot, reniform, and stunt nematodes in mixed infection. International Nematology Network Newsletter, 3, 3.

Khan, R.M. and Khan, M.W. (1990) Eco-relationships among coinhabiting plant nematodes in crop pathosystems, in Progress in Plant Nematology (eds

S.K. Saxena, M.W. Khan, A. Rashid and R.M. Khan), CBS Publishers and Distributors, New Delhi, pp. 167–93.

Khan, R.M., Khan, M.W. and Khan, A.M. (1985) Cohabitation of *Meloidogyne incognita* and *Rotylenchulus reniformis* in tomato roots and effect on multiplication and plant growth. *Nematologia Mediterranea*, **13**, 153–9.

Khan, R.M., Khan, A.M. and Khan, M.W. (1986a) Interaction between *Meloidogyne incognita*, *Rotylenchulus reniformis* and *Tylenchorhynchus brassicae* on tomato. *Revue de Nematologie*, **9**, 245–50.

Khan, R.M., Khan, M.W. and Khan, A.M. (1986b) Interactions of *Meloidogyne incognita*, *Rotylenchulus reniformis* and *Tylenchorhynchus brassicae* as cohabitants on eggplant. *Nematologia Mediterranea*, **14**, 201–6.

Khan, R.M., Khan, M.W. and Khan, A.M. (1987) Competitive interaction between *Myloidogyne incognita*, *Rotylenchulus reniformis* and *Tylenchorhynchus brassicae* on cauliflower. *Pakistan Journal of Nematology*, **5**, 73–83.

Khan, T.A. and Husain, S.I. (1989) Relative resistance of six cowpea cultivars as affected by the concomitance of two nematodes and a fungus. *Nematologia Mediterranea*, **17**, 39–41.

Khier, A.M. and Osman, A.A. (1977) Interaction of *Meloidogyne incognita* and *Rotylenchulus reniformis*. *Nematologia Mediterranea*, **5**, 113–16.

Kinloch, R.A. and Allen, M.W. (1972) Interaction of *Meloidogyne hapla* and *M. javanica* infecting tomato. *Journal of Nematology*, **4**, 7–16.

Kort, J. and Bakker, J. (1980) The occurrence of mixtures of potato cyst nematode pathotypes or species. *Nematologica*, **26**, 272–4.

Kort, J. and Jaspers, C.P. (1973) Shift of pathotypes of *Heterodera rostochiensis* under susceptible potato cultivars. *Nematologica*, **19**, 538–45.

Kraus-Schmidt, H. and Lewis, S.A. (1981) Dynamics of concomitant populations of *Hoplolaimus columbus*, *Scutellonemabrachyurum*, and *Meloidogyne incognita* on cotton. *Journal of Nematology*, **13**, 41–8.

Lawn, D.A. and Noel, G.R. (1990) Effects of temperature on competition between *Heterodera glycines* and *Pratylenchus scribneri* on soybean. *Nematropica*, **20**, 57–69.

Marshall, J.W. (1989) Changes in relative abundance of two potato cyst nematode species *Globodera rostochiensis* and *G. pallida* in one generation. *Annals of Applied Biology*, **115**, 79–87.

McGawley, E.C. and Chapman, R.A. (1983) Reproduction of *Criconemoides similis*, *Helicotylenchus pseudorobustus*, and *Paratylenchus projectus* on soybean. *Journal of Nematology*, **15**, 87–91.

McGawley, E.C. and Winchell, K.L. (1987) Greenhouse reproduction of single and combined *Meloidogyne incognita* and *M. javanica* populations on soybean. *Journal of Nematology*, **19**, 542.

Miller, L.I. (1970) Intraspecific variation in *Heterodera glycines* Ichinohe, Summaries, X *International Symposium, Society of European Nematologists*, p. 11.

Miller, L.I. (1983) Diversity of selected taxa of *Globodera* and *Heterodera* and their interspecific and intergeneric hybrids, in *Concepts in Nematode Systematics* (eds A.R. Stone, A.M. Platt and L.F. Khalil), Academic Press, New York, pp. 207–20.

Miller, L.I. (1988) Morphological comparisons of cyst cone structures of one isolate each of *Heterodera schachtii*, *H. glycines* and one of their isolates. *Journal of Nematology*, **20**, 648–9.

Miller, L.I. (1989) Morphological comparisons of second-stage juveniles of one isolate each of *Heterodera glycines*, *H. cruciferae* and one of their hybrids. *Journal of Nematology*, **21**, 574.

Miller, P.M. (1970) Rate of increase of a low population of *Heterodera tabacum* reduced by *Pratylenchus penetrans* in the soil. *Plant Disease Reporter*, 54, 25–6.

Miller, P.M. and McIntyre, J.L. (1975) *Tylenchorhynchus claytoni* feeding on tobacco root inhibits entry of *Pratylenchus penetrans*. *Journal of Nematology*, 17, 327.

Miller, P.M. and Wihrheim, S.E. (1968) Mutual antagonism between *Heterodera tabacum* and some other parasitic nematodes. *Plant Disease Reporter*, 52, 57–8.

Minz, G. and Strich-Harari, D. (1959) Inoculation experiments with a mixture of *Meloidogyne* spp. on tomato roots. *Ktavim Records Agricultural Experiment Station*, 9, 275–9.

Mishra, S.D. and Gaur, H.S. (1981) Effect of individual and concomitant inoculation with *Meloidogyne incognita* and *Rotylenchulus reniformis* on the growth of black gram (*Vigna mungo*). *Indian Journal of Nematology*, 11, 25–8.

Misra, C. and Das, S.N. (1979) Interaction of some plant parasitic nematodes on the root-knot development in brinjal. *Indian Journal of Nematology*, 7, 46–53.

Mustika, I. (1984) [Effects of nematodes and fungi on the growth of pepper and yellow disease]. *Pemberitaan Lembaga Penelitian Tanaman Industri*, Bogor, Indonesia, 8, 28–37.

Niblack, T.L. (1989) Application of nematode community structure research to agricultural production and habitat disturbance. *Journal of Nematology*, 21, 437–43.

Niblack, T.L., Hussey, R.S. and Boerma, H.R. (1986) Effects of *Heterodera glycines* and *Meloidogyne incognita* on early growth of soybean. *Journal of Nematology*, 18, 444–50.

Norton, D.C. (1969) *Meloidogyne hapla* as a factor in alfalfa decline in Iowa. *Phytopathology*, 59, 1824–8.

Norton, D.C. (1978) *Ecology of Plant-Parasitic Nematodes*, John Wiley, New York.

Norton, D.C. and Niblack, T.L. (1991) Biology and ecology of nematodes, in *Manual of Agricultural Nematology* (ed. W.R. Nickle), Marcel Dekker, New York, pp. 47–72.

O'Bannon, J.H., Radewald, J.D., Tomerlin, A.T. and Inserra, R.N. (1976) Comparative influence of *Radopholus similis* and *Pratylenchus coffeae* on citrus. *Journal of Nematology*, 8, 58–63.

Olsson, E. (1985) Morphological–taxonomical studies and pathotype classification in potato cyst nematodes. *Bulletin OEPP/EPPO Bulletin*, 15, 281–3.

Olsson, E. (1988) Hybridization experiments between *Globodera rostochiensis* and *G. pallida*. *Nematologica*, 34, 285.

Oostenbrink, M. (1966) Major characteristics of the relation between nematodes and plants. *Mededlingen Landbouwhogeschool, Wageningen*, no. 66–4, 46 pp.

Paez, N., Arcia, M.A. and Meredith, J.A. (1976) Individual and combined effects of *Meloidogyne incognita* and *M. javanica* on four tobacco (*Nicotiana tabacum* L.) cultivars in Venezuela. *Nematropica*, 6, 68–76.

Parrot, D.M., Berry, M.M. and Farrell, K.M. (1976) Competition between *Globodera rostochiensis* Ro 1 and *G. pallida* Pa3, in *Nematology Department, Rothamsted Experimental Station Report 1975* (ed. F.G.W. Jones), pp. 197–8.

Pathak, K.N., Nath, R.P. and Haider, M.G. (1985) Effect of initial inoculum level of *Meloidogyne incognita* and *Rotylenchulus reniformis* on pigeonpea and their interrelationship. *Indian Journal of Nematology*, 15, 177–9.

Pinochet, J., Raski, D.J. and Goheen, A.C. (1976) Effects of *Pratylenchus vulnus* and *Xiphinema index* singly and combined on vine growth of *Vitis vinifera*. *Journal of Nematology*, 8, 330–5.

Pitcher, R.S. (1978) Interactions of nematodes with other pathogens, in *Plant Nematology* (ed. J.F. Southey), Her Majesty's Stationery Office, London, pp. 63–77.

Potter, J.W. and Fox, J.A. (1965) Hybridization of *Heterodera schachtii* and *H. glycines. Phytopathology*, 55, 800-1.

Powell, N.T. (1971) Interactions between nematodes and fungi in disease complexes. *Annual Review of Phytopathology*, 9, 253-74.

Powell, N.T. (1979) Internal synergisms among organisms inducing disease, in *Plant Disease*, Vol. IV (eds J.G. Horsfall and E.B. Cowling), Academic Press, New York, pp. 113-33.

Price, M., Riggs, R.D. and Caveness, C.E. (1976) Races of soybean-cyst nematode compete. *Arkansas Farm Research*, 25, 16.

Price, M., Riggs, R.D. and Caveness, C.E. (1978) Hybridization of races of *Heterodera glycines. Journal of Nematology*, 10, 114-18.

Rao, B.H.K. and Prasad, S.K. (1971) Population studies on *Meloidogyne javanica* and *Rotylenchulus reniformis* occurring together and separately and their effects on the host. *Indian Journal of Entomology*, 32, 194-200.

Rao, K.T. and Seshadri, A.R. (1981) Studies on interaction between *Meloidogyne incognita* and *Rotylenchulus reniformis* on grapevine seedlings. *Indian Journal of Nematology*, 11, 101-2.

Rao, Y.S. and Prasad, J.S. (1981) Interaction of *Heterodera oryzicola* and *Meloidogyne graminicola* in artificial inoculation in rice. *Indian Journal of Nematology*, 11, 104.

Rhoades, H.L. (1985) Effects of separate and concomitant population of *Belonolaimus longicaudatus* and *Dolichodorus heterocephalus* on *Zea mays. Nematropica*, 15, 171-4.

Ross, J.P. (1964) Interaction of *Heterodera glycines* and *Meloidogyne incognita* on soybeans. *Phytopathology*, 54, 304-7.

Santo, G.S. and Bolander, W.J. (1977) Separate and concomitant effects of *Macroposthonia xenoplax* and *Meloidogyne hapla* on Concord grapes. *Journal of Nematology*, 9, 282-3.

Seinhorst, J.W. (1986) The development of individuals and populations of cyst nematodes on plants, in *Cyst Nematodes* (eds F. Lamberti and C.E. Taylor), Plenum Press, New York, pp. 101-17.

Seinhorst, J.W. and Oostrom, A. (1989) Reproduction of a mixture of pathotypes Ro 1 of *Globodera rostochiensis* and Pa 3 of *G. pallida* and its offspring on cvs. Cardinal and Irene. *Nematologica*, 35, 469-74.

Sharma, N.K. and Sethi, C.L. (1978) Interaction between *Meloidogyne incognita* and *Heterodera cajani* on cowpea. *Indian Journal of Nematology*, 6, 1-12.

Sheela, M.S. and Venkitesan, T.S. (1981) Interrelationships of infectivity between the burrowing and root-knot nematode in black pepper, *Piper nigrum. Indian Journal of Nematology*, 11, 105.

Sikora, R.A., Taylor, D.P., Malek, R.B. and Edwards, D.I. (1972) Interaction of *Meloidogyne naasi, Pratylenchus penetrans,* and *Tylenchorhynchus agri* on creeping bentgrass. *Journal of Nematology*, 4, 162-5.

Sikora, R.A., Malek, R.B., Taylor, D.P. and Edwards, D.I. (1979) Reduction in *Meloidogyne naasi* infection of creeping bentgrass by *Tylenchorhynchus agri* and *Paratrichodorus minor. Nematologica*, 25, 179-83.

Singh, N.D. (1976) Interaction of *Meloidogyne incognita* and *Rotylenchulus reniformis* on soybean. *Nematropica*, 6, 76-81.

Sohlenius, B. (1988) Interactions between two species of *Panagrolaimus* in agar cultures. *Nematologica*, 34, 208-17.

Sturhan, D. (1966) Rassen Bei Phytoparastiaren Nematoden. *Mitteilungen Biologsche Bundesanstalt Land und Forstwirtschaft (Berlin–Dahlem)*, 118, 40-53.

Sturhan, D. (1971) Biological races, in *Plant Parasitic Nematodes*, Vol. II (eds B.M. Zuckerman, W.F. Mai and R.A. Rohde), Academic Press, New York, pp. 51-71.

Taha, A.H.Y. and Kassab, A.S. (1979) The histological reactions of *Vigna sinensis* to separate and concomitant parasitism by *Meloidogyne javanica* and *Rotylenchulus reniformis*. *Journal of Nematology*, **11**, 117–23.

Taha, A.H.Y. and Kassab, A.S. (1980) Interrelations between *Meloidogyne incognita, Rotylenchulus reniformis,* and *Rhizobium* sp. on *Vigna sinensis*. *Journal of Nematology*, **12**, 57–62.

Taha, A.H.Y. and Sultan, S.A. (1977) Populations of *Rotylenchulus reniformis* and *Tylenchulus semipenetrans* on grape seedlings as influenced by coincident infestations. *Nematologia Mediterranea*, **5**, 253–7.

Taylor, A.L., Sasser, J.N. and Nelson, L.A. (1982) *Relationship of Climate and Soil Characteristics to Geographical Distribution of Meloidogyne species in Agricultural Soils,* North Carolina State University Graphics, Raleigh.

Tedford, E.C., Fortnum, B.A. and Bridges, W.C. (1986) Effects of flue-cured tobacco varieties on *Meloidogyne* spp. population shifts in South Carolina. *Journal of Nematology*, **18**, 642.

Thomas, R.J. and Clark, C.A. (1980) Interactions between *Meloidogyne incognita* and *Rotylenchulus reniformis* on sweet potato. *Journal of Nematology*, **12**, 239.

Thomas, R.J. and Clark, C.A. (1981) *Meloidogyne incognita* and *Rotylenchulus reniformis* interactions in a sweet potato field. *Phytopathology*, **71**, 908.

Thomas, R.J. and Clark, C.A. (1983a) Effects of concomitant development on reproduction of *Meloidogyne incognita* and *Rotylenchulus reniformis*. *Journal of Nemaology*, **15**, 215–21.

Thomas, R.J. and Clark, C.A. (1983b) Population dynamics of *Meloidogyne incognita* and *Rotylenchulus reniformis* alone and in combination, and their effects on sweet potato. *Journal of Nematology*, **15**, 204–11.

Triantaphyllou, A.C. (1991) Further studies on the role of polyploidy in the evolution of *Meloidogyne*. *Journal of Nematology*, **23**, 249–53.

Triantaphyllou, A.C. and Esbenshade, P.R. (1990) Demonstration of multiple mating in *Heterodera glycines* with biochemical markers. *Journal of Nematology*, **22**, 452–6.

Turner, D.R. and Chapman, R.A. (1971) Infection of alfalfa and red clover by concomitant populations of *Meloidogyne incognita* and *Pratylenchus penetrans*. *Journal of Nematology*, **3**, 332.

Turner, D.R. and Chapman, R.A. (1972) Infection of seedling of alfalfa and red clover by concomitant populations of *Meloidogyne incognita* and *Pratylenchus penetrans*. *Journal of Nematology*, **4**, 280–6.

Upadhyay, K.D. and Swarup, G. (1981) Growth of maize plants in the presence of *Tylenchorhynchus vulgaris* Upadhyay *et al.*, 1973, singly and in combination with *Pratylenchus zeae* Graham, 1951, and *Fusarium moniliforme* Sheld. *Indian Journal of Nematology*, **1**, 29–34.

Vaishnav, M.U. and Sethi, C.L. (1978) Pathogenicity of *Meloidogyne incognita* and *Tylenchorhynchus vulgaris* on bajra and their interrelationship. *Indian Journal of Nematology*, **8**, 1–8.

Van Gundy, S.D. and Kirkpatrick, J.D. (1975) Nematode–nematode interactions on tomato. *Journal of Nematology*, **7**, 330.

Wallace, H.R. (1983) Interactions between nematodes and other factors on plants. *Journal of Nematology*, **15**, 221–7.

Webster, J.M. (1967) The significance of biological races of *Ditylenchus dipsaci* and their hybrids. *Annals of Applied Biology*, **59**, 77–83.

Winoto, S. and Lim, T.K. (1972) Interaction of *Meloidogyne incognita* and *Rotylenchulus reniformis* on tomato. *Malaysian Agricultural Research*, **1**, 6–13.

Yang, H., Powell, N.T. and Barker, K.R. (1976) Interactions of concomitant species of nematodes and *Fusarium oxysporum* f. sp. *vasinfectum* on cotton. *Journal of Nematology*, **8**, 74–80.

8

Nematode interactions with root-nodule bacteria

Abd-El-Samie H.Y. Taha

Definitions of the words 'symbiosis' and 'parasitism' have generally been more or less antithetical. Lewin (1982) defined symbionts as 'lodgers who pay rent', and parasites as 'freeloaders who do not'. *Rhizobium*-induced root nodules are a prime example of symbiosis between legume host and bacteria (Vance, 1983). For an effective nitrogen-fixing association, legume roots must become associated with *Rhizobium* bacteria. The plant (the macrosymbiont) provides an energy source and an ecological niche for the bacteria (the microsymbiont) and the bacteria provide a source of fixed nitrogen for the plant (Vance and Johnson, 1981). In contrast, plant–nematode relationships can be characterized by host efficiency (degree of nematode reproduction) and host sensitivity (degree of yield loss) (Cook, 1974).

The association of rhizobia with plant nematodes in the rhizosphere, and the beneficial effect of rhizobial symbiosis on plant nutrition and growth, led to investigations into the potential role of nematode parasitism on nodulation, and consequently on symbiotic nitrogen fixation. Interactions of nematodes with rhizobia have been reviewed recently by Huang (1987). This chapter sheds light on prerequisite information required for the right perspective of nematodes, root-nodule bacteria and host plant relations.

8.1 LEGUMES AND NON-LEGUMES

Symbiotic nitrogen fixation is not confined to the legumes, and neither are nitrogen-fixing root nodules limited to Leguminosae. Some non-leguminous angiosperms that form root nodules when invaded by soil actinomycetes fix atmospheric nitrogen at rates comparable to legumes (Callaham *et al.*, 1978). The general assumption that nodulating ability is a universal characteristic of leguminous plants was not borne out in a survey by Allen and Allen (1947). They showed that 11.8% of Mimosoideae, 3.3% of Caesalpinioideae and 84.9% of Papilionoideae had nodules. Current estimates place the nodulation status of legumes in the range of 25–30% for Caesalpinioideae, 60–70% for Mimosoideae and 90–95% for Papilionoideae (Lim and Burton, 1982).

Actinorhizal symbiosis is the best known non-leguminous nitrogen-fixing interaction and has been reviewed by Bond (1958), Bond and Wheeler (1980) and Akkermans and Van Dijk (1981). The term 'actinorhiza' has been applied to root nodules formed by *Frankia*, the actinomycete (Stowers, 1987). At present, almost 170 non-legume species, all trees or shrubs belonging to 16 genera, are known to bear root nodules tenanted by a nitrogen-fixing endophyte, with the exception of *Parasponia* spp. The nodules of all three plant species are of the same basic type in their morphology and the nature of endophytes (Bond and Wheeler, 1980). Lechevalier (1983) has catalogued 21 actinorhizal plants with *Frankia* strains. Nodulation of *Parasponia* (Ulmaceae), a non-legume (Akkermans *et al.*, 1978), by a slow-growing strain of *Rhizobium* sp., later named as *Bradyrhizobium*, is the only completely validated instance of nodulation by a member of the Rhizobiaceae outside the Leguminosae (Jordan, 1984).

8.2 THE MICROSYMBIONT

The ability of legumes and non-legumes to draw atmospheric nitrogen is dependent on the presence in the soil of microorganisms of particular groups, viz. Rhizobiaceae (Jordan, 1984) and *Frankia* (Lechevalier and Lechevalier, 1989). They stimulate the plants to form nodules on the roots within which they grow, multiply and convert atmospheric nitrogen into some available form.

The Rhizobiaceae includes two genera of root-nodule bacteria, viz. *Rhizobium* Frank, 1889 (fast-growing rhizobia) and *Bradyrhizobium* Jordan, 1982 (slow-growing rhizobia), which are normally involved in fixing atmospheric nitrogen through a process called *Rhizobium* symbiosis. Jordan (1984) described in detail their morphological and physiological characteristics in *Bergey's Manual of Systematic Bacteriology*, Vol. 1 (1984). Members of the genus *Rhizobium* produce fast growth on yeast extract–mannitol agar and acid reaction in mineral salts–mannitol medium. Conversely, members of the genus *Bradyrhizobium* produce slow growth and alkaline reaction in the above-mentioned medium. Rhizobia are aerobic rods, non-spore forming, motile and facultative symbionts.

Callaham *et al.* (1978) made the first successful isolation of *Frankia* Brunchorst 1886, in in vitro culture from *Comptonia peregrina* (L.) Coult. (Myricaceae) root nodules. They were able to reinfect *Comptonia* seedlings with the isolate, obtained nitrogen-fixing root nodules, and reisolated the causal actinomycete from the induced root nodules.

A full description of the genus *Frankia* has been given by Lechevalier and Lechevalier (1989) in *Bergey's Manual of Systematic Bacteriology*, Vol. 4. Suitable plant hosts are readily infected by most frankiae, giving rise to nitrogen-fixing (effective) nodules on the roots (Callaham *et al.*, 1978; Lalonde, 1978; Bond and Becking, 1982; Berry and Torrey, 1983).

8.3 LEGUME-*RHIZOBIUM* ASSOCIATIONS

Rhizobia are facultative symbionts able to live in soil in the temporary absence of their hosts. They are more numerous in the rhizosphere than in the surrounding soil (Rovira, 1965a, 1965b). Legumes stimulate rhizobia much more than other rhizosphere microorganisms (Nutman, 1965). The ratio of numbers of *Rhizobium* in the rhizosphere and in soil is rarely smaller than 10^2 and often exceeds 10^6.

Actively growing plant roots exert a distinct selective action on soil microorganisms through root exudate, the composition of which varies both qualitatively and quantitatively (Nutman, 1957; Rovira, 1965a). As rhizobia differ in their ability to metabolize various compounds present in root exudates, growth of some rhizobia is promoted in the rhizosphere, while that of others is retarded. Rhizobia that are stimulated by the plant secretions successfully form infection threads. The plant-elaborated rhizosphere exerts selectivity over the legume–rhizobia interaction in this manner (Broughton, 1978).

The ability of a given strain of *Rhizobium* to overcome antagonism from other microorganisms and competition from other *Rhizobium* strains in nodule formation is one of the major factors in deciding the success of the strain in nodule formation. The bacteria in soil respond much more to the presence of plant roots than do the actinomycetes, fungi, algae and protozoa (Rovira, 1965b). Certain actinomycete strains inhibit growth of *Rhizobium* species in the rhizosphere of several plants (Van Schreven, 1964).

In the rhizosphere, competition for substrates seems to favour, at least a little, the rhizobia which are stimulated and attain high numbers close to the roots of both legumes and non-legumes (Trinick, 1982). Competition between effective and ineffective strains may sometimes greatly influence the results of infection and the fixation of nitrogen (Burton and Allen, 1950).

8.3.1 Legume–*Rhizobium* specificity

Symbiotic nitrogen fixation is a complex process involving physiological and biochemical properties of both the bacterium and the host plant (Stacey *et al.*, 1980). The first step is the selective attachment and penetration of the plant by the bacterium. A considerable degree of specificity is often shown between rhizobia and legume host which is the basis for cross-inoculation classification in legume production and for species differentiation in the genus *Rhizobium* (Bohlool and Schmidt, 1974; Sequeira, 1978).

Baldwin *et al.* (1927) used serological methods for classifying leguminous species into cross-inoculation groups. A cross-inoculation group refers to a collection of leguminous species which develop nodules when exposed to rhizobia obtained from the nodules of any member of that particular plant group. Consequently, a single cross-inoculation group ideally includes all host species which are infected by an individual bacterial strain.

The validity of the cross-inoculation system has not gone unchallenged, because many legumes are nodulated by rhizobia of other host-rhizobial groups. The rhizobial strains which invade legumes outside their particular class and plants which are thus infected are examples of a phenomenon termed 'symbiotic promiscuity' (Alexander, 1961).

How is the host able to recognize its *Rhizobium* symbiont among the multitude of soil microorganisms? Specific contact between microbe elicitor and host receptor is required for recognition (Albersheim and Anderson-Prouty, 1975; Sequeira, 1978). Albersheim and Anderson-Prouty (1975) suggested that the most likely molecular determinants of plant–microorganism recognition would be host plant lectins and carbohydrate components of microbial cell surfaces.

Carbohydrates on cell surfaces may be important in intercellular recognition. Certain plants contain substances capable of agglutinating erythrocytes (Toms and Western, 1971). Although in most species of plants the highest concentration of these 'phytohaemagglutinins' is present in seeds, they may also be present to a lesser extent in the leaves, roots and stems. Boyd and Sharpleigh (1954) introduced the term 'lectin', derived from the Latin *legere* (to choose or select), as a term for plant proteins with haemagglutinating activity.

Although the term lectin is commonly used as a synonym for phytohaemagglutinin, its original definition needs to be broadened in order to include those substances which may not be of plant origin. Since the proteins capable of specific interaction with carbohydrate-containing substances are found not only in plants but also in invertebrates (e.g. nematodes), fungi and bacteria, the term lectin should be used as a generic term to denote all sugar-specific proteins, with a designation as 'phytolectin', 'zoolectin' or 'mycolectin', depending on whether they are of plant, animal or fungal origin, respectively (Liener, 1976).

Some kind of *Rhizobium* surface polysaccharide is perhaps important in establishing the symbiotic relationship (Bohlool and Schmidt, 1974; Bhuvaneswari *et al.*, 1977; Dazzo and Brill, 1979). Several different kinds of *Rhizobium* cell surface polysaccharides have been described: exopolysaccharides, capsular polysaccharides, lipopolysaccharides, 2-linked glucans, 3-, 6-, 3,6-linked glucans, and β-4-linked glucans cellulose (Bauer, 1981).

8.3.2 Entry of rhizobia

Rhizobia enter the root via the root hairs. In peanut, however, mode of entry is through ruptured tissue at the site of lateral rootlet emergence.

Rhizobia entering the roots at these sites occupy the space between the 'root hair' wall and the adjoining epidermal and cortical cells (Chandler, 1978). Penetration of root hairs appears to be under control of both the bacterium and the plant and only certain discrete portions of the root are susceptible to infection (Thornton, 1952; Nutman, 1962; Bauer, 1981; Bhuvaneswari *et al.*, 1980, 1981).

Multiplication of rhizobia in the rhizosphere is very rapid (Lochhead, 1952). Fewer than 100 rhizobia in the whole rhizosphere are sufficient to start infection. The number of infections and the density of rhizobia are simply related, such that doubling rhizobial density produces twice as many infections. As soon as the first nodule is formed more rhizobia are required for each infection; the additional number of bacteria required is greatest for those species that are least abundantly infected (Nutman, 1965).

After the rhizobia bind to the root hair surface, the next step in the invasion process involves marked root hair curling. Infected hairs are usually, but not invariably, deformed; the growing tip of the hair (or of the lateral branch of the hair) curls in a characteristic 'shepherd's crook' manner. Only a small proportion of all hairs are deformed and not all deformed hairs are visibly infected, i.e. contain infection threads (see below) (Nutman, 1965). Curling, however, does not seem to be a sole requirement for infection since both heterologous and homologous strains can induce curling (Vance, 1983), nor is it an absolute requirement since infections can occur directly through the epidermis and straight root hairs (Yao and Vincent, 1976). The curling response in any particular host is induced not only by the limited number of rhizobial strains that can form infection threads in its root hairs, but also by rhizobia that infect other host species (Nutman, 1965).

Although the exact mechanism of root hair penetration has not been resolved, three hypotheses have been offered to explain the phenomenon: invagination, cell wall degrading enzymes, and lectin–enzyme theory (Vance, 1983). Nutman (1962) and Bauer (1981) discussed in detail the evidence suggesting that infection thread formation and proliferation is self-regulated by the plant. Only a few root hairs become infected, and in most of those that do become infected infection threads abort before nodules are formed. The root cells susceptible to infection are limited primitively to the root zone just below the smallest emergent root hair and just above the zone of rapid cell elongation. Development of the first nodule population inhibits subsequent infection, whereas removal of nodules, nodule tips and root tips stimulates infection. Application of nitrogen reduces penetration and infection thread formation and stimulates abortion of infection threads (Vance, 1983).

The infected root hair can be recognized readily by the formation within it of an infection thread or tube containing the rhizobia (Nutman, 1958). The rhizobia grow and divide within the infection thread and become aligned along the axis of growth, often in a single row (Meijer, 1982).

The rhizobia produce and become embedded in a mucopolysaccharide-containing matrix, termed 'zoogleal matrix' by Dart (1977). As the infection thread grows towards the root cortex, it crosses already existing cell walls, becoming attached to and continuous with them (Dart, 1977).

Within the cortex, extensive branching of the thread occurs, resulting in the infection of many cells by the same infection thread (Nutman, 1958; Dart, 1977). A close association between host cell nucleus in the root hair and the infection thread has been observed in the passage of the thread through the cortex. Within the root hair, the nucleus is usually found near the site of infection, and it stays close to the tip of the infection thread during its growth towards the base of the root-hair cell (Dart, 1977). These observations seem to suggest that the nucleus, which is enlarged in infected root-hair cells, directs the growth of the infection thread (Meijer, 1982).

Usually rhizobia are released shortly after the penetration of the cortical cells by the infection thread (Meijer, 1982). Prior to release, disintegration of the thread wall takes place, often at the tip, but also along the sides (Goodchild and Bergersen, 1966). Soon after the rhizobia are released into the cytoplasm, the host cells and also the adjacent non-infected ones undergo rapid cell division (Allen and Allen, 1953). As the rhizobia are released, they localize in the peripheral cytoplasmic layer around the vacuoles of the host cells where they undergo multiplication. Eventually, many cells become entirely filled with rhizobia and the cell nuceli become distorted. In the early stages of infection the rhizobia are typical rods ($c.$ $1.0 \times 2.5 \, \mu m$) which stain intensely and entirely. In more mature stages the rods become vacuolated, banded, swollen, and exhibit bacteroid forms characteristic of certain host associations (Allen and Allen, 1953).

Infection threads penetrate the occasional tetraploid cells of the cortex. These cells are stimulated into meristematic activity and with nearby diploid cells constitute the initial nodule primordium. This characteristic cytological situation resulting from the early association of infection threads with disomatic cells has been confirmed for a large number of host species and even for artificially induced polyploidy. The tetraploid cells are known to be present in the root before infection occurs and are found also in uninoculated plants; they often occur in close proximity to lateral root initials. The tetraploid centres appear to be the primary foci of infection; they are preformed and only become activated by infection (Nutman, 1958).

In an alternate mode of infection observed in peanut, root hairs and nodules develop only around the base of emergent lateral roots. No infection threads are found in either the hairs or the nodules. Infections are initiated at the basal junction between the axillary hair cells and adjacent epidermal cells, and then progress through the middle lamellar region. Movement of the rhizobia from the middle lamella into the host cell cytoplasm appears to involve alteration or disintegration of the host wall and enclosure of the bacteria in membrane envelopes. The site of nodule initiation appears to be in the cortex of the lateral root, not in the cortex of the primary root (Chandler, 1978).

8.4 THE NODULE

Root nodules are desirable hypertrophies for symbiotic nitrogen fixation. The nodule is merely a protective tissue, and bacteroids are the ultimate stages in the life cycle of *Rhizobium* (Subba Rao, 1988). The term 'mature nodule', in a histological sense, refers to that stage when all tissues essential for nodule function are well defined as to structure. Nodule maturity is perhaps best characterized by increased volume and secondary thickening of vascular elements which accentuate accompanying anatomical features (Allen and Allen, 1953).

8.4.1 The histology of nodule

The histology of nodules from widely different species is remarkably similar. Within certain limits this is also true for the histology of effective and ineffective nodules, although the pattern of the former is accepted as the norm (Allen and Allen, 1953). Structural similarity between effective and ineffective nodules has also been indicated by several workers (Bergersen, 1957; Jordan, 1974; Vance and Johnson, 1981), but ineffective nodules do not fix nitrogen. On the other hand, some observations indicate that ineffective nodules differ from effective nodules in both structure and physiology (Pankhurst *et al.*, 1972; Viands *et al.*, 1979; Vance and Johnson, 1983).

Four distinct zones of tissue differentiation are conspicuous in longitudinal sections of nodules found on all species, viz. nodule cortex, meristematic zone, vascular system and bacteroid zone (Allen and Allen, 1953).

The nodule cortex

The outermost tissue of the nodule consists of four to ten layers of non-infected parenchyma which make up the nodule cortex. These cells are derived from the nodule meristem and are readily distinguished from root cortex by compactness and smaller size. In early developmental stages the young nodule is embedded in the root cortex. As the mass of proliferating cells enlarges, the loosely packed outer cortical parenchyma becomes stretched and broken. Nodules do not emerge or digest their way out of the root cortex as do rootlets, but, as exemplified by large nodules in many species, the outer root cortex is eventually sloughed away.

Meristematic zone

Underneath the cortical layers of the nodule lies the conspicuous nodule meristem composed of numerous small, compact, densely stained cells. This is the region of active nuclear division. The cells do not contain infection threads or rhizobia and provide the nodule with a continuous supply of young cells which soon differentiate towards the outside into vascular or cortical tissues, or become invaginated towards the inside for the continuance of symbiotic function.

The prominence of the meristematic zone varies according to the host species. Its size and duration of activity depend on the effectiveness of the rhizobial strain. Effective nodules are divided into three major groupings based on shape and meristematic activity (Allen and Allen, 1953; Vance, 1983).

1. Apical meristems produce the elongated, cylindrical nodules common on alfalfa and clover (Fig. 8.1(a)). Growth occurs only in the distal extremity of the nodule. Unequal growth or bifurcation of this type of meristem produces nodules with branched tips characteristic of alfalfa.
2. Hemispherical or bowl-shaped meristems produce spherical nodules, e.g. those of bean, pea, soya bean and cowpea (Fig. 8.2(a)). In nodules of this type an equal rate of cell division throughout the peripheral zone produces radial enlargement. Uneven or interrupted meristematic activity in certain areas later results in lobed or irregular shaped nodules.
3. Divided lateral meristems are responsible for the development of collar-type nodules found on lupin. Horizontal growth in opposite directions within these nodules produces a hypertrophy which tends to surround the parent root. The initiation and development of collar type nodules are less well understood (Dart, 1977).

The vascular system

The efficiency of the symbiotic relationship between the leguminous plant and rhizobia is dependent upon an adequate vascular system. Normally, vascular tissue differentiation is discernible within a week after nodule initiation. Radial divisions which give rise to provascular strands occur in cells between the periphery of the inner infected tissue and the nodule cortex, while the nodule is still largely meristematic.

The development of the vascular system keeps pace with the growth of the nodule. Nodules of alfalfa and clover usually contain only two or four vascular bundles. In large spherical nodules of cowpea, soya bean and *Sesbania* sp., the vascular strands branch repeatedly to form a network of conductive tissue around the bacteroid zone. The strands never come into direct contact with the bacteroidal tissue. A few layers of non-infected parenchyma always separate the bundle sheath from the inner zone (Fig. 8.2(c)). Diffusion of nutrients across the parenchymatous layer gradually becomes restricted due to the suberization of the bundle endodermis.

The bacteroid zone

This area, which is the heart of the nodule, comprises all the host cells enclosed within the tissues previously described. Only bacteroidal cells (*Rhizobium*-infected) and interstitial cells (*Rhizobium*-free) are contained within the bacteroid zone (Allen and Allen, 1953; Tu, 1975). The size of these cells varies within wide limits due to hypertrophy accompanying the infection. In general, the mature infected cells are four to eight times greater in size than the neighbouring rhizobia-free ones. The number of infected cells is dependent on the plant species and also upon effectiveness of the rhizobial strain.

The cells in the bacteroid zone are loosely organized, having numerous large intercellular spaces. The ventilation of the cells may be provided mainly by the intercellular spaces. The interstitial cells may serve in diffusion and translocation of materials to and from the bacteroidal cells (Tu, 1975). The bacteroid zone represents from 16% to over 50% of the total dry weight of nodule (Bond, 1941) and its mean relative volume is much greater in effective nodules than in ineffective nodules (Chen and Thornton, 1940).

Once liberated from the infection thread into the root cytoplasm, the rhizobia assume a peculiar morphology, a cellular form that has been termed the 'bacteroid'. In morphology, the bacteroids found within the nodule are swollen and irregular, frequently appearing in star, clubbed or branched shapes.

The number and shape of bacteroids in a cell vary with the bacterial strain and the host plant cultivar (Subba Rao, 1988). In the elongated cylindrical nodules, the infected cells do not divide, but are increased in number by the continued ramification of infection threads into new tissue. In spherical nodules, the central, roughly spherical masses of bacteroid–containing cells are formed by continued division of infected cells (Chen and Thornton, 1940; Dart, 1977).

The bacteroids are arranged in groups within the cell, each group of four or six bacteroids being enclosed in a system of double membranes. These membranes appear to originate from the host cells. Bergersen (1960) proved that the membrane envelopes are the site of the primary reactions of nodule nitrogen fixation.

The effective functioning of bacteroid tissue of legume root nodules in symbiotic nitrogen fixation is dependent on two vital prerequisites, viz. the bacteroids and the leghaemoglobin (Lb) (Chopra and Subba Rao, 1967). Lb is the major soluble protein found in legume nodules formed after legume roots are invaded by rhizobia (Sidloi-Lumbroso and Schulman, 1977). Although the presence of Lb is confined to the nodules, genetic evidence indicates that the type of Lb is a characteristic of the plant host rather than the infecting *Rhizobium* (Dilworth, 1969; Broughton and Dilworth, 1971; Cutting and Schulman, 1971; Baulcombe and Verma, 1978).

Bergersen (1971) concluded that Lb lies within the membrane envelope which surrounds the bacteroids in soya bean nodules. The function of Lb is to deliver oxygen to the aerobic endosymbiotic bacteroids at an oxygen tension sufficiently low so as not to inhibit the oxygen-sensitive bacteroid nitrogenase (Nadler and Avissar, 1977). Lb does not have a direct role in the nitrogen-fixing system (Bergersen and Turner, 1967). The capacity of Lb to bind oxygen depends on the presence of a non-polypeptide unit, namely a haem group (Fe-protoporphyrin IX) (Stryer, 1981). The Lb consists of apoprotein (globin) and haem. The globin portion of Lb has been shown to be produced by the host plant (Dilworth, 1969; Cutting and Schulman, 1971), while haem is synthesized by the bacteroids (Cutting and Schulman, 1969; Nadler and Avissar, 1977). The haem gives Lb its distinctive colour; it is markedly similar to blood haemoglobin. Allen and Allen (1953) identified four well-defined pigments in the bacteroid zone: Lb (red), legcholeglobin (green), legmethaemoglobin (brown) and coproporphyrin (brown). The Lb content and the extent of bacteroid tissue in nodules have a direct correlation with the amount of nitrogen fixed by legumes (Bergersen and Briggs, 1958; Chopra and Subba Rao, 1967).

The enzyme systems necessary for nitrogen fixation by legume root nodules are located within the bacteroids (Kennedy *et al.*, 1966; Bergersen and Turner, 1967).The nitrogenase complex, which carries the biological process of nitrogen fixation, consists of two kinds of protein components: a reductase, which provides electrons with high reducing power, and a nitrogenase, which uses these electrons to reduce nitrogen to NH_4^+. Each component is an iron–sulphur protein. The nitrogenase component of the complex also contains one or two molybdenums and so it has been known as the MoFe protein. The reductase is also called the Fe protein. In the nitrogenase complex, one or two Fe proteins are associated with a MoFe protein (Stryer, 1981).

The conversion of nitrogen into NH_4 by the nitrogenase complex requires ATP and a powerful reductant. In most nitrogen-fixing microorganisms the source of high-potential electrons in this six-electron reduction is reduced ferrodoxin, an electron carrier. Whether reduced ferrodoxin is then regenerated by photosynthetic or oxidative processes depends on the particular species (Stryer, 1981).

8.4.2 Nodule senescence

Sutton (1983) stated that nodule senescence commences when nitrogen fixation rates begin to decline.The process is soon accompanied by cellular disintegration. The earliest indication of senescence in effective nodules is evidenced by the change in colour of the bacteroid zone from red (Lb) to green (legcholeglobin). Nodule senescence has a threefold practical importance: (1) the amount of fixed nitrogen made available to the host plant obviously depends on the duration as well as the volume and activity of the symbiotic tissue (Chen and Thornton, 1940); (2) nodule degeneration may allow the release of at least some host-specific rhizobia back into the rhizosphere (Thornton, 1930); and (3) nodule decay may result in an underground transfer of combined nitrogen from legumes to other plants in the ecosystem (Butler and Bathurst, 1956).

8.5 PLANT-PARASITIC NEMATODES

Several interactions between plant-parasitic nematodes and other soil-inhabiting pathogens are a source of plant diseases. The associations of beneficial organisms, viz. mycorrhizal fungi (Smith, 1987) and rhizobia (Huang, 1987) with plant-parasitic nematodes, and the effect of mycorrhizal and rhizobial symbioses on plant growth, have been investigated.

8.5.1 Nematodes in the rhizosphere

The rhizosphere is a dynamic environment where the relationships among nematodes, plant and environment are often of a chemical nature. The number of nematodes in the rhizosphere was greater than in the non-rhizosphere soil of wheat, barley, oats, soya bean, peas (Henderson and Katznelson, 1961), okra and eggplant (Khan *et al.*, 1975).The abundance of parasitic and non-parasitic nematodes around roots may be due to 'attractants' produced by root microflora as well as to root exudates (Katznelson and Henderson, 1962, 1963). The root microflora exerts an influence completely unrelated to its food value by producing substances which attract or repel nematodes and/or by creating conditions which are favourable or inimical to them (Katznelson and Henderson, 1962).

O'Brien and Prentice (1930) were the first to show that the cysts of potato nematodes (*Heterodera schachtii*: now *Globodera rostochiensis*) hatched in the presence of root washings of potato but not the washing of beet, rape, lupin, mustard or oat roots. Dependence on root diffusates for hatching, however, varies with the nematode species. Root diffusates are not required for hatching of root-knot nematodes, although they may increase the number of emerging juveniles (Wallace, 1966). Linford and Oliveira (1940) found that newly hatched juveniles of *Rotylenchulus reniformis*, held in water without plant roots, soon passed through three superimposed moults to become young infective females and males. The exudates also influenced the moulting of pre-adult juveniles of the pin nematode, *Paratylenchus* spp. (Rhoades and Linford, 1959). Red clover (*Trifolium pratense* L.) root diffusates induced moulting in *P. projectus* but not in *P. dianthus*, although red clover is a suitable host for both species.

Root exudates may also inhibit egg hatching or may repel or kill nematodes. The control of lesion nematode populations in soil and around roots of other plants by cultivation of marigolds, *Tagetes patula* L. and *T. erecta* L. (Oostenbrink *et al.*, 1957), and control of root-knot nematodes by *Crotalaria spectabilis* Roth. (McBeth and Taylor, 1944), represent direct action by growing plants.

8.5.2 Nematode feeding habits

The group of root-parasitic nematodes is large and diverse. Some are ectoparasites that remain outside the host tissue except for the stylet; some are semi-endoparasites that bury the anterior body portion in the host tissue, leaving the posterior exposed; and others are endoparasites that are entirely embedded in host tissue while feeding. Many parasitic nematodes show preference for feeding or entry at a specific stage in root ontogeny (Kirkpatrick *et al.*, 1964). Some species prefer the meristem, some the zone of elongation, and others the region of maturation. Species preference, however, may change with the plant parasitized. Some nematodes feed ectoparasitically along the entire length of the root, with no preferred site (Sutherland, 1967; Cohn, 1970).

Nematodes not only show varying degrees of specificity for certain root regions, but they also show preference for tissues and cells within a root region, i.e. apical meristem, epidermis, cortex, stele, etc. (Kirkpatrick *et al.*, 1964). They damage plant cells to various degrees by removing cell contents, or by inducing cell lysis. A few parasites, however, induce the host cell to provide specific feeding sites containing specialized nutritive cells or syncytia upon which the nematodes feed (Dropkin, 1969; Endo, 1975). Although many nematodes are capable of inflicting severe above-ground damage (stunting, wilting, chlorosis, reduced yield), they most commonly cause direct root damage (galls, lesions, root-pruning, stubby root) which is indirectly responsible for the above-ground symptoms.

8.6 NEMATODE–RHIZOBIA INTERACTIONS

The environment greatly influences not only the growth and longevity of nodule bacteria in the soil, but also the production and behaviour of the nodules and the development of the host plant, and for all these reasons it also influences uptake of nitrogen from the soil and from the air (Van Schreven, 1958). One of the biological factors affecting nodule formation or dysfunction of existing nodules is the presence of nematodes in the rhizosphere. Dysfunction of the symbiotic process in legumes also occurs with viral and fungal infections (Tu *et al.*, 1970; Orellana *et al.*, 1976, 1978; Bowen, 1978).

8.6.1 Effect on nodulation

Nutman's review (1965) of his extensive studies on red clover and strains of *Rhizobium trifolii* showed that the number and size of nodules, the presence and absence of nodules, and the factors influencing the effectiveness of the nodules in nitrogen fixation, are controlled through the genetic make-up inherent to the plant. Evidence that nodules of different kinds do in fact differentially inhibit further nodule formation was obtained in experiments on the effects of nodule excision and delayed inoculation. The excision of the growing nodule and delayed inoculation cause an immediate increase in the rate of nodule formation. Inhibition of nodule formation is related to the activity of the nodular meristem, upon which the size of the nodule depends. The inverse relationship between nodule size and abundance varies in its expression between species (Nutman, 1958).

Several plant-parasitic nematodes with different modes of parasitism cause a reduction in nodulation on leguminous plants. Ectoparasites – *Trichodorus christiei* and *Criconemoides curvatum* on hairy vetch, *Belonolaimus longicaudatus* on soya bean grown in the greenhouse and *Scutellonema cavenessi* on soya bean; semi-endoparasites – *Rotylenchulus reniformis* on cowpea; and endoparasites – *Meloidogyne javanica* on hairy vetch, alfalfa and mung bean, *M. incognita* on mung bean, chickpea, cowpea, Wando pea and green gram, *M. hapla* on white clover, *Meloidogyne* spp. on horse bean, lupin, clover and peas, *Heterodera glycines* on soya bean, *H. goettingiana* on peas, *H. cajani* on cowpea, *H. trifolii* on white clover and *Pratylenchus sfeansis* on soya bean – all inhibit nodulation. *M. hapla* on hairy vetch completely inhibited nodulation. When soya bean plants were inoculated simultaneously with a high density of soya bean cyst nematode race 1 (12 500 juveniles per plant) and rhizobia, complete inhibition of nodule development occurred (Huang and Barker, 1983).

Stimulation of nodule formation on leguminous plants by plant-parasitic nematodes has been observed. *Meloidogyne incognita*, *M. hapla* and *Pratylenchus penetrans* on soya bean grown in the greenhouse, and *P. penetrans* and *Belonolaimus longicaudatus* on soya bean grown in the phytotron (Hussey and Barker, 1976), and *M. incognita* on pea and black bean (Verdejo *et al.*, 1988) were reported to stimulate nodule formation. However, nodules on plants with increased nodulation were smaller than nodules formed on plants free of nematode.

Other reports indicated that nematode infection had no significant effect on number and size of nodules. *Meloidogyne javanica* and *Heterodera trifolii* on white clover (Taha and Raski, 1969), and *M. incognita* on soya bean (Caroppo and Pelagatti, 1988) did not affect nodule formation. Introduction of nematodes one week before, simultaneously, or one week following inoculation with rhizobia did not hinder nodule formation (Taha and Raski, 1969). Only nodules formed on nematode galls were significantly smaller than healthy or infected nodules not on galls. Individual cases occurred in which nodules heavily infected with *H. trifolii* were significantly smaller than healthy nodules (Taha and Raski, 1969).

Races of *Heterodera glycines* on soya bean differed in their influence on nodulation. Race 1 significantly reduced nodulation, especially when added simultaneously with rhizobia, while races 2 and 4 did not (Lehman *et al.*, 1971; Barker *et al.*, 1972). Race 3 did not reduce nodule mass on resistant or susceptible soya bean cultivars compared to nematode-free plants, except for one instance where nodule mass decreased on a resistant variety (McGinnity *et al.*, 1980). Results with race 4 were variable, i.e. reducing nodule mass in one instance but increasing it in three instances. Differential response of races 3 and 4 on nodule mass compared to race 1 was attributed to the differences in the reproduction rates of *H. glycines* race.

The cause of reduced nodulation has been postulated. According to Masefield (1958), nematode galls on the roots may affect nodulation by causing nutrient deficiency in host plants and by occupying space on the root system, a suggestion which was supported later by Malek and Jenkins (1964). A competition phenomenon between nematode juveniles and root-nodule bacteria was also postulated as a cause of reduction (Epps and Chambers, 1962). Devitalization of root tips by *Trichodorus christiei* and the resulting lack of formation of root hairs were primarily responsible for reduced nodulation (Malek and Jenkins, 1964). Other work, however, provided no support for the hypothesis of site occupation or the competition phenomenon (Taha and Raski, 1969). Infection by *Meloidogyne javanica* on white clover (Taha and Raski, 1969), and *Heterodera incognita* on pea and black bean (Verdejo *et al.*, 1988) reduced the size of the root system and the number or weight of nodules per gram of root was not affected. Barker and Hussey (1976) used an inoculum level of *M. incognita* 14 times the size of that used by Verdejo *et al.* (1988) to suppress nodulation of Wando pea.

Nutman (1958) noted that rhizobial infection is not a haphazard process but is firmly controlled by influences within the root. *Rotylenchulus reniformis* infection may deplete root hairs through which rhizobial infection could take place (Taha and Kassab, 1980). Also, suppression of lateral root formation by *R. reniformis* (Oteifa and Salem, 1972) may cause reduction in the number of sites for nodule initiation, since lateral roots bear their own sites.

Interference of the nematode with soya bean lectin metabolism was determined as a factor that reduces binding of rhizobia to *Heterodera glycines*-infected soya bean roots and thus suppresses nodule formation (Huang *et al.*, 1984). In contrast, nematode infection caused an increase in numbers of root hairs and therefore the surface area of total root system. The binding between rhizobia and soya bean root hairs is generally believed to be mediated by the rhizobial lipopolysaccharide and the 120 000 Da soya bean lectin (Bohlool and Schmidt, 1974; Maier and Brill, 1978; Huang *et al.*, 1984). Presence of saccharides, viz. *N*-acetylglucosamine, galactose, *N*-acetylgalactosamine and mannose and/or glucose, on the cuticle surface of plant-parasitic nematodes (Spiegel *et al.*, 1982; Zuckerman and Jansson, 1984; Forrest and Robertson, 1986), according to Zuckerman and Jansson (1984) may play an important role in the interaction between nematode and their hosts. Huang *et al.* (1984) found that binding of rhizobia to nematode-free roots was inhibited only after pretreatment with N-acetyl-D-galactosamine or D-galactose, the haptens of the soya bean lectin, but not glucose. Therefore, very few rhizobia were observed on root-hair surfaces of nematode-infected plants in comparison with root-hair surfaces of nematode-free plants.

Histology of nodules

Nodule invasion by nematodes occurs on a number of legumes. Invasion of nodules by *Meloidogyne* spp. like *M. javanica* in velvet bean, cowpea, alfalfa and white clover, *M. incognita* in soya bean, peanut, cowpea, gram, clover, horse bean, lupin and black bean, *M. hapla* in alfalfa, and *Meloidogyne* spp. in common bean varieties has been reported. Other nematodes like *Heterodera trifolii* in white clover, *H. glycines* in soya bean, *H. cajani* in cowpea, *H. daverti* in clover, *Rotylenchulus reniformis* in pigeonpea and cowpea, *Pratylenchus globulicola* in alfalfa, clover and pea, and *P. penetrans* in Wando pea invade rhizobial nodules.

Minimal damage of nodules by *Pratylenchus penetrans* on soya bean and peanut was considered to be related to the fact that the nematode colonizes only cortical tissues (Barker and Hussey, 1976). The capacity of this nematode to penetrate nodules of Wando pea and its association with microbivorous nematodes such as *Acrobeloides buetschlii* may result in considerable damage in the field situation on this and certain other legumes. High numbers of *A. buetschlii* found in the central portion of Wando pea nodules were apparently feeding and reproducing on bacteroids (Westcott and Barker, 1976). Damage of Wando pea and peanut nodules by the ectoparasite *Belonolaimus longicaudatus* was usually limited to cortical tissues. Nodules tended to break down after extensive feeding. The severe damage normally encountered in the field on soya bean and peanut parasitized by *B. longicaudatus* results from a suppression of root growth rather than damage to nodular tissues which does occur with *Heterodera glycines* (Barker and Hussey, 1976).

The histological changes in legume nodular tissues induced by the infection of *Melodiogyne* spp. occurs in velvet bean (Robinson, 1961), white clover (Taha and Raski, 1969), soya bean (Barker and Hussey, 1976), peanut (Taha and Yousif, 1976), common bean (Taha *et al.*, 1977), cowpea (Taha and Kassab, 1979; Ali *et al.*, 1981), clover, horse bean, lupin and pea (Yousif 1979). The formation of giant cells inside the vascular bundles was observed, while the structure of nodular tissues was not disturbed (Fig. 8.1(b)). Heavily infected nodules formed on nematode galls had a small amount of bacteroids (Fig. 8.2(b)). A female was found inside the bacteroidal tissue of cowpea nodule, inducing syncytial formation in the cortex (Fig. 8.2(f)) (Taha and Kassab, 1979). Infected nodules degenerated earlier than uninfected ones. The rapid degeneration of *M. javanica*-infected white clover nodules cannot be attributed entirely to the presence of nematodes inside the nodules and/or

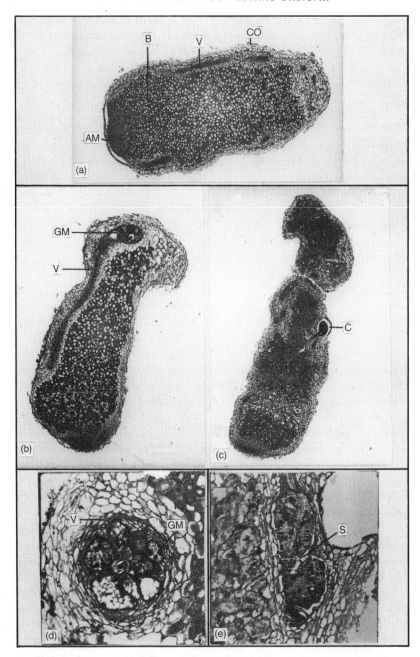

Figure 8.1 (a) Longitudinal section of healthy white clover nodule. (b) Longitudinal section of *Meloidogyne javanica*-infected white clover nodule showing the giant cells inside the vascular bundle. (c) Longitudinal section of *Heterodera trifolii*-infected white clover nodule showing the cyst extending to the cortex. (d) Transverse section of *M. javanica* giant cells inside the vascular bundle. (e) Longitudinal section of

to the large number of larvae reinfecting the nodules (Taha and Raski, 1969). The destruction of the root system by the nematode is also important. *M. hapla* was not detected in nodules of peanut, and had no apparent effect on nodules that developed on galls (Barker and Hussey, 1976).

The presence of syncytia extending into the vascular bundles was observed in infection by *Heterodera trifolii* on white clover (Taha and Raski, 1969) and *H. daverti* on Egyptian clover (Massoud and Ghorab, 1988). Taha and Raski (1969) found *H. trifolii* cyst in the nodular cortex (Fig. 8.1(c)), and Massoud and Ghorab (1988) observed hypertrophied cells in the bacteroidal tissues. The giant cells of *Meloidogyne javanica* (Fig. 8.1(d)) and the syncytia of *H. trifolii* (Fig. 8.1(e)) are structurally different, and they resemble their respective counterparts in the roots. Nodular tissue of soya bean, however, was unfavourable for the development of *H. glycines* larvae of either race 1 or 4, although race 1 penetrated nodular tissues at a much greater rate than race 4 (Barker *et al.*, 1972). Most nematodes that matured in nodules were males, and although a few matured cysts developed on nodules, most infections caused considerable tissue necrosis, and most syncytia did not develop fully. Endo (1964) indicated that *H. glycines*, especially males, often resulted in necrosis and degeneration of syncytia as the nematodes matured.

Soya bean nodules emerging from plants infected with race 1 of *Heterodera glycines* were poorly organized, with less distinct zones of nodular tissues and early appearance of vascular elements and sclerenchyma layers. However, the most conspicuous features in the infected plants were the massive accumulation of starch granules and crystalline arrays of phytoferritin in the plastids of cells in the nodular tissues. Although small starch granules occasionally occurred in similar tissues of check plants, phytoferritin was not observed. Starch and phytoferritin are, respectively, energy and iron reserves. Their accumulation in nodules of nematode-infected soya beans suggests that the metabolism of carbohydrates and iron-containing compounds is affected by the presence of the cyst nematode (Ko *et al.*, 1985).

Rotylenchulus reniformis appears to be a pericycle feeder. Taha and Kassab (1979) found cowpea nodules infected with more than one *R. reniformis* female. Although *R. reniformis* is a semi-endoparasite, it was found embedded in nodular cortex of cowpea, inducing hypertrophied pericycle cells. Coincidence of giant cells and hypertrophied pericycle cells (Fig. 8.2(c)) was observed in separate vascular bundles of nodules infected with both *Meloidogyne javanica* (Fig. 8.2(d)) and *R. reniformis* (Fig. 8.2(e)), respectively. Furthermore, giant cells and hypertrophied pericycle cells in a single vascular strand were revealed by successive sections of cowpea nodule containing both nematodes. Both types of feeding site

---

H. trifolii syncytium. (AM, apical meristem; B, bacteroidal tissue; C, cyst; CO, cortex; GM, giant cells; V, vascular bundle.) Figure reproduced with permission of the Society of Nematologists.)

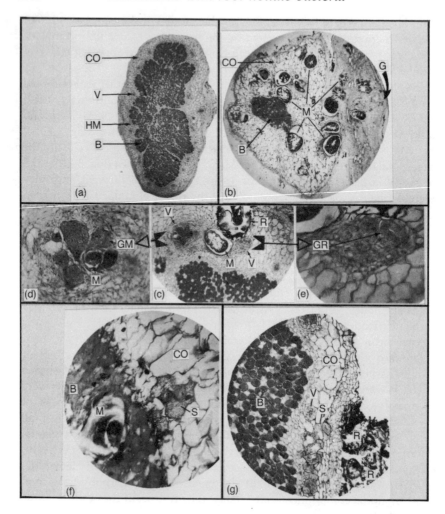

Figure 8.2 (a) Cross-section of cowpea healthy nodule. (b) Longitudinal section of cowpea nodule heavily infected with *Meloidogyne javanica* on galled root. (c) Cross-section of cowpea nodule co-inhabited by *M. javanica* and *Rotylenchulus reniformis*, showing *M. javanica* giant cells in one vascular bundle (d) and *R. reniformis* hypertrophied pericycle cells in another (e). (f) Cross-section of cowpea nodule infected with *M. javanica*, showing embedded female in bacteroidal tissue, and a syncytium in the cortex. (g) Cross-section of cowpea nodule infected with *R. reniformis*, showing a syncytium consisting of cortical cells with wall gaps. (B, bacteroidal tissue; CO, cortex; G, nematode gall; GM, giant cell; GR, hypertrophied pericycle cells; HM, hemispherical meristerm; M, *M. javanica*; R, *R. reniformis*; S, syncytium; V, vascular bundle.) (Figure reproduced with permission of the Society of Nematologists.)

appeared together at a point between separate ones. Each nematode species produced its own histopathological response in either root or nodular tissue, and the response to one species was not affected by the presence of the other species. *R. reniformis* also induced a syncytium consisting of cortical cells (Fig. 8.2(g)).

8.6.2 Effect on nitrogen fixation

The effective functioning of bacteroid tissue of legume root nodules in symbiotic nitrogen fixation is dependent on two vital prerequisites, viz. the bacteroids and Lb (Chopra and Subba Rao, 1967). *Acrobeloides buetschlii* inhibited nitrogen fixation in nodular tissue of Wando pea because of their apparent feeding and reproduction on bacteroids (Westcott and Barker, 1976). Infected nodules revealed cavities that developed after nodules had been invaded by *A. buetschlii*.

Reduction in number and size of nodules and early degeneration of nodules because of nematode infection caused reduced nitrogen fixation in legumes (Romaniko, 1961; Taha and Raski, 1969; Taha *et al.*, 1974; Yeates *et al.*, 1977; Ali *et al.*, 1981; Chahal and Chahal, 1988). Chen and Thornton (1940) named two quantitative characters closely correlated with effectiveness of nodules: the combined volume of the nodular tissues and the length of time that elapses before they collapse and disintegrate. Taha and Raski (1969) found that the nitrogen-fixation efficiency of nematode-infected nodules was not impaired; however, earlier disintegration of nodules as a result of *Meloidogyne javanica* infection ultimately deprived the plants of nitrogenous materials.

In soya bean root nodules, Lb accounts for 25–30% of the total soluble protein but is not detected in other tissues (Baulcombe and Verma, 1978). The Lb from soya bean nodules may be chromatographically fractionated into four components, Lba, Lbb, Lbc and Lbd, on a DEAE–cellulose column (Appleby *et al.*, 1975). Lb content per gram of nodule was lower in soya bean plants infected with *Heterodera glycines* race 1 than in the nodules of check plants (Huang and Barker, 1983). Lba from nematode-infected and check plants had similar ultraviolet and visible light spectra and gel electrophoresis profiles, as did Lbb, Lbc and Lbd. The ratio of Lbc/Lba, however, was higher from nematode-infected soya bean than from check plants (Huang and Barker, 1983).

Different Lb components are coded for by different plant mRNAs, and the relative levels of these mRNAs change during root nodule development. Analysis of the in vitro translation products of mRNA from nodules of different ages showed that Lbc was synthesized at a higher rate than Lba in young nodules, and the reverse was true in mature nodules (Verma *et al.*, 1979). The ratio of Lbc/Lba in soya bean root nodules therefore remained high in the early stages of soya bean growth and decreased during flowering and fruiting (Fuchsman *et al.*, 1976). Since nodules from nematode-infected soya beans had a higher Lbc/Lba ratio, this suggested that nodule development was impaired with nematode infection (Huang and Barker, 1983). The significant reduction in overall Lb content, however, indicated that nodules from nematode-infected plants were senescent. Although infection of soya bean by the cyst nematode limited nodule size (Lehman *et al.*, 1971), the cause, whether due to impairment of nodule development or acceleration of nodule senescence, remains to be determined (Huang and Barker, 1983).

Root-knot nematode, *Meloidogyne incognita*, adversely affected the functioning of chickpea nodules as reflected by nitrogenase activity (Chahal and Chahal, 1988). Higher inoculum levels (1000 and 2000 juveniles per pot) caused significantly more reduction in the nitrogenase activity, Lb and bacteroid content of nodules than did lower levels (250 and 500 juveniles per pot). A similar trend in total nitrogen uptake by the plant tops was also observed. In contrast. *M. incognita* did not seem to alter the function of pea and black bean nodules that developed (Verdejo *et al.*, 1988). Nitrogenase activity was increased in nodules that developed on black bean infected by *M. incognita*, but on pea it was only increased when *Rhizobium* and the nematode were introduced together. Moreover, an increase in Lb in nodules on pea, and decrease in nodules on black bean, was found. Verdejo *et al.* (1988) suggested that pea nodules are indeterminate (apical) and therefore able to grow and form young tissue rich in Lb when the plant is short of nitrogen. Black bean nodules are determinate (spherical) and once formed are less able to develop so that young tissue can only form in new nodules.

McGinnity *et al.* (1980) indicated that nodulation and nitrogen fixation are influenced by cultivar × *Rhizobium* strain and soya bean cyst nematode race interactions and that the nematode is not equally deleterious to all susceptible cultivars.

8.7 CONCLUSION

The continuous drain on the nitrogen resources of the soil and the necessity for higher crop yields have led to an ever-increasing emphasis on means of conserving the limited supply of the element. As only a fraction of the total agricultural need for nitrogen comes from synthetic and natural fertilizers, the remaining portion must be satisfied from the soil reserves and through the biological fixation of atmospheric nitrogen (Alexander, 1961). Symbiotic fixation of nitrogen is not confined to the legumes, and neither are nitrogen-fixing root nodules confined to Leguminosae.

The contribution of legumes to the maintenance of soil fertility and to the spiralling need for pulses and oil seeds for human consumption, and forage crops for animals, has been of peculiar concern to investigators in various fields of science. The wide interest in the fundamental process of symbiotic nitrogen fixation has been shown by the voluminous work on factors affecting the *infectiveness*, the ability of a rhizobial strain to nodulate a given host, and the *effectiveness*, the relative capacity of the symbiotic system, once established, to assimilate molecular nitrogen.

The importance of heritable host factors in legume symbiosis, long suspected from comparison of cultivars, has now been demonstrated in various leguminous crops but attempts to exploit this source of variability to increase nitrogen fixation through breeding programmes appear to be reported only for species of *Trifolium* and *Medicago sativa* (Nutman and Riley, 1981). Interference of nematodes in the legume–*Rhizobium* association creates an antagonistic force by which nematodes withdraw food from the plant, leading to impoverishment of the plant and, eventually, yield reduction. This could be alleviated by applying nematicides.

Most studies on the genetics of host–nematode interactions deal primarily with inheritance of plant resistance. However, the introduction of cultivars resistant to a pathogen often results in changes in the pathogen population which enable the pathogen to overcome host resistance. Such changes have occurred with cyst nematodes, including the soya bean cyst nematode, *Heterodera glycines* (Riggs *et al.*, 1977). Nevertheless, research of this type should be followed by teamwork between nematologists and plant breeders, and in cooperation with other scientists who are interested in legumes, but looking at them from different points of view. This programme may offer the best approach to certain unusually difficult problems.

REFERENCES

Akkermans, A.D.L. and Van Dijk, C. (1981) Non-leguminous root-nodule symbioses with actinomycetes and *Rhizobium*, in *Nitrogen Fixation, Vol. I: Ecology* (ed. W.J. Broughton), Clarendon Press, Oxford, pp. 57–103.

Akkermans, A.D.L., Abdulkader, S. and Trinick, M.J. (1978) Nitrogen-fixing root nodules in Ulmaceae. *Nature*, **247**, 190.

Albersheim, P. and Anderson-Prouty, A.J. (1975) Carbohydrates, proteins, cell surfaces, and the biochemistry of pathogenesis. *Annual Review of Plant Physiology*, **26**, 31–52.

Alexander, M. (1961) *Introduction to Soil Microbiology*. John Wiley, New York.

Ali, M.A., Trabulsi, I.Y. and Abd-Elsamea, M.E. (1981) Antagonistic interaction between *Meloidogyne incognita* and *Rhizobium leguminosarum* on cowpea. *Plant Disease*, **65**, 432–5.

Allen, D.N. and Allen, E.K. (1947) A survey of nodulation among leguminous plants. *Proceedings of Soil Science Society of America*, **12**, 203–8.

Allen, D.N. and Allen, E.K. (1953) Morphogenesis of the leguminous root nodule, in *Abnormal and Pathological Plant Growth*. Brookhaven Symposia in Biology no. 6, Brookhaven National Laboratory, Associated Universities, Upton, New York, pp. 209–34.

Appleby, C.A., Nicola, N.A., Hurrell, J.G.R. and Leach, S.J. (1975) Characterization and improved separation of soybean leghaemoglobins. *Biochemistry*, **14**, 4444–50.

Baldwin, I.L., Fred, E.B. and Hastings, E.G. (1927) Grouping of legumes according to biological reactions of their seed proteins: Possible explanation of phenomenon of cross inoculation. *Botanical Gazette*, **83**, 217–43.

Barker, K.R. and Hussey, R.S. (1976) Histopathology of nodule tissues of legumes infected with certain nematodes. *Phytopathology*, **66**, 851–5.

Barker, K.R., Huisingh, D. and Johnson, S.A. (1972) Antagonistic interactions between *Heterodera glycines* and *Rhizobium japonicum* on soybean. *Phytopathology*, **62**, 1201–5.

Bauer, W.D. (1981) Infection of legumes by rhizobia. *Annual Review of Plant Physiology*, **32**, 407–49.

Baulcombe, D. and Verma, D.P.S. (1978) Preparation of a complementary DNA for leghaemoglobin and direct demonstration that leghaemoglobin is encoded by the soybean genome. *Nucleic Acid Research*, **5**, 4141–53.

Bergersen, F.J. (1957) The structure of ineffective root nodules of legumes: An unusual new type of the ineffectiveness, and an appraisal of present knowledge. *Australian Journal of Biological Science*, **10**, 233–42.

Bergersen, F.J. (1960) Biochemical pathways in legume root nodule nitrogen fixation. *Biochemical Review*, **24**, 246–50.

Bergersen, F.J. (1971) Biochemistry of symbiotic nitrogen fixation in legumes. *Annual Review of Plant Physiology*, **22**, 121–40.

Bergersen, F.J. and Briggs, M.J. (1958) Studies on the bacteroid component of soybean root nodules: Cytology and organization in the host tissue. *Journal of General Microbiology*, **19**, 482–90.

Bergersen, F.J. and Turner, G.L. (1967) Nitrogen fixation by the bacteroid fraction of breis of soybean root nodules. *Biochimica et Biophysica Acta*, **141**, 507–15.

Berry, A.M. and Torrey, J.G. (1983) Root hair deformation in the infection process of *Alnus rubra*. *Canadian Journal of Botany*, **61**, 2863–76.

Bhuvaneswari, T.V., Pueppke, S.G. and Bauer, W.D. (1977) Role of lectins in plant–microorganism interactions. I. Binding of soybean lectin to rhizobia. *Plant Physiology*, **60**, 486–91.

Bhuvaneswari, T.V., Turgeon, B.G. and Bauer, W.D. (1980) Early events in the infection of soybean (*Glycine max* L. Merr.) by *Rhizobium japonicum*: Localization of infectible root cells. *Plant Physiology*, **66**, 1027–31.

Bhuvaneswari, T.V., Bhagwat, A. and Bauer, W.D. (1981) Transient susceptibility of root cells in four common legumes to nodulation by rhizobia. *Plant Physiology*, **68**, 1144–9.

Bohlool, B.B. and Schmidt, E.L. (1974) Lectins: A possible basis for specificity in the *Rhizobium*–legume root nodule symbiosis. *Science*, **185**, 269–71.

Bond, G. (1941) Symbiosis of leguminous plants and nodule bacteria. I. Observations on respiration and on the extent of utilization of host carbohydrates by the nodule bacteria. *Annals of Botany*, **5**, 313–37.

Bond, G. (1958) Symbiotic nitrogen fixation by non-legumes, in *Nutrition of the Legumes* (ed. E.G. Hallsworth), Butterworths Scientific Publications, London, pp. 216–31.

Bond, G. and Becking, J.H. (1982) Root nodules in the genus *Colletia*. *New Phytologist*, **90**, 57–65.

Bond, G. and Wheeler, C.T. (1980) Non-legume nodule systems, in *Method for Evaluating Biological Nitrogen Fixation* (ed. F.J. Bergersen), John Wiley, Chichester, pp. 185–211.

Bowen, G.D. (1978) Dysfunction and shortfalls in symbiotic responses, in *Plant Disease: An Advanced Treatise*, Vol. III (eds J.G. Horsfall and E.B. Cowling), Academic Press, New York, pp. 231–6.

Boyd, W.C. and Sharpleigh, E. (1954) Specific precipitating activity of plant agglutinins (lectins). *Science*, **119**, 419.

Broughton, W.J. (1978) A review control of specificity in legume–*Rhizobium* associations. *Journal of Applied Bacteriology*, **45**, 165–94.

Broughton, W.J. and Dilworth, M.J. (1971) Control of leghaemoglobin synthesis in snake beans. *Biochemical Journal*, **125**, 1075–80.

Burton, J.C. and Allen, O.N. (1950) Inoculation of crimson clover (*Trifolium incarnatum* L.) with mixtures of rhizobia strains. *Proceedings of Soil Science Society of America*, **14**, 191–5.

Butler, G.W. and Bathurst, N.O. (1956) The underground transference of nitrogen from clover to associated grass. *Proceedings Seventh International Grassland Conference*, pp. 168–78.

Callaham, D., Tredici, P.D. and Torrey, J.G. (1978) Isolation and cultivation *in vitro* of the actinomycete causing root nodulation in *Comptonia*. *Science*, **199**, 899–902.

Caroppo, S. and Pelagatti, O. (1988) The effects of the infestation by *Meloidogyne incognita* (Kofoid & White) Chitwood on the symbiotic nitrogen fixation in *Glycine max* (L.) Merr. *Redia*, **71**, 1–10.

Chahal, P.P.K. and Chahal, V.P.S. (1988) Effects of different population levels of *Meloidogyne incognita* on nitrogenase activity, leghaemoglobin and bacteroid contents of chickpea (*Cicer arietinum* L.) nodules formed by *Rhizobium* spp. *Zentralblatt für Mikrobiologie*, **143**, 63–5.

Chandler, M.R. (1978) Some observations on infections of *Arachis hypogea* L. by *Rhizobium*. *Journal of Experimental Botany*, **29**, 749–55.

Chen, H.K. and Thornton, H.G. (1940) The structure of ineffective nodules and its influence on nitrogen fixation. *Proceedings of the Royal Society of London, Series B*, **129**, 208–29.

Chopra, C.L. and Subba Rao, N.S. (1967) Mutual relationships among bacteroids, leghaemoglobin and nitrogen content of Egyptian clover (*Trifolium alexandrinum*) and gram (*Cicer arietinum*). *Archiv für Mikrobiologie*, **58**, 71–6.

Cohn, E. (1970) Observations of the feeding and symptomatology of *Xiphinema* and *Longidorus* on selected host roots. *Journal of Nematology*, **2**, 167–73.

Cook, R. (1974) Nature and inheritance of nematode resistance in cereals. *Journal of Nematology*, **6**, 165–73.

Cutting, J.A. and Schulman, H.M. (1969) The site of heme synthesis in soybean root nodules. *Biochimica et Biophysica Acta*, **192**, 486–93.

Cutting, J.A. and Schulman, H.M. (1971) The determinant in the *R*–legume symbiosis for leghemoglobin specificity. *Biochimica et Biophysica Acta*, **229**, 58–62.

Dart, P.J. (1977) Infection and development of leguminous nodules, in *A Treatise on Dinitrogen Fixation, Section 3: Biology* (eds R.W.F. Hardy and W.S. Silver), Wiley–Interscience, New York, pp. 367–72.

Dazzo, F.B. and Brill, W.J. (1979) Bacterial polysaccharide which binds *R. trifolli* to clover root hairs. *Journal of Bacteriology*, **137**, 1362–73.

Dilworth, M.J. (1969) The plant as the genetic determinant of leghaemoglobin production in the legume root nodule. *Biochimica et Biophysica Acta*, **184**, 432–41.

Dropkin, V.H. (1969) Cellular responses of plants to nematode infections. *Annual Review of Phytopathology*, **7**, 101–22.

Endo, B.Y. (1964) Penetration and development of *Heterodera glycines* in soybean roots and related anatomical changes. *Phytopathology*, **54**, 79–88.

Endo, B.Y. (1975) Pathogenesis of nematode-infected plants. *Annual Review of Phytopathology*, **13**, 213–38.

Epps, J.M. and Chambers, A.Y. (1962) Effect of seed inoculation, soil fumigation, and cropping sequences on soybean nodulation in soybean-cyst nematode-infested soil. *Plant Disease Reporter*, **46**, 48–51.

Forrest, J.M.S. and Robertson, W.M. (1986) Characterization and localization of saccharides on the head region of four populations of the potato cyst nematode *Globodera rostochiensis* and *G. pallida*. *Journal of Nematology*, **18**, 23–6.

Fuchsman, W.H., Barton, C.R., Stein, M.M., Thomson, J.T. and Willett, R.M. (1976) Leghaemoglobin: Different roles for different components. *Biochemical and Biophysical Research Communications*, **68**, 387–92.

Goodchild, D.J. and Bergersen, F.J. (1966) Electron microscopy of the infection and subsequent development of soybean nodule cells. *Journal of Bacteriology*, **92**, 204–13.

Henderson, V.E. and Katznelson, H. (1961) The effect of plant roots on the nematode population of the soil. *Canadian Journal of Microbiology*, 7, 163–7.

Huang, J.S. (1987) Interactions of nematodes with rhizobia, in *Vistas on Nematology* (eds J.A. Veech and D.W. Dickson), Society of Nematologists, Hyattsville, pp. 301–6.

Huang, J.S. and Barker, K.R. (1983) Influence of *Heterodera glycines* on leghemoglobins of soybean nodules. *Phytopathology*, 73, 1002–4.

Huang, J.S., Barker, K.R. and Van Dyke, C.G. (1984) Suppression of binding between rhizobia and soybean roots by *Heterodera glycines*. *Phytopathology*, 74, 1381–4.

Hussey, R.S. and Barker, K.R. (1976) Influence of nematodes and light sources on growth and nodulation of soybean. *Journal of Nematology*, 8, 48–52.

Jordan, D.C. (1974) Ineffectiveness in the *Rhizobium* leguminous plant association. *Proceedings of the Indian National Science Academy*, 40, 713–40.

Jordan, D.C. (1984) Family III. Rhizobiaceae Conn 1938, in *Bergey's Manual of Systematic Bacteriology*, Vol. 1 (eds N.R. Krieg and J.G. Holt), Williams and Wilkins, Baltimore, pp. 234–56.

Katznelson, H. and Henderson, V.E. (1962) Studies on the relationships between nematodes and other soil microorganisms. I. Influence of actinomycetes and fungi on *Rhabditis* (*Cephaloboides*) *oxycerca* de Man. *Canadian Journal of Microbiology*, 8, 875–82.

Katznelson, H. and Henderson, V.E. (1963) Ammonium as an 'attractant' for a soil nematode. *Nature*, 198, 907–8.

Kennedy, I.R., Parker, C.A. and Kidby, D.K. (1966) The probable site of nitrogen fixation in root nodules of *Orinthopus sativa*. *Biochimica et Biophysica Acta*, 130, 517–19.

Khan, M.W., Khan, A.M. and Saxena, S.K. (1975) Studies on the rhizosphere effect in relation to fungi and nematodes of eggplant and okra. *Acta Botanica Indica*, 3, 114–20.

Kirkpatrick, J.D., Van Gundy, S.D. and Mai, W.F. (1964) Interrelationships of plant nutrition, growth, and parasitic nematodes. *Plant Analysis and Fertilizer Problems*, 4, 189–225.

Ko, M.P., Huang, P.Y., Huang, J.S. and Barker, K.R. (1985) Accumulation of phytoferritin and starch granules in developing nodules of soybean roots infected with *Heterodera glycines*. *Phytopathology*, 75, 159–64.

Lalonde, M. (1978) Confirmation of the infectivity of a freeliving actinomycete isolated from *Comptonia peregrina* root nodules by immunological and ultra-structural studies. *Canadian Journal of Botany*, 56, 2621–35.

Lechevalier, M.P. (1983) Cataloging *Frankia* strains. *Canadian Journal of Botany*, 61, 2964–7.

Lechevalier, M.P. and Lechevalier, H.A. (1989) Genus *Frankia* Brunchorst 1886, 174, in *Bergey's Manual of Systematic Bacteriology*, Vol. 4 (ed. S.T. Williams), Williams and Wilkins, Baltimore, pp. 2410–17.

Lehman, P.S., Huisingh, D. and Barker, K.R. (1971) The influence of races of *Heterodera glycines* on nodulation and nitrogen-fixing capacity of soybean. *Phytopathology*, 61, 1239–44.

Lewin, R.A. (1982) Symbiosis and parasitism: Definitions and evaluations. *Bioscience*, 32, 254–60.

Liener, I.E. (1976) Phytohemagglutinins (phytolectins). *Annual Review of Plant Physiology*, 27, 291–319.

Lim, G. and Burton, J.C. (1982) Nodulation status of the Leguminosae, in *Nitrogen Fixation, Vol. 2: Rhizobium* (ed. W.J. Broughton), Clarendon Press, Oxford, pp. 1–34.

Linford, M.B. and Oliveira, J.M. (1940) *Rotylenchulus reniformis* nov. gen. n. sp., a nematode parasite of roots. *Proceedings of the Helminthological Society of Washington, 7,* 35–42.

Lochhead, A.G. (1952) Soil microbiology. *Annual Review of Microbiology, 6,* 185–206.

Maier, R.J. and Brill, W.J. (1978) Involvement of *Rhizobium japonicum* O-antigen in soybean nodulation. *Journal of Bacteriology, 133,* 1292–9.

Malek, R.B. and Jenkins, W.R. (1964) Aspects of the host–parasite relationships of nematodes and hairy vetch. *New Jersey Agricultural Experiment Station, Bulletin 813,* 31 pp.

Masefield, G.B. (1958) Some factors affecting nodulation in the tropics, in *Nutrition of the Legumes* (ed. E.G. Hallsworth). Butterworths Scientific Publications, London, pp. 202–15.

Massoud, S.I. and Ghorab, A.I. (1988) Parasitism of *Heterodera daverti* on clover root *Rhizobium* nodules in Egypt. *Egyptian Journal of Phytopathology,* 20–78.

McBeth, C.W. and Taylor, A.L. (1944) Immune and resistant cover crops valuable in root-knot infested peach orchards. *Proceedings American Society of Horticultural Science, 45,* 158–66.

McGinnity, P.J., Kapusta, G. and Mayers, D., Jr (1980) Soybean cyst nematode and *Rhizobium* finances on soybean nodulation and N_2-fixation. *Agronomy Journal, 72,* 785–9.

Meijer, E.G.M. (1982) Development of leguminous root nodules, in *Nitrogen Fixation, Vol. 2: Rhizobium* (ed. W.J. Broughton), Clarendon Press, Oxford, pp. 311–31.

Nadler, K.D. and Avissar, Y.J. (1977) Heme synthesis in soybean root nodules. I. On the role of bacteroid δ-aminolevulinic acid synthase and δ-aminolevulinic acid dehydrase in the synthesis of the heme of leghemoglobin. *Plant Physiology, 60,* 433–6.

Nutman, P.S. (1957) Studies on the physiology of nodule formation. V. Further experiments on the stimulating and inhibitory effects of root secretions. *Annals of Botany, 21,* 321–37.

Nutman, P.S. (1958) The physiology of nodule formation, in *Nutrition of the Legumes* (ed. E.G. Hallsworth), Butterworths Scientific Publications, London, pp. 87–107.

Nutman, P.S. (1962) The relation between root hair infection by *Rhizobium* and nodulation in *Trifolium* and *Vicia. Proceedings of the Royal Society of London, Series B, 156,* 137.

Nutman, P.S. (1965) The relation between nodule bacteria and the legume host in the rhizosphere and in the process of infection, in *Ecology of Soil-borne Plant Pathogens, Prelude to Biological Control* (eds K.F. Baker and W.C. Snyder), University of California Press, Berkeley, pp. 231–47.

Nutman, P.S. and Riley, J. (1981) Breeding of nodulated red clover (*Trifolium pratense*) for high yield. *Annals of Applied Biology, 98,* 319–31.

O'Brien, D.G. and Prentice, E.G. (1930) An eelworm disease of potatoes caused by *Heterodera schachtii. Scottish Journal of Agriculture, 13,* 415–32.

Oostenbrink, M., Kuiper, K. and Jacob, J.J. (1957) *Tagetes* als feindpflanzen von *Partylenchus*-arten. *Nematologica* (Suppl.), *2,* 424–33.

Orellana, R.G., Sloger, C. and Miller, V.L. (1976) *Rhizoctonia–Rhizobium* interactions in relation to yield parameters of soybean. *Phytopathology, 66,* 464–7.

Orellana, R.G., Fan, F. and Sloger, C. (1978) Tobacco ring spot virus and *Rhizobium* interaction in soybean: Impairment of leghemoglobin accumulation and nitrogen fixation. *Phytopathology, 68,* 577–82.

Oteifa, B.A. and Salem, A.A. (1972) Biology and histopathogenesis of the reniform nematode, *Rotylenchulus reniformis,* on Egyptian cotton, *Gossypium barbadense.*

Proceedings of the Third Congress of the Mediterranean Phytopathological Union, Oeiras, Portugal, pp. 299–304.

Pankhurst, C.E., Schwinghamer, E.A. and Bergersen, F.J. (1972) The structure of acetylene reducing activity of root nodules formed by a riboflavin-requiring mutant of *Rhizobium trifolii. Journal of General Microbiology*, 70, 161–77.

Rhoades, H.L. and Linford, M.B. (1959) Molting of preadult nematodes of the genus *Paratylenchus* stimulated by root diffusates. *Science*, 130, 1476–7.

Riggs, R.D., Hamblen, M.L. and Rakes, L. (1977) Development of *Heterodera glycines* pathotypes as affected by soybean cultivars. *Journal of Nematology*, 9, 312–18.

Robinson, P.E. (1961) Root-knot nematodes and legume nodules. *Nature*, 189, 506–7.

Romaniko, V.I. (1961) Injury and economic damage to legumes caused by *Pratylenchus globulicola* n. sp., in *Questions in Phytonematology*. Izdat. Akad. Nauk SSSR. English translation of selected Russian and Polish papers in Phytonematology, Bulletin 546 (eds M.W. Brzeski and B.M. Zuckerman), Agricultural Experimental Station, University of Massachusetts, pp. 161–74.

Rovira, A.D. (1965a) Plant root exudates and their influence upon soil microorganisms, in *Ecology of Soil-Borne Plant Pathogens, Prelude to Biological Control* (eds K.F. Baker and W.C. Snyder), University of California Press, Berkeley, pp. 170–84.

Rovira, A.D. (1965b) Interactions between plant roots and soil microorganisms. *Annual Review of Microbiology*, 19, 241–66.

Sequeira, L. (1978) Lectins and their role in host–pathogen specificity. *Annual Review of Phytopathology*, 16, 453–81.

Sidloi-Lumbroso, R. and Schulman, H.M. (1977) Purification and properties of soybean leghaemoglobin messenger RNA. *Biochimica et Biophysica Acta*, 476, 295–302.

Smith, G.S. (1987) Interactions of nematodes and mycorrhizal fungi, in *Vistas on Nematology* (eds J.A. Veech and D.W. Dickson), Society of Nematologists, Hyattsville, pp. 292–300.

Spiegel, Y., Cohn, E. and Spiegel, S. (1982) Characterization of sialyl and galactosyl residues on the body wall of different plant parasitic nematodes. *Journal of Nematology*, 14, 33–9.

Stacey, G., Paau, A.S. and Brill, W.J. (1980) Host-recognition in the *Rhizobium*–soybean symbiosis. *Plant Physiology*, 66, 609–14.

Stowers, M.D. (1987) Collection, isolation, cultivation and maintenance of *Frankia*, in *Symbiotic Nitrogen Fixation Technology* (ed. G.H. Elkan), Marcel Dekker, New York, pp. 29–53.

Stryer, L. (1981) *Biochemistry*, W.H. Freeman, San Francisco.

Subba Rao, N.S. (1988) Biological nitrogen fixation: Potentials, prospects and limitations, in *Biological Nitrogen Fixation, Recent Developments* (ed. N.S. Subba Rao), Oxford & IBH Publishing Co., New Delhi, pp. 1–19.

Sutherland, J.R. (1967) Parasitism of *Tylenchus emarginatus* on conifer seedling roots and some observations on the biology of the nematode. *Nematologica*, 13, 191–6.

Sutton, W.D. (1983) Nodule development and senescence, in *Nitrogen Fixation, Vol. 3: Legumes* (ed. W.J. Broughton), Clarendon Press, Oxford, pp. 144–212.

Taha, A.H.Y. and Kassab, A.S. (1979) The histopathological reactions of *Vigna sinesis* to separate and concomitant parasitism by *Meloidogyne javanica* and *Rotylenchulus reniformis. Journal of Nematology*, 11, 117–23.

Taha, A.H.Y. and Kassab, A.S. (1980) Interactions between *Meloidogyne javanica*, *Rotylenchulus reniformis*, and *Rhizobium* sp. on *Vigna sinensis. Journal of Nematology*, 12, 57–62.

Taha, A.H.Y. and Raski, D.J. (1969) Interrelationships between root-nodule bacteria, plant-parasitic nematodes and their leguminous host. *Journal of Nematology*, 1, 201–11.

Taha, A.H.Y. and Yousif, G.M. (1976) Histology of peanut underground parts infected with *Meloidogyne incognita*. *Nematologia Mediterranea*, 4, 175–81.

Taha, A.H.Y., Yousif, G.M. and El-Hadidy, T.T. (1974) Interaction of root-knot nematode infection and symbiotic nitrogen fixation in leguminous hosts. *Annals of Agricultural Science, Faculty of Agriculture, Ain Shams University, Cairo*, 19, 33–8.

Taha, A.H.Y., Yousif, G.M. and Oteifa, B.A. (1977) Strategies of nematode control management in newly reclaimed irrigated lands of Egypt. II. Relative suscepti-bility of common bean varieties and their nodules to root-knot nematode. *Journal of the Association for the Advancement of Agricultural Sciences in Africa*, 4, 1–8.

Thornton, H.G. (1930) The influence of the host plant in inducing parasitism in lucerne and clover nodules. *Proceedings of the Royal Society of London, Series B*, 106, 110–22.

Thornton, H.G. (1952) The symbiosis between *Rhizobium* and leguminous plants and the influence of this on the bacterial strain. *Proceedings of the Royal Society of London, Series B*, 139, 171–6.

Toms, G.C. and Western, A. (1971) Phytohaemagglutinins, in *Chemotaxonomy of the Leguminosae* (eds J.B. Harborne, D. Boulter and B.L. Turner), Academic Press, London, pp. 367–462.

Trinick, M.J. (1982) Competition between rhizobial strains for nodulation, in *Nitrogen Fixation in Legumes* (ed. J.M. Vincent), Academic Press, New York, pp. 229–38.

Tu, J.C. (1975) Rhizobial root nodules of soybean as revealed by scanning and transmission electron microscopy. *Phytopathology*, 65, 447–54.

Tu, J.C., Ford, R.E. and Quiniones, S.S. (1970) Effect of soybean mosaic virus and/or bean pod mottle virus infection on soybean nodulation. *Phytopathology*, 60, 518–23.

Vance, C.P. (1983) Rhizobium infection and nodulation: A beneficial plant disease? *Annual Review of Microbiology*, 37, 399–424.

Vance, C.P. and Johnson, L.E.B. (1981) Nodulation: A plant disease perspective. *Plant Disease*, 65, 118–24.

Vance, C.P. and Johnson, L.E.B. (1983) Plant determined ineffective nodules in alfalfa (*Medicago sativa*): Structural and biochemical comparison. *Canadian Journal of Botany*, 61, 93–106.

Van Schreven, D.A. (1958) Some factors affecting the uptake of nitrogen by legumes, in *Nutrition of the Legumes* (ed. E.G. Hallsworth), Butterworths Scientific Publications, London, pp. 137–63.

Van Schreven, D.A. (1964) The effect of some actinomycetes on rhizobia and *Agrobacterium radiobacter*. *Plant and Soil*, 21, 283–302.

Verdejo, M.S., Green, C.D. and Podder, A.K. (1988) Influence of *Meloidogyne incognita* on nodulation and growth of pea and black bean. *Nematologica*, 34, 88–97.

Verma, D.P.S., Ball, S., Guerin, C. and Wanekaker, L. (1979) Leghemoglobin biosynthesis in soybean root nodules. Characterization of the nascent and released peptides and the relative rate of synthesis of the major leghemoglobin. *Biochemistry*, 18, 476–83.

Viands, D.R., Vance, C.P., Heichel, G.H. and Barnes, D.K. (1979) An ineffective nitrogen fixation trait in alfalfa (*Medicago sativa* L.). *Crop Science*, 19, 905–8.

Wallace, H.R. (1966) Factors influencing the infectivity of plant parasitic nematodes. *Proceedings of the Royal Society of London, Series B*, 164, 592–614.

Westcott, S.W., III and Barker, K.R. (1976) Interaction of *Acrobeloides buetschlii* and *Rhizobium leguminosarum* on Wando pea. *Phytopathology*, **66**, 468–72.

Yao, P.Y. and Vincent, J.M. (1976) Factors responsible for the curling and branching of clover root hairs by *Rhizobium*. *Plant and Soil*, **45**, 1–16.

Yeates, G.W., Ross, D.J., Bridger, B.A. and Visser, T.A. (1977) Influence of the nematodes *Heterodera trifolii* and *Meloidogyne hapla* on nitrogen fixation by white clover under glasshouse conditions. *New Zealand Journal of Agricultural Research*, **20**, 401–13.

Yousif, G.M. (1979) Histological responses of four leguminous crops infected with *Meloidogyne incognita*. *Journal of Nematology*, **11**, 395–401.

Zuckerman, B.M. and Jansson, H.B. (1984) Nematode chemotaxis and possible mechanisms of host/prey recognition. *Annual Review of Phytopathology*, **22**, 95–113.

9

Interactions of nematodes with mycorrhizae and mycorrhizal fungi

L.J. Francl

Plant-parasitic nematodes often encounter roots that have been transformed structurally and physiologically by mycorrhizal fungi. These symbiotic associations of fungi with higher plants are broadly termed mycorrhizae (or mycorrhizas, sing., mycorrhiza; the literal translation is 'fungus root'). The fungal partner obtains carbon compounds from the plant and the plant partner effectively extends its root volume, thereby improving its access to possibly scarce inorganic soil nutrients, especially phosphorus, and soil moisture. The interaction of nematodes and these beneficial symbionts is unlike those between parasitic nematodes and pathogenic organisms that are described in other chapters of this book.

9.1 MYCORRHIZAE AND MYCORRHIZAL FUNGI

Mycorrhizae are a diverse group that have been classified according to structures formed in the root and families of plants that are infected. The two most widespread and important groups are the ectomycorrhiza and the endomycorrhiza. An intermediate form, the ectendomycorrhiza, has been described and endomycorrhizae can be further classified into vesicular-arbuscular, ericoid and orchid mycorrhizae, with the latter two occurring in association with the Ericaceae and Orchidaceae (Harley and Smith, 1983). One should be careful to make the distinction between the fungi that cause mycorrhiza to form and the mycorrhiza itself. This chapter will concentrate upon interactions of nematodes with ectomycorrhizae, vesicular-arbuscular endomycorrhizae and the fungal symbionts, after an introduction to mycorrhizal biology.

Virtually all plants infected by ectomycorrhizal fungi are woody perennials. Most are temperate forest species, such as members of the Pinaceae, Fagaceae and Betulaceae, but there are important tropical families represented as well, such as the Dipterocarpaceae. Ectomycorrhizal fungi are predominantly higher Basidiomycetes and reproduce by forming the familiar fleshy fruiting structures. A few taxa in the Ascomycetes and Zygomycetes are also represented to a lesser extent (Harley and Smith, 1983). Ectomycorrhizal fungi promote rapid seedling growth and have been found to be essential to produce woody perennials economically in nurseries (Wilde, 1954; Maronek *et al.*, 1981; Kropp and Langlois, 1990). Thus, in fumigated soils and areas previously devoid of ectomycorrhizal tree species, there is an impetus to inoculate soils with ectomycorrhizal fungi for increased productivity.

Ectomycorrhizal fungi penetrate juvenile roots and develop a hyphal network called a Hartig net in the middle lamellae between root epidermal and cortical cells. Eventually, a sheath or mantle of fungal tissue, which is interconnected with the Hartig net, surrounds the exterior of the root. There is apparently little or no penetration of plant cells by the hyphae. The ectomycorrhizae are typically stubby in appearance and as such are distinctly different morphologically from uninfected roots.

Endomycorrhizae that form vesicles and arbuscules within roots are commonly called VAM fungi. Hyphae grow within the root cortex (inter- and intracellularly) and arbuscules penetrate cortical cell walls and interface with the cell's plasma membrane, much like haustoria of obligately parasitic fungal species (Brown and King, 1987). In fact, VAM fungi have yet to be cultured away from their host (Hepper, 1987). Arbuscules are regarded as the primary avenue of bidirectional transfer of materials between the two symbionts. Vesicles are globose to irregularly shaped and contain large amounts of lipid (Holley and Peterson, 1979). Vesicular–arbuscular mycorrhizae are usually similar in appearance to uninfected roots, so histological preparation is necessary to observe the internal morphological structures. Sometimes the size and density of vesicles are great enough in older roots that the mycorrhizae take on an uneven, torose appearance.

Most families of Angiospermae contain species that are symbionts with VAM fungi. The economically important exceptions occur in the Chenopodiaceae and Brassicaceae. Vesicular–arbuscular endomycorrhizal fungi form symbiotic associations with many woody perennial species not colonized by ectomycorrhizae, including many gymnosperms outside the Pinaceae (Harley and Smith, 1983). Both tropical and temperate forests have a high incidence of infection by VAM fungi (Baylis, 1961; Alwis and Abeybayake, 1980; Thapar and Khan, 1985). Vesicular–arbuscular mycorrhizal fungi are classified as Zygomycetes in the family Endogonaceae. Taxonomic placement into species is involved because most, if not all, forms are obligate symbionts and the sexual state is unknown (Trappe and Schenck, 1982).

Chapters in *Methods and Principles of Mycorrhizal Research* (Schenck, 1982) can be consulted for further information on procedures for extraction and handling of ectomycorrhizal and VAM fungi. In addition, an important resource available to researchers is the International Culture Collection of VA Mycorrhizal Fungi, or INVAM. INVAM correspondence can be addressed to the attention of Joseph B. Morton, West Virginia University, Morgantown, WV 26506-6057 (USA). The goals of INVAM are to serve as a repository of fungal specimens, to train individuals in the identification of VAM fungi, to distribute reliable starter cultures, and to provide taxonomic verification of VAM fungi. Schenck and Pérez (1986) have published an INVAM manual for identification of VAM taxa.

It should be apparent from the description above that the mycorrhizal condition is the rule in higher plants rather than the exception. Aside from species in the families mentioned above, plants that are not mycorrhizal are predominantly found in wetlands, presumably because reduced soil oxygen is detrimental to mycorrhizal fungi (Read *et al.*, 1976). Because plant-parasitic nematodes also obtain sustenance from plant roots and mycophagous nematodes can feed upon a number of fungal species, there is potential for interaction and competition. Research interest has focused upon whether VAM can ameliorate damage to plants caused by the more pathogenic species of nematodes. The interactions detailed in this chapter are but a sampling of those undoubtedly occurring among nematodes and mycorrhizae–mycorrhizal fungi.

There have been several excellent reviews on the subject of VAM–nematode interactions (Hussey and Roncadori, 1982; Smith, 1987; Ingham, 1988). Smith (1987) should be read closely for hypotheses and consequent methodologies for testing interactions between mycorrhizae and nematodes. Also recommended for its emphasis on experimental technique is a treatment of nematode interactions with other organisms (Sikora and Carter, 1987). Interactions between ectomycorrhizae and nematodes were reviewed some years ago (Marx, 1972; Ruehle, 1973). The effects of fungal-feeding nematodes on endomycorrhizae were considered by Ingham (1988). Selected research results will be summarized again here for the reader with limited access to the literature.

9.2 PARAMETERS OF INTERACTIONS AMONG ORGANISMS

Plant-parasitic nematodes may enhance or depress colonization of roots and sporulation of mycorrhizal fungi. Likewise, mycorrhizae may decrease or increase nematode penetration, development and reproduction. If nematode reproduction is decreased, the plant and its fungal symbiont can be described as having induced resistance or as a poorer host to the parasitic nematode than the non-mycorrhizal plant. Mycorrhizal fungi also are capable of directly interacting with sedentary states of plant-parasitic nematodes. Similarly, some species of nematodes are capable of feeding upon mycorrhizal fungi. The reaction of plant growth to these interactions may be positive, negative or neutral. Yield, therefore, is dependent dynamically upon how much damage the nematode is doing and how much benefit the plant is deriving from the fungus. If growth is enhanced, the plant (again with its fungal symbiont) can be described as more tolerant to the nematode (Hussey and Roncadori, 1982).

Research has shown little consistency in the outcome of interacting organisms in this system. Surely, many seemingly conflicting results encountered in the literature can be attributed to the influence of soil fertility, specificity of plant (mycorrhiza)–nematode–mycorrhizal fungus combinations, population densities of nematode and fungus, and timing of inoculations of nematode and fungus. Experimental designs that included soil fertility as a variable, particularly the role of phosphorus, were rare in earlier work and most experiments have been conducted exclusively in the greenhouse, customarily using phosphorus-poor soil (Smith, 1988). The possible number of combinations of species and species mixtures is astronomical and organism composition seems to contribute materially to results found in the literature. Consider also that populations of any of the three species may behave significantly differently from other populations. The consequences of this variability will be discussed in the chapter summary after reviewing literature accounts.

9.3 EFFECT OF FUNGAL-FEEDING NEMATODES ON MYCORRHIZAL FUNGI

Mycophagous or, more strictly, mycetophagous nematodes (to distinguish rhabditids that feed on yeast fungi) are commonly isolated from natural and agricultural ecosystems (Ruehle, 1967; Nematode Geographical Distribution Committee of the Society of Nematologists, 1984). Species in the genera *Aphelenchus*, *Aphelenchoides*, *Bursaphelenchus* and *Ditylenchus* are mycetophagous and potentially can reduce the growth of mycorrhizal fungi. Grazing of hyphae external to the root may reduce the potential yield of mycorrhizal plants and such reductions may be induced by naturally occurring population densities of nematodes (Finlay, 1985; Ingham, 1988). Conceivably, even limited feeding can break the chain of nutrient translocation in linear stretches of hyphae.

Riffle (1971) found that *Aphelenchus cibolensis* fed and reproduced on 53 of 58 ectomycorrhizal species tested under culture conditions. Thirty-two fungal species had their radial growth reduced by more than 20% and aerial hyphal growth was reduced from 21% to 100% in 15 species. Sutherland and Fortin (1968) had previously found that *A. avenae* reduced the radial growth of seven ectomycorrhizal species in vitro, including two species not tested by Riffle (1971). In another in vitro study, *Aphelenchoides bicaudatus* severely decreased the growth of all five ericoid endomycorrhizal fungi tested (Shafer *et al.*, 1981).

Aphelenchus avenae did not affect sporulation by *Gigaspora margarita* or *Glomus etunicatum* in pot culture with cotton (*Gossypium hirsutum*) (Hussey and Roncadori, 1981). Plant shoot growth was reduced only at extraordinarily high population levels of *A. avenae*. However, with soya bean (*Glycine max*), concomitant addition of *A. avenae* and a *Glomus* sp. to soil resulted in poor sporulation by the fungus and reduced plant growth (Salawu and Estey, 1979). Nodulation of soya bean roots was significantly reduced, possibly because the plants were not able to supply sufficient nutrients to support the bacterial symbiont. Reproduction of *Aphelenchoides composticola* on *Agaricus bisporus* and *Glomus clarum* was associated with reduced yield and a lower shoot phosphorus content in VAM red clover (*Trifolium pratense*), although there was no significant effect on the level of mycorrhizal infection (Giannakis and Sanders, 1990). However, mycetophagous nematodes did not affect the yield of non-mycorrhizal plants, supporting the conclusion that nematodes limited the contribution to plant growth by mycelia in the soil.

Further studies on the effect of nematode mycetophagy on mycorrhizal fungi in the soil system are necessary to determine important components of this relationship. Direct observations of nematodes feeding on hyphae in soil have yet to be made, although rhizotrons and other techniques (Schüepp *et al.*, 1987) may make this possible. Furthermore, there are apparently no reports concerning the effect of nematode mycetophagy on the ectomycorrhizal mantle.

9.4 EFFECT OF PLANT-PARASITIC NEMATODES ON MYCORRHIZAE AND MYCORRHIZAL FUNGI

Sporulation on and percentage infection of the root have been measured to show an influence of plant-parasitic nematodes on mycorrhizal fungi. In addition, researchers have described in detail vesicles, arbuscules and hyphae within mycorrhizae parasitized by nematodes. Mycelia external to the mycorrhizae have not yet been assayed directly.

Meloidogyne incognita-induced galls and immediately adjacent areas in roots of soya bean were devoid of vesicles and arbuscules of *Glomus macrocarpum* (Kellam and Schenck, 1980). However, neither sporulation nor percentage infection were affected by the presence of *M. incognita*. *M. incognita* had no effect on soya bean root colonization by VAM fungi, but at low phosphorus levels the nematode increased sporulation by *G. margarita* and decreased sporulation by *G. etunicatum* (Carling *et al.*, 1989). Clearly, specific organisms involved in the association can interact differently. Reduced root infection and sporulation of *G. fasciculatum* and *G. epigaeum* resulted from inoculation of *Heterodera cajani* on cowpea (*Vigna unguiculata*) (Jain and Sethi, 1987). Mycorrhizal structures developed normally in feeder roots of *Citrus limon* (symbiont: *G. mosseae*) in the absence of *Tylenchulus semipenetrans*, but when the nematode was present vesicles were not formed (O'Bannon *et al.*, 1979). Vesicle formation and mycelial growth of *G. etunicatum* were reduced in mycorrhizal roots of *C. limon* infected by *Radopholus similis* when compared to VAM without nematodes (O'Bannon and Nemec, 1979). Cortical tissue destruction by

R. similis presumably left fewer habitable sites for *G. etunicatum*. Similarly, Umesh *et al.* (1988) noted decreased colonization of banana (*Musa acuminata*) by *G. fasciculatum* in roots infected by *R. similis* but there was no effect on sporulation.

An effect of nematodes on mycorrhizae or possibly a direct effect on VAM fungi was suggested when increased VAM root infection followed treatment of soil with nematicides (Bird *et al.*, 1974; Germani *et al.*, 1980; Menge, 1982; Ingham *et al.*, 1986). No VAM were seen after treatment of soil with the biocide methyl bromide (Bird *et al.*, 1974; Linderman, 1987).

Apparently, mycorrhizal fungi are sensitive to transformations of roots induced by sedentary nematodes (galls, syncytia and nurse cells) and, without a doubt, are deterred from colonizing roots severely crippled by migratory endoparasites. Ectoparasitic nematodes might be expected to interact with VAM in a manner similar to migratory endoparasites. The first organism to arrive at a habitable site on the root would appear to have some advantage; moreover, population density will play a key role in the relationship. More needs to be known about the mechanisms by which nematode parasitism influences colonization by ectomycorrhizal fungi.

9.5 EFFECT OF MYCORRHIZAE AND MYCORRHIZAL FUNGI ON PLANT-PARASITIC NEMATODES

The interaction between ectoparasitic nematodes and ectomycorrhizae was described by Mancini *et al.* (1983) for *Helicotylenchus digonicus*, *Suillus luteus* and three *Pinus* spp. *H. digonicus* increased its population on pines with mycorrhizae but reproduction on non-mycorrhizal pines was not measured. Shoot growth in soil with a combination of nematode and ectomycorrhizal fungus was similar to that in sterile soil but less than ectomycorrhizal pines without nematodes.

The protective effect of VAM and ectomycorrhizae against penetration and development of migratory plant-parasitic nematodes appears to be negligible in some associations but not in others. *Hoplolaimus galeatus* readily entered ectomycorrhizae and parasitized morphologically altered cells as well as non-mycorrhizal roots of two *Pinus* spp. (Ruehle and Marx, 1969). *Pratylenchus penetrans* was uninhibited in its invasion of ectomycorrhizae and root-lateral primordia of *Pinus radiata*. Nematodes moved freely within the cortical tissue, completing their life cycle entirely within the roots (Marks *et al.*, 1987). Fewer *Pratylenchus brachyurus* were found per gram of cotton root colonized by *Glomus margarita* than in non-colonized roots; however, total nematode number per plant was unaffected by VAM (Hussey and Roncadori, 1978). Numbers of *Radopholus similis* in banana roots were significantly lower when the endomycorrhizal fungus *G. fasciculatum* was applied simultaneously or seven days prior to nematode inoculation (Umesh *et al.*, 1988).

Sedentary plant-parasitic nematodes do not seem to fare as well as migratory nematodes, but results are again mixed. No significant differences were observed in juvenile penetration of *Meloidogyne incognita* between tomato (*Lycopersicon esculentum*) seedlings infected with *Glomus fasciculatum* and non-mycorrhizal seedlings (Suresh *et al.*, 1985). Penetration of *M. incognita* was similar in cotton roots inoculated with *G. intraradices* and non-mycorrhizal roots after seven days but the population was reduced after 28 days in VAM. The rate of development of second-stage juveniles to ovipositing females was unaffected by *G. intraradices* or phosphorus when *M. incognita* was added at planting, but was delayed when *M. incognita* was added 28 days after planting in soil containing *G. intraradices* (Smith *et al.*, 1986a). Similarly, the time required for inoculated second-stage juveniles of *M. hapla* to mature to adults was greater by 1000 degree hours (base = 9°C) in VAM than in non-mycorrhizal onion (*Allium cepa*) plants supplemented with phosphorus (MacGuidwin *et al.*, 1985).

Roncadori and Hussey (1977) found that *Meloidogyne incognita* reproduced more on a per plant basis on mycorrhizal cotton (symbiont: *Glomus margarita*) compared to non-colonized roots, presumably because the root system was larger. Numbers of *M. incognita* on cotton, however, were reduced at various densities of *G. fasciculatum*, both on a per plant and per gram of root basis (Saleh and Sikora, 1984). Sitaramaiah and Sikora (1982) found that maturation of females of *Rotylenchulus reniformis* was delayed and the nematode reproduced less on endomycorrhizal tomato roots (symbiont: *G. fasciculatum*) compared to non-colonized roots. Significantly reduced galling and populations of *M. incognita* were found when *Piper nigrum* plants were pre-inoculated with *G. fasciculatum* or *G. etunicatum* (Sivaprasad *et al.*, 1990). Mycorrhizae induced by *G. fasciculatum* on cowpea decreased cyst production and reproduction of *Heterodera cajani*, but *G. epigaeum* tended to exhibit a reverse trend (Jain and Sethi, 1987). The reproduction of *M. javanica* on chickpea (*Cicer arietinum*) was differentially suppressed by VAM fungi, being most pronounced with *G. manihotis*, less with *G. margarita*, only slightly with *Gigaspora gigantea* and not at all with *Entrophospora colombiana* (Diederichs, 1987). Egg production by *M. incognita* (eggs per root system and eggs per gram of root) on soya bean inoculated with *Glomus margarita* or *G. etunicatum* was suppressed at low phosphorus as well as by increased phosphate fertilization, suggesting indirectly that the induced resistance was due to better phosphorus nutrition (Carling *et al.*, 1989).

Biochemical alteration of the plant root has been hypothesized as one explanation of reduced nematode infection (Smith, 1987). Phenolic compounds have long been thought to play a role in disease resistance (Goodman *et al.*, 1967) and they have been shown to be formed after infection by mycorrhizal fungi (Sylvia and Sinclair, 1983). Phytoalexin production is greater in mycorrhizal than non-mycorrhizal soya bean roots (Morandi, 1987) and phytoalexins previously have been associated with an incompatible root-knot–soya bean combination (Kaplan *et al.*, 1980). Morandi (1987) found that populations of *Meloidogyne javanica* were significantly reduced in roots of tomato by pre-inoculation with *Glomus fasciculatum*. In addition, increases in lignins and phenols in mycorrhizal roots were associated with reduced reproduction of *M. javanica* on tomato (Singh *et al.*, 1990) and *Radopholus similis* on banana (Umesh *et al.*, 1988). The effect of biochemical changes in mycorrhizae on nematode parasitism needs to be more thoroughly explored.

Mycorrhizal fungi also can affect sedentary nematodes directly. On several occasions, samples from field soils infested with *Heterodera avenae* contained cysts stuffed full of VAM spores (Willcox and Tribe, 1974; Graham and Stone, 1975). Tribe (1977), in a review of fungal parasites of *Heterodera* and *Globodera* spp., reported the observations of Arnold Steele on a *Glomus* sp. and *H. schachtii* cultured in a greenhouse on tomato. Cysts of *H. schachtii* were intensively occupied by chlamydospores and eggs within were apparently crushed. *Glomus* sp. also was found on and within young females. Chlamydospores of *G. fasciculatum* were found occupying cysts of *H. glycines* collected from several geographical locations from Missouri, USA (Francl and Dropkin, 1985). Co-inoculation of soya bean with the Missouri VAM fungal isolates or commercially purchased isolates of *G. fasciculatum* and second-stage juveniles of *H. glycines* resulted in nematode egg parasitism. In addition to this, runner hyphae (VAM hyphae growing along the surface of the mycorrhiza, typically in clusters) were observed to penetrate the cuticle of a sedentary female. The percentage of eggs parasitized under experimental conditions was too low to have much effect on population increase by *H. glycines* (Francl and Dropkin, 1985). VAM fungal spores are capable of occupying many types of soil voids, including insect carcasses, leaf litter and seed testae (Rabatin and Rhodes, 1982; Taber, 1982). Presumably, nematode parasitism by *G. fasciculatum* and possibly other VAM fungi is opportunistic and dependent on carbon nutrition from the autotrophic symbiont, rather than representative of a true host–parasite relationship. Therefore, few encounters between nematodes and VAM fungi may be expected to result in parasitism.

9.6 EFFECTS OF NEMATODE–MYCORRHIZAL INTERACTIONS ON PLANT GROWTH

Concomitant effects of nematodes and mycorrhizae on plant growth are measured relative to growth of plants in media free of nematodes or mycorrhizal fungi. If mycorrhizae increase plant growth over that in nematode-infested soil, then tolerance is said to have been increased. Enhanced resistance is indicated by decreased nematode reproduction and was covered implicitly in the previous section. Generally speaking, plants with mycorrhizae and nematode parasites yield less than mycorrhizal plants without nematodes and more than non-mycorrhizal plants with nematodes.

Including several levels of soil fertility allows an evaluation of the role of mycorrhizae over and above that of improving plant nutrition. The role of phosphorus was taken into account in the research reports that follow.

The interaction between *Glomus intraradices, Meloidogyne incognita* and cantaloupe (*Cucumis melo*) was studied at three soil phosphorus levels (Heald *et al.*, 1989). In soil amended with $50\,\mu g$ $P\,g^{-1}$ soil, *M. incognita* suppressed the growth of non-mycorrhizal plants by 84% compared to the control but growth of mycorrhizal plants inoculated with *M. incognita* was retarded by only 21%. A similar trend occurred in plants grown in soil with $100\,\mu g$ $P\,g^{-1}$ soil and plant growth was reduced if no phosphorus was added to soil. Similarly, mycorrhizal infection (by a mixture of VAM fungal species) improved growth of tamarillo (*Cyphomandra betacea*) and suppressed reproduction and development of *M. incognita* (Cooper and Grandison, 1987). The beneficial effects of VAM on plant vigour could not be matched by adding phosphate fertilizer. A mixed VAM fungal inoculum was able to increase soya bean tolerance to *Heterodera glycines* in greenhouse experiments and outdoor microplots (Tylka *et al.*, 1991). The addition of VAM fungi to microplot soil infested with *H. glycines* increased yields of two soya bean cultivars by an average of 23% over phosphate fertilization. Smith *et al.* (1986b) found that *G. intraradices* increased tolerance of cotton to *M. incognita* in microplots and fertilization with phosphate actually decreased yields. Deficient levels of zinc were found in plant tissue from fertilized microplots but not in plots with mycorrhizae, leading Smith *et al.* (1986b) to conclude that a nutrient imbalance had occurred. In contrast to these results, Smith and Kaplan (1988), working with *Radopholus citrophilus*, *G. intraradices* and *Citrus limon*, found increased plant growth that appeared to have resulted from improved phosphorus nutrition rather than antagonism between fungus and nematodes.

In most instances, therefore, phosphate fertilizer alone does not duplicate the effect of mycorrhizae on plant vigour in soils infested with nematodes. Clearly, greenhouse experiments should be verified by field experimentation whenever possible. More subtle beneficial effects of mycorrhizae, mainly enhanced micronutrient absorption (zinc, copper, etc.), balanced nutrition and increased drought tolerance, are suitable subjects for research. In addition, selection of tolerance to high soil phosphorus in VAM fungi (Porter *et al.*, 1978; Davis *et al.*, 1984) introduces another permutation for researchers to consider in their experimental designs.

9.7 CONCLUSIONS AND FUTURE PROSPECTS

Genetic transformation of ectomycorrhizal fungi has been accomplished for at least one species (Barrett *et al.*, 1990). The fungal symbiont may soon be genetically engineered in several ways to benefit the plant. Inhibition of nematode parasitism by incorporation of the capability to produce nematicidal or nematostatic gene products should be one of those objectives.

Endomycorrhizal fungi will be much more difficult to manipulate genetically because of their obligate symbiont status and because large-scale culture of inoculum is impossible at present for the same reason. Improved productivity of greenhouse plants and transplants therefore are logical subjects for applied research.

The wide variety of outcomes for interactions between mycorrhizae, mycorrhizal fungi and nematodes demonstrates that generalizations about particular plant–nematode–mycorrhizal fungus combinations are difficult to make. More data on parameters associated with the use of mycorrhizal fungi in nematode-infested soil and their mode of action are required. Situations where mycorrhizal fungi are operating in a suboptimal manner need to be explored and remedial actions identified. Likewise, circumstances that have engendered a more optimal response also need to be described. In the final analysis, the primary economic benefit of mycorrhizal fungi is to promote soil nutrient and water absorption in their capacity of symbiont. Biological protection of roots from nematodes and other pathogens is essentially a secondary role. Increased tolerance may be a more appropriate and achievable goal in many cases; nonetheless, there are possibilities for improvement of biological control of plant-parasitic nematodes by VAM by selecting the most effective species and strain for specific situations. Once this is accomplished, the prospects of mycorrhizal fungi as biological control agents can be fully assessed.

REFERENCES

Alwis, D.P. and Abeybayake, K. (1980) A survey of mycorrhizae in some forest trees of Sri Lanka, in *Tropical Mycorrhizal Research* (ed. P. Mikola), Cambridge University Press, Cambridge, pp. 135–55.

Barrett, V., Dixon, R.K. and Lemke, P.A. (1990) Genetic transformation of a mycorrhizal fungus. *Applied Microbiology and Biotechnology*, 33, 313–16.

Baylis, G.T.S. (1961) The significance of mycorrhizas and root nodules in New Zealand vegetation. *Proceedings of the Royal Society of New Zealand*, 89, 45–50.

Bird, G.W., Rich, J.R. and Glover, S.U. (1974) Increased endomycorrhizae of cotton roots in soil treated with nematicides. *Phytopathology*, 64, 48–51.

Brown, M.F. and King, E.J. (1987) Morphology and histology of vesicular-arbuscular mycorrhizae: Anatomy and cytology, in *Methods and Principles of Mycorrhizal Research* (ed. N.C. Schenck), American Phytopathological Society, St Paul, pp. 15–21.

Carling, D.E., Roncadori, R.W. and Hussey, R.S. (1989) Interactions of vesicular-arbuscular mycorrhizal fungi, root-knot nematode, and phosphorus fertilization on soybean. *Plant Disease*, 73, 730–3.

Cooper, K.M. and Grandison, G.S. (1987) Effects of vesicular–arbuscular mycorrhizal fungi on infection of tamarillo (*Cyphomandra betacea*) by *Meloidogyne incognita* in fumigated soil. *Plant Disease*, 71, 1101–6.

Davis, E. A., Young, J.L. and Rose, S.L. (1984) Detection of high-phosphorus tolerant VAM-fungi colonizing hops and peppermint. *Plant and Soil*, 81, 29–36.

Diederichs, C. (1987) Interaction between five endomycorrhizal fungi and the root-knot nematode *Meloidogyne javanica* on chickpea under tropical conditions. *Tropical Agriculture*, **64**, 353–5.

Finlay, R.D. (1985) Interactions between soil micro-arthropods and endomycorrhizal associations of higher plants, in *Ecological Interactions in Soil: Plants, Microbes and Animals* (eds A.H. Fitter, D. Atkinson, D.J. Read and M.B. Usher), Blackwell Scientific Publications, Oxford, pp. 319–31.

Francl, L.J. and Dropkin, V.H. (1985) *Glomus fasciculatum*, a weak pathogen of *Heterodera glycines*. *Journal of Nematology*, **17**, 470–5.

Germani, G., Diem, H.G. and Dommergues, Y.R. (1980) Influence of 1,2-dibromo-3-chloropropane fumigation on nematode population, mycorrhizal infection, N_2 fixation and yield of field-grown groundnut. *Revue de Nematologie*, **3**, 75–9.

Giannakis, N. and Sanders, F.E. (1990) Interactions between mycophagous nematodes, mycorrhizal and other soil fungi. *Agriculture, Ecosystems and Environment*, **29**, 163–7.

Goodman, R.N., Király, Z. and Zaitlin, M. (1967) *The Biochemistry and Physiology of Infectious Plant Disease*, D. Van Nostrand, Princeton.

Graham, C.W. and Stone, L.E.W. (1975) Field experiments on the cereal cyst nematode (*Heterodera avenae*) in south-east England, 1967–1972. *Annals of Applied Biology*, **80**, 61–73.

Harley, J.L. and Smith, S.E. (1983) *Mycorrhizal Symbiosis*, Academic Press, New York.

Heald, C.M., Bruton, B.D. and Davis, R.M. (1989) Influence of *Glomus intraradices* and soil phosphorus on *Meloidogyne incognita* infecting *Cucumis melo*. *Journal of Nematology*, **21**, 69–73.

Hepper, C.M. (1987) VAM spore germination and hyphal growth *in vitro*: Prospects for axenic culture, in *Mycorrhizae in the Next Decade, Practical Applications and Research Priorities*, Proceedings of the 7th North American Conference on Mycorrhizae (eds D.M. Sylvia, L.L. Hung and J.H. Graham), Institute of Food and Agricultural Sciences, University of Florida, Gainesville, pp. 172–4.

Holley, J.D. and Peterson, R.L. (1979) Development of a vesicular–arbuscular mycorrhiza in bean roots. *Canadian Journal of Botany*, **57**, 1960–78.

Hussey, R.S. and Roncadori, R.W. (1978) Interaction of *Pratylenchus brachyurus* and *Gigaspora margarita* on cotton. *Journal of Nematology*, **10**, 16–20.

Hussey, R.S. and Roncadori, R.W. (1981) Influence of *Aphelenchus avenae* on vesicular–arbuscular endomycorrhizal growth response in cotton. *Journal of Nematology*, **13**, 48–52.

Hussey, R.S. and Roncadori, R.W. (1982) Vesicular–arbuscular mycorrhizae may limit nematode activity and improve plant growth. *Plant Disease*, **66**, 9–14.

Ingham, E.R., Trofymow, J.A., Ames, R.N., Hunt, H.W., Morley, C.R., Moore, J.C. and Coleman, D.C. (1986) Trophic interactions and nitrogen cycling in a semi-arid grassland soil. Part II. System responses to removal of different groups of soil microbes or fauna. *Journal of Applied Ecology*, **23**, 615–30.

Ingham, R.E. (1988) Interactions between nematodes and vesicular–arbuscular mycorrhizae. *Agriculture, Ecosystems and Environment*, **24**, 169–82.

Jain, R.K. and Sethi, C.L. (1987) Pathogenicity of *Heterodera cajani* on cowpea as influenced by the presence of VAM fungi, *Glomus fasciculatum* or *G. epigaeus*. *Indian Journal of Nematology*, **17**, 165–70.

Kaplan, D.T., Keen, N.T. and Thomason, I.J. (1980) Association of glyceollin with the incompatible response of soybean roots to *Meloidogyne incognita*. *Physiological Plant Pathology*, **16**, 309–18.

Kellam, M.K. and Schenck, N.C. (1980) Interactions between a vesicular–arbuscular mycorrhizal fungus and root-knot nematode on soybean. *Phytopathology*, **70**, 293–6.

Kropp, B.R. and Langlois, C.G. (1990) Ectomycorrhizae in reforestation. *Canadian Journal of Forest Research*, **20**, 438–51.

Linderman, R.G. (1987) Response of shade tree seedlings, VA mycorrhizal fungi, and *Pythium* to soil fumigation with Vapam or methyl bromide, in *Mycorrhizae in the Next Decade, Practical Applications and Research Priorities*, Proceedings of the 7th North American Conference on Mycorrhizae (eds D.M. Sylvia, L.L. Hung and J.H. Graham), Institute of Food and Agricultural Sciences, University of Florida, Gainesville, p. 30.

MacGuidwin, A.E., Bird, G.W. and Safir, G.R. (1985) Influence of *Glomus fasciculatum* on *Meloidogyne hapla* infecting *Allium cepa*. *Journal of Nematology*, **17**, 389–95.

Mancini, G., Cotroneo, A. and Moretti, F. (1983) Response of three pines to parasitism by *Helicotylenchus digonicus* (Nematoda: Hoplolaimidae). *European Journal of Forest Pathology*, **13**, 245–50.

Marks, G.C., Winoto-Suatmadji, R. and Smith, I.W. (1987) Effects of nematode control on shoot, root and mycorrhizal development of *Pinus radiata* seedlings growing in a nursery soil infested with *Pratylenchus penetrans*. *Australian Forest Research*, **17**, 1–10.

Maronek, D.M., Hendrix, J.W. and Kierman, J. (1981) Mycorrhizal fungi and their importance in horticultural crop production. *Horticultural Review*, **3**, 172–213.

Marx, D.H. (1972) Ectomycorrhizae as biological deterrents to pathogenic root infections. *Annual Review of Phytopathology*, **10**, 429–54.

Menge, J.A. (1982) Effect of soil fumigants and fungicides on vesicular–arbuscular fungi. *Phytopathology*, **72**, 1125–32.

Morandi, D. (1987) VA mycorrhizae, nematodes, phosphorus and phytoalexins on soybean, in *Mycorrhizae in the Next Decade, Practical Applications and Research Priorities*, Proceedings of the 7th North American Conference on Mycorrhizae (eds D.M. Sylvia, L.L. Hung and J.H. Graham), Institute of Food and Agricultural Sciences, University of Florida, Gainesville, p. 212.

Nematode Geographical Distribution Committee of the Society of Nematologists (1984) *Distribution of Plant-Parasitic Nematode Species in North America*, Society of Nematologists, Hyattsville, 205 pp.

O'Bannon, J.H. and Nemec, S. (1979) The response of *Citrus limon* seedlings to a symbiont, *Glomus etunicatus*, and a pathogen, *Radopholus similis*. *Journal of Nematology*, **11**, 270–5.

O'Bannon, J.H., Inserra, R.N., Nemec, S. and Volvas, N. (1979) The influence of *Glomus mosseae* on *Tylenchulus semipenetrans*-infected and uninfected *Citrus limon* seedlings. *Journal of Nematology*, **11**, 247–50.

Porter, W.M., Abbott, L.K. and Robson, A.D. (1978) Effect of rate of application of superphosphate on populations of vesicular arbuscular endophytes. *Australian Journal of Experimental Agriculture and Animal Husbandry*, **18**, 573–8.

Rabatin, S.C. and Rhodes, L.H. (1982) *Acaulospora bireticulata* inside oribatid mites. *Mycologia*, **74**, 859–61.

Read, D.J., Koucheki, H.K. and Hodgson, J. (1976) Vesicular–arbuscular mycorrhiza in natural vegetation systems. I. The occurrence of infection. *New Phytologist*, **77**, 641–53.

Riffle, J.W. (1971) Effect of nematodes on root-inhabiting fungi, in *Mycorrhizae*, Proceedings of the First North American Conference on Mycorrhizae (ed. E. Hacskaylo), Miscellaneous Publication 1189, US Department of Agriculture, Washington, pp. 97–113.

Roncadori, R.W. and Hussey, R.S. (1977) Interaction of the endomycorrhizal fungus *Gigaspora margarita* and root-knot nematode on cotton. *Phytopathology*, **67**, 1507–11.

Ruehle, J.L. (1967) *Distribution of Plant-Parasitic Nematodes Associated with Forest*

Trees of the World, US Forest Service, Southeastern Forest Experiment Station.

Ruehle, J.L. (1973) Nematodes and forest trees: Types of damage to tree roots. *Annual Review of Phytopathology*, **11**, 99–118.

Ruehle, J.L. and Marx, D.H. (1969) Parasitism of pine mycorrhizae by lance nematodes. *Journal of Nematology*, **1**, 303.

Salawu, E.O. and Estey, R.H. (1979) Observations on the relationships between a vesicular–arbuscular fungus, a fungivorous nematode, and the growth of soybeans. *Phytoprotection*, **60**, 99–102.

Saleh, H. and Sikora, R.A. (1984) Relationship between *Glomus fasciculatum* root colonization of cotton and its effect on *Meloidogyne incognita*. *Nematologica*, **30**, 230–7.

Schenck, N.C. (ed.) (1982) *Methods and Principles of Mycorrhizal Research*, American Phytopathological Society, St Paul.

Schenck, N.C. and Pérez, Y. (1986) *Manual for the Identification of VA Mycorrhizal Fungi*, INVAM, Gainesville.

Schüepp, H., Miller, D.D. and Bodmer, M. (1987) A new technique for monitoring hyphal growth of vesicular–arbuscular mycorrhizal fungi through soil. *Transactions of the British Mycological Society*, **89**, 429–35.

Shafer, S.R., Rhodes, L.H. and Riedel, R.M. (1981) *In-vitro* parasitism of endomycorrhizal fungi of ericaceous plants by the mycophagous nematode *Aphelenchoides bicaudatus*. *Mycologia*, **73**, 141–9.

Sikora, R.A. and Carter, W.W. (1987) Nematode interactions with fungal and bacterial plant pathogens: Fact or fantasy, in *Vistas on Nematology* (eds J.A. Veech and D.W. Dickson), Society of Nematologists, Hyattsville, pp. 307–12.

Singh, Y.P., Singh, R.S. and Sitaramaiah, K. (1990) Mechanism of resistance of mycorrhizal tomato against root-knot nematode, in *Trends in Mycorrhizal Research*, Proceedings of the National Conference on Mycorrhiza (eds B.L. Jalali and H. Chand), Haryana Agricultural University, Hissar, pp. 96–7.

Sitaramaiah, K. and Sikora, R.A. (1982) Effect of the mycorrhizal fungus *Glomus fasciculatus* on the host–parasite relationship of *Rotylenchulus reniformis* in tomato. *Nematologica*, **28**, 412–19.

Sivaprasad, P., Jacob, A., Nair, S.K. and George, B. (1990) Influence of VA mycorrhizal colonization on root-knot nematode infestation in *Piper nigrum* L., in *Trends in Mycorrhizal Research*, Proceedings of the National Conference on Mycorrhiza (eds B.L. Jalali and H. Chand), Haryana Agricultural University, Hissar, pp. 100–1.

Smith, G.S. (1987) Interactions of nematodes with mycorrhizal fungi, in *Vistas on Nematology* (eds J.A Veech and D.W. Dickson), Society of Nematologists, Hyattsville, pp. 292–300.

Smith, G.S. (1988) The role of phosphorus nutrition in interactions of vesicular–arbuscular mycorrhizal fungi with soil-borne nematodes and fungi. *Phytopathology*, **78**, 371–4.

Smith, G.S. and Kaplan, D.T. (1988) Influence of mycorrhizal fungus, phosphorus and burrowing nematode interactions on growth of rough lemon citrus seedlings. *Journal of Nematology*, **20**, 539–44.

Smith, G.S., Hussey, R.S. and Roncadori, R.W. (1986a) Penetration and postinfection development of *Meloidogyne incognita* on cotton as affected by *Glomus intraradices* and phosphorus. *Journal of Nematology*, **18**, 429–35.

Smith, G.S., Roncadori, R.W. and Hussey, R.S. (1986b) Interaction of endomycorrhizal fungi, superphosphate and *Meloidogyne incognita* on cotton in microplot and field studies. *Journal of Nematology*, **18**, 208–16.

Suresh, C.K., Bagyaraj, D.J. and Reddy, D.D.R. (1985) Effect of vesicular–arbuscular mycorrhiza on survival, penetration and development of root-knot nematode in tomato. *Plant and Soil*, **87**, 305–8.

Sutherland, J.R. and Fortin, J.A. (1968) Effect of the nematode *Aphelenchus avenae* on some ectotrophic, mycorrhizal fungi and on a red pine mycorrhizal relationship. *Phytopathology*, **58**, 519–23.

Sylvia, D.M. and Sinclair, W.A. (1983) Phenolic compounds and resistance to fungal pathogens induced in primary roots of Douglas-fir seedlings by the ectomycorrhizal fungus *Laccaris laccata*. *Phytopathology*, **73**, 390–7.

Taber, R.A. (1982) Occurrence of *Glomus* spores in weed seeds in soil. *Mycologia*, **74**, 515–20.

Thapar, H.S. and Khan, S.N. (1985) Distribution of VA mycorrhizal fungi in forest soils in India. *Indian Journal of Forestry*, **8**, 5–7.

Trappe, J.M. and Schenck, N.C. (1982) Taxonomy of the fungi forming endomycorrhizae, in *Methods and Principles of Mycorrhizal Research* (ed. N.C. Schenck), American Phytopathological Society, St Paul, pp. 1–9.

Tribe, H.T. (1977) Pathology of cyst-nematodes. *Biological Review*, **52**, 477–507.

Tylka, G.L., Hussey, R.S. and Roncadori, R.W. (1991) Interactions of vesicular-arbuscular mycorrhizal fungi, phosphorus, and *Heterodera glycines* on soybean. *Journal of Nematology*, **23**, 122–33.

Umesh, K.C., Krishnappa, K. and Bagyaraj, D.J. (1988) Interaction of burrowing nematode, *Radopholus similis* (Cobb, 1893) Thorne 1949 and VA mycorrhiza, *Glomus fasciculatum* (Thaxt) Gerd and Trappe in banana (*Musa acuminata* Colla.). *Indian Journal of Nematology*, **18**, 6–11.

Wilde, S.A. (1954) Mycorrhizal fungi: Their distribution and effect on tree growth. *Soil Science*, **78**, 23–31.

Willcox, J. and Tribe, H.T (1974) Fungal parasitism in cysts of *Heterodera*. I. Preliminary investigations. *Transactions of the British Mycological Society*, **62**, 585–94.

10

Nematode–virus interactions

Bernhard Weischer

The interactions between plant-parasitic nematodes and plant viruses can be divided into two groups: (1) specific interrelationships between certain ectoparasitic nematode species and some viruses which they transmit; (2) more general effects of plant viruses on various plant-parasitic nematodes via the host plant. The first have attracted much attention since their discovery in 1958 and constitute the main part of this chapter. The second type of interaction has been investigated to a much lesser extent.

10.1 SPECIFIC INTERACTIONS

After many years of speculation, the first reliable proof of transmission of a plant virus by a plant-parasitic nematode was given when Hewitt *et al.* (1958) succeeded in transmitting grapevine fanleaf nepovirus by *Xiphinema index*. This discovery incited an intensive search for nematode vectors of other soil-borne viruses and led to research on many aspects of the taxonomy, biology and ecology of both the nematode vectors and the viruses (Lamberti *et al.*, 1975). Practically all nematodes feeding on virus-infected plants ingest virus particles but out of an estimated total of 2600 nominal species of plant-parasitic nematodes only about 30 species are virus vectors. They belong to the families Longidoridae and Trichodoridae of the order Dorylaimida. Both longidorid and trichodorid nematodes are ectoparasites on roots of annual and perennial plants. In addition to virus transmission, they can cause severe direct damage to crops. Within the Longidoridae seven out of 172 valid species of the genus *Xiphinema* and seven out of 83 species of *Longidorus* are known to be virus vectors. Similarly 13 out of 50 species of *Trichodorus* and *Paratrichodorus* can transmit plant viruses. However, these figures are subject to continuous changes due to progress in taxonomy and improvement in transmission techniques.

Nematode-transmitted viruses belong to two different groups, nepo- and tobraviruses respectively. *Nepo* is an acronym of *ne*matode-transmitted *po*lyhedral, and *tobra* of *to*bacco *ra*ttle. Although it has been claimed that other viruses can also be transmitted experimentally by nematodes, a biologically unequivocal role of nematodes as vectors has been ascertained only for nepoviruses/longidorids and tobraviruses/trichodorids.

10.1.1 Nematodes transmitting plant viruses

Longidorid nematodes are characterized by an axial stylet measuring up to 200 μm long consisting of an elongated spear (odontostyle) plus an extension (odontophore) usually half as long as the odontostyle; by a typical dorylaimid oesophagus with the stylet connected to the cylindrical, muscular and glandular bulb by a slender food canal; and by their 1.5–13 mm long slender body. Juveniles are recognized by having both a functional and a replacement odontostyle, for the next stage, situated in the wall of the anterior oesophagus. The tail is usually short, hemispheroid or conoid, rarely elongate or filiform. Of the five genera comprising the family, only some members of the genera *Xiphinema* and *Longidorus* are virus vectors. Species of *Longidorus* are characterized by a single guide ring around the anterior part of the odontostyle; by a simple junction between odontostyle and odontophore; and by an odontophore usually without basal flanges. In *Xiphinema* the guiding sheath is around the posterior part of the odontostyle. The base of the odontostyle is forked at the junction with the odontophore and the odontophore has prominent basal flanges.

Trichodorid nematodes are readily recognized by their ventrally curved mouth spear (onchiostyle), which is an elongated dorsal tooth without a lumen and without basal knobs or flanges. The oesophagus consists of a narrow anterior part swelling posteriorly to form an elongate or pear-shaped basal bulb. The body length in adults ranges from 0.5 to 1 mm, the females appearing plump and cigar shaped. The cuticle appears thick, particularly in dead individuals. Virus vectors occur in the genera *Trichodorus* and *Paratrichodorus*. Both are similar in appearance, the most evident difference being the abnormally strong swelling of the cuticle in *Paratrichodorus* when heat-relaxed or fixed in acid fixatives. Furthermore, males of *Paratrichodorus* have a bursa, which is absent in *Trichodorus*.

10.1.2 Viruses transmitted by nematodes

The plant viruses transmitted by nematodes form a heterogeneous collection that can be differentiated on the basis of particle morphology and relationship with vectors into two groups, nepo- and tobraviruses, according to the virus classification scheme by Harrison *et al.* (1971). There are at present 36 known nepoviruses, of which 11 have a recognized nematode vector. All nepoviruses have isodiametric particles with icosahedral symmetry and are 23–30 nm in diameter. Their genome is bipartite, with two functional ribonucleic acids (RNA-1 and RNA-2), separately encapsidated. Common biological properties are host responses and seed/pollen transmission. Some physicochemical properties and the type of intracellular behaviour are typical for the group (Martelli, 1975; Stace-Smith, 1977; Martelli and Taylor, 1989). There are, however, some differences in physicochemical characters and hydrodynamic behaviour between members of this group. Also the geographical distribution, the vectors, natural means of spread and serological properties differ. Depending on the criteria used (serology, physical properties, RNA analysis) the nepoviruses can be divided into two, three or four subgroups. Since some nepoviruses are still incompletely characterized, a final grouping is not yet possible (Martelli and Taylor, 1989). The number of recognized nepoviruses though has increased from eight to 36 during the last twenty years; the number of tobraviruses has increased from two to three only, namely tobacco rattle, pea early-browning and pepper ringspot tobraviruses. They have rigid, rod-shaped particles of three different lengths: very short, ±45 nm; short 50–110 nm; and long, 185–200 nm. The bipartite genome has two functional RNA species: a larger RNA-1 and a smaller RNA-2. The short particles encapsidate one molecule of RNA-2, the long particles one molecule of RNA-1. The coat protein consists of one type of subunit with a molecular weight of 21 000–23 000 Da. The three tobraviruses are distinguished genetically, i.e. by the sequence homology of RNA-1 (Robinson and Harrison, 1985).

10.1.3 Nematode feeding and virus acquisition

After a period of exploration of the root surface, longidorid nematodes penetrate the root tissue by rapid thrusting and twisting of the stylet usually two or three cell layers deep. Before the cell contents are ingested, secretions from the oesophageal glands are discharged through the stylet into the cell. During the feeding process, periods of pumping (ingestion) and of quiescence (salivation) alternate and a nematode may remain at a feeding site for hours (Weischer and Wyss, 1976). The odontostyle can be thrust several cell layers deep into the root tissue and eventually reach the vascular bundle. Trichodorid nematodes explore the root surface similarly in search for a suitable feeding site. They feed on epidermal cells and root hairs only. The cell walls are penetrated by rapid thrusts of the onchiostyle. Then secretions of the oesophageal glands are used to form a feeding tube through the cell wall and subsequently are injected into the cell and homogenize the cytoplasm. This is then rapidly ingested and the nematodes move to another cell (Wyss, 1977). Both nematode groups acquire viruses when ingesting cytoplasm of virus-infected plants. The time needed for successful acquisition is short, e.g. *Xiphinema index* has been shown to acquire grapevine fanleaf nepovirus in less than 5 minutes (Alfaro and Goheen, 1974). However, the number of successful transmissions, i.e. the number of viruliferous nematodes, increases with increasing access periods.

10.1.4 Virus retention

After the ingestion of virus particles during the feeding process, nematode vectors retain their corresponding viruses for some time. Generally *Xiphinema* spp. are said to retain the viruses they transmit for 10–12 months and most *Longidorus* vectors for shorter periods, usually shorter than three months. However, the English strain of raspberry ringspot nepovirus was detected in its vector *Longidorus macrosoma* after 60 months of starvation (Buser, 1990) and *X. rivesi* transmitted tomato ringspot nepovirus to bait plants up to two years after storage in soil without host plants (Bitterlin, 1986). These findings indicate that virus–vector relations are not the same for all combinations and that generalizations can be misleading. It was further shown that retention periods can vary with temperature. Mulberry ringspot nepovirus was retained by its vector *L. martini* in host-free soil for more than 18 months at 0–9°C, for 13 months at room temperature and for three months at 20–24°C (Yagita, 1977). Tobacco rattle tobravirus has been shown to persist in trichodorid vectors for periods of up to two years (Van Hoof, 1970). Such long retention periods have been observed solely in populations without access to host plants and apply mainly to adults. Nematodes lose some virus particles whenever they feed (salivate).

Nepo- and tobraviruses are retained in their appropriate vectors at distinct sites. In *Longidorus* vectors the virus particles associate with the interior surface of the lumen of the odontostyle and with the guiding sheath. In *Xiphinema* they are adsorbed to the cuticular lining of the lumen of the odontostyle and of the oesophagus, and in *Trichodorus* and *Paratrichodorus* they are attached to the cuticle lining the oesophagus but not to the onchiostyle. All surfaces serving as retention sites are shed during moulting, together with the outer cuticle of the body. Consequently, the adhering virus particles are lost and cannot be transferred from one developmental stage to the following. As far as can be said at present the viruses retained in their vectors do not multiply nor are they involved in the vector metabolism. No direct influence of plant viruses on nematodes has been proven. The mechanism of retention is still incompletely known. Evidence shows that nematode-transmitted viruses are selectively and specifically adsorbed at the retention sites in their associated vectors when cell contents of virus-infected plants are ingested. Other viruses that may be ingested but not transmitted by these nematodes are not retained and pass into the intestine, where they are unavailable for transmission. Experiments indicate that in some *Xiphinema* species the specific retention is based on a recognition process between lectin-like molecules associated with the virus protein coat and carbohydrates present in the cuticular lining of the oesophagus. However, the mechanism of recognition may be different in detail in different virus–vector combinations (Brown and Robertson, 1990).

Since the properties of the virus protein coat are genetically encoded for by the RNA-2, this part of the bipartite viral genome determines the specificity and transmissibility. In nepoviruses vector specificity is most strongly developed where the viruses are most distantly related serologically. In most cases a given virus is transmitted by one nematode species only (Tables 10.1–10.3). Where a nematode vector can transmit more than one virus these are serologically unrelated. Thus *Xiphinema diversicaudatum* transmits arabis mosaic nepovirus and the serologically unrelated strawberry latent ringspot nepovirus. Even serologically distant strains of the same virus can have different vectors. The Scottish strains (serotypes) of tomato blackring nepovirus and the unrelated raspberry ringspot nepovirus are both transmitted by *Longidorus elongatus*, whereas the English serotypes of these viruses are transmitted by *L. attenuatus* and *L. macrosoma* respectively. Experimental results showed that in such cases the nematodes ingested the virus to which they were exposed, but individuals of non-transmitting populations had very few or no virus particles retained at the typical sites (Martelli and Taylor, 1989). With tobraviruses evidence of a high specificity between virus and vector has only recently been identified (Brown and Ploeg, unpublished). The ability to transmit a virus must also be considered in respect to biological differences between populations of one species. This is particularly true when the nominal species is widely distributed. Local populations of a vector species are usually most efficient in transmitting local virus isolates. Thus, geographical isolation can lead to a high specificity, as shown for strawberry latent ringspot nepovirus from Italy, which is only transmitted by *X. diversicaudatum* from Italy (Brown, 1985).

After the specific retention, the dissociation of virus particles from their sites of retention is a prerequisite for successful transmission. It occurs when nematode saliva passes from the basal bulb anteriorly through the oesophagus and the stylet into the plant cell during the feeding process. Although the mechanism is not yet known in detail it is assumed that the saliva modifies the pH within the lumen and alters the surface charge of the virus particles (Martelli and Taylor, 1989). This leads to a detachment of a number of particles. The long retention periods observed show that not all particles adsorbed to the retention site are released at the same time, i.e. during a single feeding act. It has been shown above that the main cause of specific transmission is the specific association of virus particles at the sites of retention. However, the mechanism of dissociation can also in some instances determine specificity. *Longidorus macrosoma* transmits the English strain of raspberry ringspot nepovirus efficiently, but only rarely the Scottish strain. However, although particles of both strains become attached to the typical sites, only the English strain dissociates, whereas the Scottish strain does not (Taylor and Robertson, 1973).

10.1.5 Virus transmission

Virus transmission in its strict sense occurs when the dissociated particles are inoculated with the stream of saliva into plant cells during feeding. For a successful infection the virus particles must be viable and the plant cell must not be seriously damaged by the nematode attack. Otherwise the virus is not able to replicate and to pass into neighbouring cells. This is particularly true for tobraviruses and their trichodorid vectors. It takes only about 3 minutes to complete the feeding cycle on an individual cell, from the first stylet thrust to pierce the cell wall until leaving the emptied cell, of which the ingestion of the cytoplasm only takes 30 seconds. Therefore, a successful inoculation can only occur when the feeding cycle is not completed and the cytoplasm not entirely removed. In longidorid vectors the feeding periods are much longer than in trichodorids. Individuals have been observed staying at one feeding site for hours or even days. When a longidorid nematode stops feeding at a cell, the stylet is inserted into the next deeper cell so that a column of cells fed upon by the nematode remains, progressing from the epidermis into deeper cell layers. Hence there is ample time for virus particles to become established (Weischer and Wyss, 1976).

10.1.6 Geographical distribution of vectors and viruses

The present distribution of nematodes is the result of a natural development based on geological events and of dissemination caused by man's activities. The natural occurrence and distribution can best be studied in undisturbed habitats. In cultivated land nematode species may have been introduced with propagation material or soil used as ships' ballast, even over larger distances. Due to the lack of thorough surveys over larger areas reliable information on the geographical distribution of phytonematodes is scarce, with the exception of Europe, where centrally conducted surveys have been done (Alphey and Taylor, 1987; Brown *et al.*, 1990b). Historical biogeography is a new field in nematology (Ferris, 1983) which can give interesting information about the origin of nematodes and viruses and the evolutionary development of their associations. Recently Coomans (1984/85) studied Longidoridae under the aspects of phylogeny and biogeography. Of the two genera with virus-transmitting species *Xiphinema* is a cosmopolitan genus originating from Gondwanaland, from where it spread to Laurasia before the break-up of Pangaea some 180 million years ago. The main speciation subsequently occurred in Africa, where the majority of species are found. For *Longidorus* the situation is less clear. The probable origin lies in South-East Africa and India, from where it spread to Laurasia, with the main speciation occurring in Europe. *Longidorus* spp. are found mainly in the Northern Hemisphere. Little is known about the historical biogeography of trichodorid nematodes. At present they are most frequent in Europe and North America. However, the increasing number of reports, e.g. from Africa, Brazil, India, Japan and New Zealand, indicates a widespread occurrence, single species often having a regional or local distribution. Nepoviruses are primarily pathogens of wild plants and depend for survival and natural spread on dissemination by their vectors and by host seeds (Murant, 1981). Due to this limited natural mobility the actual geographical distribution (Tables 10.1 and 10.2) corresponds most probably to their areas of origin or differentiation. Generally, nepoviruses with a wide host range have a wide distribution, whereas viruses with few hosts are more restricted. An exception is grapevine fanleaf nepovirus, which is present in all major grape-growing areas of the world. Although a highly specialized virus associated with *Vitis vinifera*, it has been widely disseminated with infected propagating material over the centuries. In a similar way arabis mosaic nepovirus may have reached New Zealand, and the American tomato ringspot nepovirus Australia (Martelli and Taylor, 1989). Of the three recognized tobraviruses tobacco rattle tobravirus has a worldwide distribution, whereas pea early-browning tobravirus is restricted to Europe and pepper ringspot tobravirus to South America. Within these virus 'species' there are serologically distinct strains having different nematode vectors and a different distribution (Table 10.3).

10.1.7 Methodology

When Hewitt *et al.* (1958) reported the successful transmission of grapevine fanleaf nepovirus by *Xiphinema index*, research activities related to the problem of nematode-virus interactions increased rapidly. In the following years a great number of such associations were reported. However, many reports were contradictory and it became evident that the methods applied were often inappropriate. Main problems were difficulties in species determination and the lack of adequate standardized methods for sampling and transmission experiments. Many of the older species descriptions are incomplete and do not allow an exact identification. *Xiphinema americanum* Cobb, 1913 is a good example. It has been considered to be one species for about 50 years. Now it is regarded as a group of closely related species, at present 38 (Lamberti and Carone, 1991). All former records of *X. americanum* s. 1. as vector need to be rechecked. In order to obtain reliable results while studying nematode–virus interactions under natural and experimental conditions certain requirements must be fulfilled. Brown and Boag (1988) and Brown *et al.* (1990a) described the sampling techniques for virus-vector nematodes and their viruses. It was shown that the distribution of nematode-transmitted viruses is highly aggregated and that sampling for their detection should be based on a regular grid lattice with 7 m between sampling points. The optimum sampling depth is between 10 and 40 cm, depending on nematode species, host plant root system, soil characteristics and climate.

Table 10.1 *Xiphinema* vector nematodes and transmitted nepoviruses

Nematode	Virus	Host range of virus*	Geographical distribution of association
X. americanum sensu lato	Tomato ringspot	Fruits, vegetables, ornamentals	N. America
	Cherry rasp leaf	Fruits	N. America
X. americanum sensu stricto	Tobacco ringspot	Fruits, vegetables, ornamentals	N. America
	Peach rosette mosaic	Fruits	N. America
X. californicum	Tomato ringspot	Fruits, vegetables, ornamentals	N. America
	Grapevine yellow vein strain	Grapevine	N. America
X. diversicaudatum	Arabis mosaic	Fruits, vegetables, ornamentals	Europe, Mediterranean
	Strawberry latent ringspot	Fruits, vegetables, ornamentals	Europe, Mediterranean
X. index	Grapevine fanleaf	Grapevine	Worldwide in all major vinegrowing areas
X. italiae	Grapevine fanleaf	Grapevine	Europe, Mediterranean
X. rivesi	Tomato ringspot	Fruits, vegetables, ornamentals	N. America

* Nepoviruses are recovered from a wide range of naturally infected weeds but grapevine fanleaf occurs naturally only in grapevine.

Table 10.2 *Longidorus vector nematodes and transmitted nepoviruses*

Nematode	Virus	Host range of virus*	Geographical distribution of association
L. apulus	Artichoke Italian latent Italian strain	Vegetables	Mediterranean
L. attenuatus	Tomato blackring	Fruits, vegetables, ornamentals	Europe
L. diadecturus	Peach rosette mosaic	Fruits	N. America
L. elongatus	Raspberry ringspot Scottish strain	Fruits	Europe
	Tomato blackring beet ringspot strain	Sugar-beet	Europe
	Peach rosette mosaic	Fruits	N. America
L. fasciatus	Artichoke Italian latent Greek strain	Vegetables	Mediterranean
L. macrosoma	Raspberry ringspot English strain	Fruits	Europe
L. martini	Mulberry ringspot	Mulberry	Japan

* See Table 10.1

Table 10.3 Trichodorus and Paratrichodorus vector nematodes and transmitted tobraviruses

Nematode	Virus	Host range of virus	Geographical distribution of association
T. cylindricus	Tobacco rattle	Wide	Europe
T. hooperi	Tobacco rattle	Wide	Europe
T. primitivus	Tobacco rattle	Wide	Europe
	Pea early browning	Leguminosae	Europe
T. similis	Tobacco rattle	Wide	Europe
T. viruliferus	Tobacco rattle	Wide	Europe
	Pea early browning	Leguminosae	Europe
P. allius	Tobacco rattle	Wide	N. America
P. anemones	Tobacco rattle	Wide	Europe
	Pea early browning	Leguminosae	Europe
P. minor	Tobacco rattle	Wide	N. America, Japan
	Pepper ringspot	Vegetables	S. America
P. nahus	Tobacco rattle	Wide	Europe
P. pachydermus	Tobacco rattle	Wide	Europe
	Pea early browning	Leguminosae	Europe
P. porosus	Tobacco rattle	Wide	N. America
P. teres	Tobacco rattle	Wide	Europe
P. tunisiensis	Tobacco rattle	Wide	Europe

To establish the vector status of a nematode species several criteria have to be met for transmission experiments. The most important are as follows.

1. Nematode and virus must be correctly identified.
2. The nematode being tested must be the only possible vector.
3. The virus must be available to the nematode.
4. The conditions must be suitable for the transmission to take place.
5. Virus contamination of the bait plant must be avoided.
6. Bait plant tissue must be shown to contain the virus being tested.

Recently Trudgill *et al.* (1983) discussed methods and criteria for transmission experiments with longidorid nematodes. Ploeg *et al.* (1989) and Brown *et al.* (1989) did the same for trichodorid nematodes. On the basis of these criteria the authors made a critical review and came to the conclusion that only 14 out of 46 published results on virus transmission by longidorids and 13 out of 40 reports on trichodorids were supported by adequate evidence and therefore valid. They are presented here in Tables 10.1–10.3. The other reports were rejected for the following three main reasons.

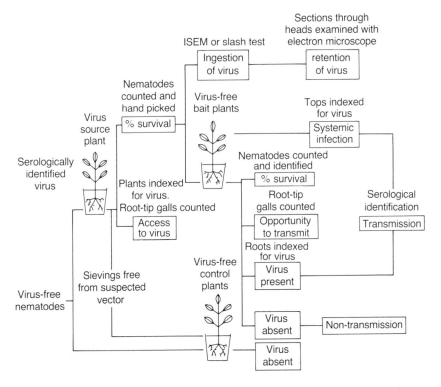

Figure 10.1 Procedure for establishing the ability of longidorid nematodes to transmit plant viruses (from Trudgill *et al.*, 1983).

1. There was no systemic infection: the viruses found could have been contaminants on the roots.
2. The description of methods used was inappropriate.
3. Vector and/or virus were not adequately identified.

Some of the rejected reports may be basically right and need only to be confirmed by further experimentation. The majority are definitely wrong.

For reliable transmission experiments with longidorid nematodes and nepoviruses, Trudgill *et al.* (1983) developed a test system that meets all requirements mentioned above (Fig. 10.1). In this test system, well-identified virus-free nematodes are placed around the roots of source plants with an identified virus. After an appropriate period of access the nematodes are extracted, counted and in small groups transferred to virus-free bait plants. Some of the nematodes are examined under the electron microscope for the presence of virus particles. Later the bait plants are tested for virus infection. Virus and vector must be re-identified at the end of the test. Controls and numerous replicates are prerequisites for these experiments. This procedure is not applicable to study the associations between trichodorid nematodes and tobraviruses. Therefore, Brown *et al.* (1989) developed a special testing system. The use of small (0.5 cm³) plastic capsules containing a single bait plant and inoculated with a single nematode allows the specific identification of each individual nematode transmitting a given virus.

10.2 GENERAL NEMATODE–VIRUS INTERACTIONS

It has already been mentioned that plant viruses have no ascertainable direct influence on carrier nematodes, vectors or non-vectors. Indirect effects via the host plant are well documented. For earlier literature on this aspect, the paper by Weischer (1975) may be referred to. The indirect host-mediated effects are based on changes in host plant metabolism caused by viruses and nematodes respectively. Development and multiplication of nematodes can be enhanced or inhibited depending on the virus, nematode or plant species. In tobacco plants stem nematode *Ditylenchus dipsaci* and leaf nematode *Aphelenchoides ritzemabosi* were both inhibited by tobacco mosaic tobamovirus and tobacco rattle tobravirus but enhanced by arabis mosaic nepovirus. Tomato blackring nepovirus enhanced *D. dipsaci* but suppressed *A. ritzemabosi* (Weischer, 1975). Root-knot nematode *Meloidogyne incognita* produced five to ten times more individuals on cardamom plants infected with 'Katte' mosaic virus than on healthy plants (Ali, 1988). Similarly plants of *Zinnia elegans* infected with zinnia mosaic virus were better hosts for *M. incognita* than healthy plants (Jabri *et al.*, 1985). The population build-up of both the ectoparasite *Tylencho-rhynchus brassicae* and the semi-endoparasite *Rotylenchulus reniformis* on eggplant (*Solanum melongena*) was promoted when plants were infected with brinjal mosaic virus (Naqvi and Alam, 1975). In *Solanum khasium* the root-knot index was higher on plants inoculated with tobacco mosaic tobamovirus than on healthy plants (Ismail *et al.*, 1979). Inhibitory effects on root-knot nematode *M. javanica* were observed in zucchini (*Cucurbita pepo*) infected with watermelon mosaic potyvirus. Virus infection retarded the establishment of these nematodes in the roots as compared with healthy plants (Huang and Chu, 1984). All these effects, favourable or detrimental, occurred or were more pronounced when nematode inoculation was preceded for two to three weeks by virus infection. This points to biochemical processes in plant tissues as a decisive mechanism involved in these interactions. This is underlined by results obtained by Alam *et al.* (1990). They observed antagonistic effects between tomato mosaic tobamovirus and *M. incognita*. When virus infection preceded nematode inoculations nematodes were suppressed, and when nematodes were the first agent the virus was inhibited. The changes caused by one pathogen were detrimental to the other. The influence of the host on this type of interaction was demonstrated by Moura and Powell (1977). In two out of three tomato varieties the egg production of *M. incognita* was significantly increased by the presence of tobacco mosaic tobamovirus but in the third it was not. Similarly arabis mosaic nepovirus inhibited *D. dipsaci* in petunias (Fritzsche, 1970) but furthered it in tobacco (Weischer, 1975).

The concomitant presence of nematode and virus in most cases results in a synergistic effect that aggravates the plant damage considerably. Little is known about the physiological or biochemical basis of enhancing or inhibiting nematode development by a virus infection of the host. It has been suggested that changes in protein metabolism caused by nematodes may interact with corresponding changes caused by the virus, resulting in beneficial effects in one case and detrimental effects in another (Weischer, 1975). This view was recently supported by Showler *et al.* (1990). They showed that the changes in free amino acid levels in sugar-cane caused by sugar-cane mosaic virus were responsible for population changes in various nematodes on sugar-cane.

REFERENCES

Alam, M.M., Samad, A. and Anver, S.(1990) Interaction between tomato mosaic virus and *Meloidogyne incognita* in tomato. *Nematologia Mediterranea*, **18**, 131–3.

Alfaro, A. and Goheen, A.C. (1974) Transmission of strains of grapevine fanleaf virus by *Xiphinema index*. *Plant Disease Reporter*, **58**, 549–52.

Ali, S.S. (1988) Influence of 'Katte' mosaic virus of cardamom on the population of *Meloidogyne incognita*. *Nematologia Mediterranea*, **17**, 121–2.

Alphey, T.J.W. and Taylor, C.E. (1987) *European Atlas of Longidoridae and Trichodoridae*, Scottish Crop Research Institute, Dundee.

Bitterlin, M.W. (1986) Tomato ringspot virus: interactions with its nematode vector *Xiphinema rivesi*, studies on virus transmission to and detection in fruit trees, serological characterization, and implications for cross protection. PhD. thesis, Cornell University, Ithaca, New York.

Brown, D.J.F. (1985) The transmission of two strains of strawberry latent ringspot virus by populations of *Xiphinema diversicaudatum* (Nematoda: Dorylaimoidea). *Nematologia Mediterranea*, **13**, 217–23.

Brown, D.J.F. and Boag, B. (1988) An examination of methods used to extract virus-vector nematodes (Nematoda: Longidoridae and Trichodoridae) from soil samples. *Nematologia Mediterranea*, **16**, 93–9.

Brown, D.J.F. and Robertson, W.M. (1990) Factors involved in the acquisition, retention and release of viruses by virus-vector nematodes. *Nematologica*, **39**, 336 (Abstract).

Brown, D.J.F., Ploeg, A. and Robinson, D.J. (1989) A review of reported associations between *Trichodorus* and *Paratrichodorus* species and tobraviruses with a description of laboratory methods for examining virus transmission by trichodorids. *Revue de Nematologie*, **12**, 235–41.

Brown, D.J.F., Boag, B., Jones, A.T. and Topham, P.B. (1990a) An assessment of the soil-sampling density and spatial distribution required to detect viruliferous nematodes (Nematoda: Longidoridae and Trichodoridae) in fields. *Nematologia Mediterranea*, **18**, 153–60.

Brown, D.J.F., Taylor, C.E., Cholewa, B. and Romanenko, N.D. (1990b) The occurrence of Longidoridae (Nematoda: Dorylaimida) in Western USSR with further comments on longidorid nematodes in Europe and the Mediterranean basin. *Nematologia Mediterranea*, **18**, 199–207.

Buser, A. (1990) *Untersuchungen über die Pfeffingerkrankheit der Süsskirsche und deren Vektor Longidorus macrosoma*, Eidgenossische Technische Hochschule Zürich, dissertation no. 9194.

Coomans, A. (1984/85) A phylogenetic approach to the classification of the Longidoridae (Nematoda: Dorylaimida). *Agriculture, Ecosystems and Environment*, **12**, 335–54.

Ferris, V.R. (1983) Phylogeny, historical biogeography and the species concept in soil nematodes, in *Concepts in Nematode Systematics* (eds A.R. Stone, H.M. Platt and L.F. Khalil), Academic Press, London, pp. 143–61.

Fritzsche, R. (1970) Wechselbeziehungen zwischen Virus- und Nematodenbefall in ihrem Einfluss auf den Schädigungsgrad der Pflanzen. *Biologisches Zentralblatt*, **89**, 225–32.

Harrison, B.D., Finch, J.T., Gibbs, A.J., Hollings, M., Shepherd, R.J., Valenta, V. and Wetter, C. (1971) Sixteen groups of plant viruses. *Virology*, **45**, 356–63.

Hewitt, W.B., Raski, D.J. and Goheen, A.C. (1958) Nematode vector of soil-borne fanleaf virus of grapevines. *Phytopathology*, **48**, 568–95.

Huang, S.P. and Chu, E.Y. (1984) Inhibitory effect of watermelon mosaic virus on *Meloidogyne javanica* (Treub) Chitwood infecting *Cucurbita pepo* L. *Journal of Nematology*, **16**, 109–12.

Ismail, W., Johri, J.K., Zaidi, A.A. and Singh, B.P. (1979) Influence of root-knot nematode, tobacco mosaic virus and complex on the growth and carbohydrates of *Solanum khasium* Clarke. *Indian Journal of Experimental Biology*, **17**, 1266–7.

Jabri, M., Khan, T.A., Husain, S.I. and Mahmood, K. (1985) Interaction of zinnia mosaic virus with root-knot nematode, *Meloidogyne incognita* on *Zinnia elegans*. *Pakistan Journal of Nematology*, **3**, 17–21.

Lamberti, F. and Carone, M. (1991) A dichotomous key for the identification of species of *Xiphinema* (Nematoda: Dorylaimida) within the *X. americanum*-group. *Nematologia Mediterranea*, **19**, 341–8.

Lamberti, F., Taylor, C.E. and Seinhorst, J.W. (eds) (1975) *Nematode Vectors of Plant Viruses*, Plenum Press, New York.

Martelli, G. (1975) Some features of nematode-borne viruses and their relationship with the host plant, in *Nematode Vectors of Plant Viruses* (eds F. Lamberti, C.E. Taylor and J.W. Seinhorst), Plenum Press, New York, pp. 223–52.

Martelli, G. and Taylor, C.E. (1989) Distribution of viruses and their nematode vectors. *Advances in Disease Vector Research*, **6**, 151–89.

Moura, R.M. and Powell, N.T. (1977) Estudos sobre o complexo TMV *M. incognita* em tomate. *Sociedade Brasileira de Nematologia, Publicacao*, no. **2**, 175–81.

Murant, A.F. (1981) The role of wild plants in the ecology of nematode-borne viruses, in *Pests, Pathogens and Vegetation* (ed. J.M. Thresh), Pitman, Boston, pp. 237–48.

Naqvi, S.Q.A. and Alam, M.M. (1975) Influence of brinjal mosaic virus on the population of *Tylenchorhynchus brassicae* and *Rotylenchulus reniformis* around eggplant roots. *Geobios*, **2**, 120–1.

Ploeg, A., Brown, D.J.F. and Robinson, D. (1989) Transmission of tobraviruses by trichodorid nematodes. *EPPO Bulletin*, **19**, 605–10.

Robinson, D. and Harrison, B.D. (1985) Unequal variation in the two genome parts of tobraviruses with evidence for the existence of three separate viruses. *Journal of General Virology*, **66**, 171–6.

Showler, A., Reagan, T. and Shao, K. (1990) Nematode interactions with weeds and sugarcane mosaic virus in Louisiana sugarcane. *Journal of Nematology*, **22**, 31–8.

Stace-Smith, R. (1977) Characteristics of nematode-borne plant viruses. *American Phytopathological Society, Proceedings of the Symposium on Nematode Transmission of Viruses*, pp. 11–20.

Taylor, C.E. and Robertson, W.M. (1973) Nematology–Electronmicroscopy. *Report of the Scottish Horticultural Research Institute*, **19**, 77.

Trudgill, D., Brown, D.J.F. and McNamara, D. (1983) Methods and criteria for assessing the transmission of plant viruses by longidorid nematodes. *Revue de Nematologie*, **6**, 133–41.

Van Hoof, H.A. (1970) Some observations on retention of tobacco rattle virus in nematodes. *Netherlands Journal of Plant Pathology*, **76**, 329–30.

Weischer, B. (1975) Further studies on the population development of *Ditylenchus dipsaci* and *Aphelenchoides ritzemabosi* in virus-infected and virus-free tobacco. *Nematologica*, **21**, 213–18.

Weischer, B. and Wyss, U. (1976) Feeding behaviour and pathogenicity of *Xiphinema index* on grapevine roots. *Nematologica*, **22**, 319–25.

Wyss, U. (1977) Feeding processes of virus transmitting nematodes. *American Phytopathological Society, Proceedings of the Symposium on Nematode Transmission of Viruses*, 30–41.

Yagita, H. (1977) The life history and biology of the needle nematode, *Longidorus martini* Merny III. Studies on the host range of mulberry ringspot virus and mode of transmission by *L. martini*. *Japanese Journal of Nematology*, **7**, 15–20.

11

Nematode bacterial disease interactions

K. Sitaramaiah and K.N. Pathak

One of the most significant developments in plant pathology and plant nematology during the last three decades has been the demonstration of many complex associations between nematodes and fungi (Powell, 1963, 1971a, 1971b, 1979), bacteria (Pitcher, 1961, 1963, 1965), and viruses (Raski and Hewitt, 1963; Harrison, 1964; Weischer, 1968; Taylor, 1972) in causing plant diseases. Experiments conducted under carefully controlled conditions conclusively proved that plant-parasitic nematodes greatly increase the development of plant diseases caused by fungi and bacteria. Synergistic relationships between plant pathogenic bacteria and nematodes in increasing the severity of plant diseases have been demonstrated (Pitcher, 1963, 1965; Pitcher and Crosse, 1958). Plant-parasitic nematodes as primary pathogens favour the establishment of secondary pathogens like bacteria which otherwise cannot infect the plant under normal conditions. Primary pathogens induce changes in a host, leading to synergistic association for disease development, whereas secondary pathogens participate actively and alter the course of pathogenesis. Secondary invaders, however, merely colonize dead cells induced by the primary pathogens (Powell, 1979; Mayol and Bergeson, 1969). It is now established that two or possibly more pathogens, rather than only one, are required to cause some diseases. There is sufficient evidence to show that the damage done by two or more pathogens is much more severe than what the pathogens cause independently. There are some examples in the literature indicating that the control of nematodes protected plants from injury by secondary pathogens (Devay et al., 1967).

Different types of interactions between nematodes and bacteria may exist in nature in relation to benefit or harm given or received. The most prominent type of interaction is that in which the bacteria are able to infect a host plant which they would otherwise be incapable of infecting in the absence of nematode. Relatively there are fewer described interactions involving nematodes and bacteria when compared to fungi in plant disease complexes. It is important to consider the role of primary participating pathogen, its interrelationships with secondary pathogens and their ultimate effects upon host plants. Reviews on interactions cover bacteria (Pitcher, 1963, 1965), and bacteria and fungi (Pitcher, 1961; Mountain, 1965; Brzeski, 1970; Bergeson, 1972; Powell, 1979; Khan, 1984; Sitaramaiah and Singh, 1990). In this chapter we describe mainly the broad principles of important nematode bacterial interactions, with emphasis on wilt-inducing bacteria.

11.1 PATHOGENESIS OF WILT-INDUCING BACTERIA

Pathogenesis is the process whereby one organism causes suffering in another organism. To understand the pathogenesis in wilt-inducing bacteria a thorough knowledge about the ecology of these bacteria is essential, but unfortunately this aspect is least known among all the soil-borne pathogens. Plant-pathogenic bacterial genera have been grouped in relation to soil variously by different workers. Crosse (1968) classified them into four groups:

1. with permanent soil phase – the pathogenic bacteria in this group are capable of living indefinitely in the soil. For example, *Agrobacterium tumefaciens* has been detected for up to 40 years in the absence of susceptible host;
2. with protracted soil phase – the bacteria in this group survive in soil or in host debris for a year or more but population declines gradually. Examples are *Clavibacter michiganense* subsp. *insidiosum* (*Corynebacterium insidiosum*), *Pseudomonas solanacearum* race 1 and *Agrobacterium rhizogenes*;
3. with transitory or ephemeral soil phase – the bacteria can survive in host debris, the rhizoplane epiphytic on the roots of non-host plant. Most of the plant-pathogenic bacteria fall in this group. Species of *Xanthomonas*, *Pseudomonas* and *Corynebacterium* survive through host debris for a limited period;
4. with no soil phase – some of the plant-pathogenic bacteria have evolved in such a way that they closely follow the life cycle of the host plant or have developed a specialized relation with some insects which help them in overcoming the unfavourable weather conditions and also help in dissemination (e.g. *Xanthomonas stewartii* overwinters in corn betels).

Assigning the particular bacterial plant pathogen into different groups is not discrete. It may vary even for the different strains of the same species. The same pathogenic strains may behave differently in different geographical regions depending on cropping pattern and soil factors. To initiate an interaction the bacteria have to be in the rhizosphere and survive until interaction is consolidated. Plant-pathogenic bacteria form a very small fraction of the total population in the rhizosphere; therefore it would be effective only when there is some specific competence in the bacterium and the response from the host.

Buddenhagen and Kelman (1964) and Kelman (1953) described sequential pathogenesis in *Pseudomonas solanacearum*. The initial introduction of a relatively small number of bacterial cells into a few xylem elements is followed by multiplication in localized areas of xylem tracheae and longitudinal spread without alteration in xylem parenchyma or phloem cells; there is involvement of a high proportion of tracheal elements as systemic invasion occurs, with resulting impairment of water movement and wilting. Finally degradation of cell walls, adjacent parenchyma cells, formation of lysogenous cavities, invasion of phloem pith and cortical tissue or both, particularly in young tissue, occur. Gallegly and Walker (1949) found that the bacterial wilt pathogen *P. solanacearum* first migrates in the large xylem vessels and in succulent stems progresses into outer cellular spaces of the cortex and pith, causing lysogenous cavities. Wilt development occurs due to invasion of all bundles by bacteria, whereas epinasty results if a single lateral bundle is invaded. When the vessels are invaded there is a tendency of adventitious root formation just outside the invaded bundles (Walker, 1972). The disease cycle of *P. solanacearum* in banana was described by Buddenhagen (1965). The bacteria are dispersed primarily to fresh wounds or root-to-root spread. Once the plant is invaded a systemic vascular disease results. In the advanced stage of the disease, bacteria ooze from wounded roots and the cycle is repeated.

11.1.1 General characteristics of vascular bacterial wilt diseases

The typical wilt-inducing bacterial diseases are restricted to plants where the functional conductive elements for transporting water are the vessels. These diseases are systemic and a single infection often kills the plant and renders it unproductive. These bacteria also have the ability to grow and reproduce in xylem vessels in the dilute sap of the transportation stream (Kessler, 1966). The sap is low in sugars but contains amino acids that serve as a carbon source for the pathogen. The sap also contains inorganic nutrients absorbed by the plant and additional food is obtained from cell walls that are attacked during pathogenesis (Dimond, 1972). Once inside the plant the bacteria live and reproduce in the intercellular spaces. The bacteria may also secrete enzymes that break down the middle lamella, resulting in maceration of the tissues accompanied by cell death. When the bacteria invade vessels they reproduce, spread and carry on all their metabolic activities (Goodman *et al.*, 1967). The mode of infection of *Agrobacterium*, however, differs from those of other pathogenic soil bacteria. Instead of causing growth reduction or destruction of cellular tissue, the bacteria stimulate cell division (hyperplasia) and cell enlargement (hypertrophy) in the affected tissues. This can result in the development of overgrowths or tumours on roots, stems or other organs. Certain species of *Coryne-bacterium* can also stimulate cell division and produce overgrowths (Goodman *et al.*, 1967).

11.1.2 Physiological changes of plants infected with vascular wilt bacterial pathogens

The symptoms of vascular wilt diseases are unique because of the locus of infection and also the nature of tissues that respond to stimuli during pathogenesis. The vascular wilt bacteria produce extracellular enzymes during the infection process and these substances are transported apically through vessels and water. Husain and Kelman (1958) found that vascular occlusion is the primary cause of wilt in banana plants infected with *Pseudomonas solanacearum* and the bacterium is able to produce higher-molecular-weight polysaccharides. These polysaccharides appear to be retained in xylem vessels and increase the resistance of the stem to flow of water. This results in plugging of the pores and reduces lateral movement of water to other vessels and to parenchyma of xylem and pith. Endopoly-galacturonase activity and cellulase activity of tomato plants infected with *P. solanacearum* was also detected (Husain and Kelman, 1958; Kelman, 1953). Pegg and Sequeira (1968) reported an increase in phenylalanine, tryptophan, tyrosine, dihydoxyphenylalanine and aromatic acids in tobacco plants inoculated with *P. solanacearum*. These precursors for indole acetic acid (IAA) and other phenolic compounds lead to vascular browning and also to increased auxin content. Tobacco and banana plants infected with *P. solanacearum* build up greater amounts of auxin levels than healthy plants (Sequeira and Kelman, 1962). IAA increases the plasticity of cell walls, making them more extensible by altering the nature of the substances deposited in them. Experiments with ^{14}C-labelled tryptamine demonstrated that the host produces most of the IAA that accumulates during the early stages of pathogenesis (Sequeira, 1964, 1965; Sequeira and Williams, 1964). Formation of tyloses reduces the flow of sap through the plants (Mace and Solit, 1966) and ultimately reduces the water economy of the plant and expressed as wilt disease.

11.2 THE ROLE OF NEMATODES IN BACTERIAL DISEASE COMPLEXES

The role of nematodes in disease complexes involving bacterial pathogens is complex in nature and not completely elucidated. Each disease complex is distinct from another and largely dependent on the type of nematode parasite (whether ectoparasite/endoparasite or migratory/sedentary), the stage at which the host plants are infected, the inoculum threshold level of each participating pathogen and environmental conditions such as moisture and temperature. All these biotic and ecological factors greatly influence the level to which the disease will develop in a particular host plant.

According to the dictionary, a syndrome is a group of concurrent symptoms characterizing a plant disease. 'Syndrome' pertains to symptoms and not to causal organisms. 'Plant disease complex' refers to a group of concurrent pathogenic organisms rather than to a group of concurrent symptoms.

Theoretically, plant disease complexes involving nematodes can be grouped into two kinds, i.e. obligatory and fortuitous relationships (Wallace, 1963; Khan, 1984). In an obligatory relationship one member is completely dependent on another or directly influenced by it. Therefore, the expression of plant disease symptoms occur only when the nematode and the bacterium are present together. In a fortuitous relationship each member acts independently and is not directly influenced by the other member. In this type of relationship the presence of the nematode is not required for the expression of disease symptoms but they enhance considerably the incidence and severity of the disease. The possible roles played by the nematode in interactions with bacterial plant pathogens are discussed below, with suitable examples.

11.2.1 Nematode acts as predisposing agent by providing wounds for the entry of bacterial pathogen

In this original concept to explain the nematode pathogen complex, the role of wounds was considered to be of great importance. Feeding of nematodes causes physical damage to the host plant and also provides avenues for the quick, direct entry and passage of pathogenic bacteria, especially when the pathogen is not strong enough to break the mechanical barriers of the host. Continuous feeding of the nematode also exposes a greater surface area for colonization. The availability of necrotic cells on the root also provides quick establishment and from there the bacterial pathogens are able to invade surrounding healthy tissue. The nematode may also carry the bacterial pathogens externally to deeper tissues. Wounding of roots may have some indirect effects. For instance it may increase the root exudation pattern and positively influence the rhizosphere microflora in such a way that the bacterial pathogens are stimulated.

The bacteria enter the plants through wounds; however, wounds need to be fresh (before healing occurs) and while leakage of cellular contents provides a congenial atmosphere of high humidity and good nutritive substrate for infection and colonization. Wilt-inducing bacterial pathogens depend mainly on wound penetration to establish an infection court (Goodman *et al.*, 1967). In this type of interaction all kinds of ecto- and endoparasitic (both migratory and sedentary) nematodes are involved (Table 11.1). The stylet opening of plant-parasitic tylenchid nematodes ranges from 0.2 to 1 μm in diameter and therefore restricts the passage of pathogenic bacterial cells into intact plant cells. The plant-parasitic nematodes can either injure the plant or act as vectors favouring bacterial cells to enter the plant system. The earliest published work of Hunger (1901) showed for the first time that tomato plants were readily infected with *Pseudomonas solanacearum* in nematode-infested soil, but remained healthy in nematode-free soil. A number of bacterial pathogens normally inhabit soil and may become pathogenic on roots. Symptoms of such infections may be manifested as galling or as root decay.

Hairy root of roses caused by *Agrobacterium rhizogenes* is usually of minor importance but serious losses have been reported. This condition, apparently caused by a new strain of the bacterium (*A. rhizogenes*), is frequently correlated with infestation by the lesion nematode *Pratylenchus vulnus* (Munnecke *et al.*, 1963).

For the development of crown gall caused by *Agrobacterium tumefaciens* some sort of injury on the crown or roots is essential before the bacterium can infect the host. Crown gall of peaches (*A. tumefaciens*) becomes more severe in the presence of root-knot nematode, *Meloidogyne javanica*. At low nematode population levels, however, crown gall symptoms were no more severe than those occurring in plants inoculated with bacterium alone after wounding (Nigh, 1966). Lele *et al.* (1978) described an interaction between *Rotylenchulus reniformis* and *A. tumefaciens* in grapevine. Initial damage by the reniform nematode facilitated entry, establishment and disease development.

Normal alfalfa has a low incidence of wilt caused by *Clavibacter michiganense* subsp. *insidiosum*, but when the roots were injured by *Meloidogyne hapla*, or mechanically, wilt incidence increased (Hunt *et al.*, 1971), suggesting that the nematode produced wounds and these wounds were the favourable sites for bacterial colonization. The root-knot nematode, *M. incognita*, enhanced the interaction of *Agrobacterium radiobacter* var. *tumefaciens* in cotton (Zutra and Orion, 1982). Bookbinder *et al.* (1982) found that *M. hapla*, *Pratylenchus penetrans* and *Helicotylenchus dihystera* produced wounds in alfalfa roots which were invaded by *Pseudomonas viridiflava*, *P. carrugata* and *P. marginalis* respectively.

Stewart and Schindler (1956) studied wilting of carnation cuttings infected with *Pseudomonas caryophylli* in association with *Meloidogyne* spp., *Helicotylenchus nannus* or *Xiphinema diversicaudatum*. Treatments with and without root wounds were included as well as parallel treatments without bacteria. The results indicated that root-wounding *Meloidogyne* spp. and *H. nannus* increased the rate of wilting in the presence of bacterial pathogens. There was no effect with *X. diversicaudatum* in the presence or absence of bacteria. From these experiments they concluded that endoparasitic nematodes as well as ectoparasitic nematodes aggravate bacterial wilt by wounding the roots and allowing the bacteria to enter the plant.

Libman *et al.* (1964) tested the response of tomato plants to inoculation with *Pseudomonas solanacearum* and either *Meloidogyne hapla*, *Helicotylenchus nannus* or *Rotylenchus* spp. Both *M. hapla* and *H. nannus* increased wilt severity, whereas *Rotylenchus* spp. had no specific effect. The reason given for increased wilt incidence in the presence of nematodes was the micropunctures made by the nematodes.

The interaction of *Meloidogyne* spp. and *Pseudomonas solanacearum* has been clearly demonstrated in many crops, viz. tobacco (Johnson, 1966; Johnson and Powell, 1969; Fukudome and Sakasegawa, 1972), potato

Table 11.1 Nematode–bacterial disease interactions

Nematode	Bacterium	Host	Disease	Reference
Anguina tritici	*Clavibacter tritici* (*Corynebacterium michiganense* pv. *tritici*)	Wheat	Yellow ear-rot or 'tundu' disease	Gupta and Swarup (1972)
A. funesta	*Clavibacter* sp. (*Corynebacterium rathayi*)	Ryegrass	Annual ryegrass toxicity	Lanigan *et al.* (1976), Stynes *et al.* (1979)
Aphelenchoides ritzemabosi or *A. fragariae*	*Rhodococcus fascians* (*Corynebacterium fascians*)	Strawberry	Cauliflower disease	Crosse and Pitcher (1952), Pitcher and Crosse (1958)
Ditylenchus dipsaci	*Pseudomonas fluorescens*	Garlic	'Café au lait' bacteriosis	Caubel and Samson (1984)
D. dipsaci	*Clavibacter michiganense* subsp. *insidiosum* (*Corynebacterium insidiosum*)	Alfalfa	Wilt	Hawn (1963, 1965)
D. destructor	*Clavibacter michiganense* subsp. *insidiosum*	Rhubarb	Crown rot	Hawn (1963)
Ditylenchus spp.	*Pseudomonas caryophylli*	Carnation	Wilt	Schuster (1959)
Helicotylenchus dihystera	*P. marginalis*	Alfalfa	Root-rot/wilt	Bookbinder *et al.* (1982)
H. nannus	*P. caryophylli*	Carnation	Wilt	Stewart and Schindler (1956)
H. nannus	*P. solanacearum*	Tomato	Wilt	Libman *et al.* (1964)
Meloidogyne hapla	*Pseudomonas viridiflava*	Alfalfa	Root rot/wilt	Bookbinder *et al.* (1982)

Nematode	Bacterium	Host	Disease	Reference
M. hapla	P. solanacearum	Tomato	Wilt	Libman et al. (1964)
M. incognita	Curtobacter (Corynebacterium) flaccumfaciens	Beans	Wilt	Schuster (1959)
M. incognita	P. solanacearum	Eggplant	Wilt	Reddy et al. (1979)
M. incognita	Agrobacterium radiobacter var. tumefaciens	Peach	Crown gall	Zutra and Orion (1982)
M. incognita	P. solanacearum	Tomato	Wilt	Napiere and Quinio (1980)
M. incognita acrita	P. solanacearum	Potato	Wilt	Jatala and Martin (1977a, 1977b)
M. incognita acrita	P. solanacearum	Tobacco	Wilt	Lucas et al. (1955)
M. javanica	P. solanacearum biotype 3	Eggplant	Wilt	Sitaramaiah and Sinha (1984a, 1984b)
M. javanica	Agrobacterium tumefaciens	Peach	Crown gall	Nigh (1966)
M. javanica	P. marginata	Gladiolus	Scab	El-Goorani et al. (1974)
Meloidogyne spp.	P. caryophylli	Carnation	Wilt	Stewart and Schindler (1956)
Meloidogyne spp.	Clavibacter michiganense	Tomato	Canker	De Moura et al. (1975)
Meloidogyne spp.	Clavibacter spp.	Tomato	Wilt	Hunt et al. (1971)
Pratylenchus penetrans	P. carrugata	Alfalfa	Root-rot/wilt	Bookbinder et al. (1982)
P. vulnus	Agrobacterium rhizogenes	Roses	Hairy root	Munnecke et al. (1963)
Rotylenchulus reniformis	A. tumefaciens	Grapevine	Crown gall	Lele et al. (1978)

(Feldmesser and Goth, 1970; Nirula and Paharia, 1970; Jatala and Martin, 1977b), tomato (Temiz, 1968; Duarte and De Albuquerque, 1971; Davide, 1972; Jenkins, 1972; Napiere, 1980; Sellam *et al.*, 1982; Thakur, 1985) and eggplant (Jenkins, 1972; Sitaramaiah and Sinha, 1984a, 1984b). Interaction of *M. incognita* with wilt bacteria *Curtobacterium* (*Corynebacterium*) *flaccumfaciens* in beans (Schuster, 1959) has also been reported. In all these studies, mechanical root injury – root wounding of plant cells by the plant-parasitic nematodes – is an important predisposing factor for direct introduction, establishment and multiplication of bacterial plant pathogens into the host tissue.

11.2.2 Nematode increases susceptibility by modification of host tissue

Endoparasitic nematodes such as *Meloidogyne* spp. become sedentary in their host, and usually induce surrounding cells to undergo profound changes in structure and physiology. Results of histochemical tests on root-knot nematode galls demonstrated that giant cell walls contain cellulose and pectin but no lignin, suberin, starch or ninhydrin-positive substances. However, giant cell protoplasm contains carbohydrates, fat, RNA and a large amount of protein. Starch disappears in galled tissues while cellulose, sugars, phosphorylated intermediates, keto acids, free amino acids, protein, nucleic acids, phosphorus and nitrogen increases when compared to healthy tissues. Levels of several metals and sulphur remain the same as in comparable healthy tissue. These findings, as well as results from histochemical studies, indicated that *Meloidogyne*-infected roots are very active metabolically (Krusberg, 1963; Sitaramaiah, 1989).

Based on chromatographic and biological activity assays, galled plant tissues were reported to contain more auxin or indole compounds than healthy tissues. During the feeding process *Meloidogyne* spp. secrete a proteolytic enzyme which releases IAA or tryptophan (Endo, 1971; Bird, 1974). The tryptophan is metabolized to IAA, resulting in a hormonal imbalance which is not confined only to the site of action but also affects other tissues. These hormonal imbalances and other physiological changes in the plants infected with nematodes as primary pathogen can influence positively the pathogenesis of bacterial pathogens. Due to a change in host physiology the mechanism of inhibiting production of toxins by bacterial wilt pathogen may be destroyed, thus enabling the wilt bacteria to break down host resistance.

In the interactions involving nematodes and bacteria, disease severity was always greater when the plants were infected with nematodes several days or weeks prior to exposure to bacterial pathogens than when the plants were infected with both the organisms simultaneously (Johnson and Powell, 1969; Sitaramaiah and Sinha, 1984a, 1984b). This shows that although physical injury plays an important role, it may not completely explain the predisposition of plants to bacterial pathogens. However, ectoparasitic nematodes are unable to cause any modification of host tissue.

Modification of host tissue by endoparasitic nematodes may be localized or systemic in nature. Powell and Nusbaum (1960) were the first to demonstrate that modification in the substrate due to nematode infestation provides an advantage to fungal pathogens. This altered tissue also favours the establishment of bacterial pathogens. Johnson and Powell (1969) showed that root-knot nematodes act as modifiers of infested tissue in such a way that the infested tissue and surrounding cells become more suitable for bacterial colonization. They found that plants inoculated with the nematodes three to four weeks prior to bacterial inoculation developed bacterial wilt symptoms to a greater extent than plants inoculated with nematodes and bacteria simultaneously.

Roots infected with *Meloidogyne* spp. are nutritionally more rich than the non-infected roots. Root galls have been found to contain more amino acids, auxins, growth-promoting substances, RNA, DNA and phosphorus than healthy tissue. These substances promote growth and establishment of bacterial pathogens. Histological observations revealed the presence of bacteria-like inclusions within giant cells in those plants which received nematodes three to four weeks before inoculation with bacteria. Histological observations of eggplant roots inoculated with *M. javanica* two or three weeks before *Pseudomonas solanacearum* biotype 3 inoculation showed extensive cavities of broken cortical and endodermal cells (Sitaramaiah and Sinha, 1984a). The nematode-inoculated roots contained pronounced hyperplastic and hypertrophic regions characterized by 'giant cells' predisposed to bacterial invasion and colonization.

11.2.3 Nematode breaks resistance to bacterial pathogens

Failure of some bacterial disease-resistant crop cultivars in the presence of plant-parasitic nematodes has drawn the keen attention of both plant breeders and nematologists to breed crop cultivars resistant to nematodes and also to discover the mechanism of resistance. The increase in growth of disease-resistant crop cultivars in fumigated soils over those in non-fumigated soils can be great, indicating clearly the importance of breeding cultivars resistant to nematodes. Resistance breaking to bacterial pathogens implies that there is a possibility that plant-parasitic nematodes could be able to supply something which is obviously lacking in the resistant cultivar in the absence of nematode invasion, or the nematode could alter cell permeability, making a particular cultivar susceptible to bacterial colonization (Pitcher, 1978).

It has been observed in several instances that a particular cultivar resistant to some bacterial pathogens becomes susceptible in the presence of *Meloidogyne* spp., and *Meloidogyne* spp. as primary pathogens can bring about physiological changes favouring the bacterial pathogens to overcome the resistance. Reddy *et al.* (1979) reported that when the eggplant cultivar

Pusa purple cluster (highly resistant to *Pseudomonas solanacearum*) was inoculated with a combination of the bacterium and *M. incognita*, a greater number of plants wilted when the organisms were inoculated simultaneously than when one was inoculated before the other. Field resistance in potato to *P. solanacearum* was broken down when the plants were infected with *M. incognita acrita* (Jatala and Martin, 1977a, 1977b). Resistance to the nematode *M. incognita acrita* in cultivated *Solanum chacoeuse* and *S. sparsipillum* could confirm resistance to *P. solanacearum*. The root-knot nematode *M. incognita acrita* predisposed the moderately resistant tobacco variety Dixie Bright 101 to *P. solanacearum* (Lucas *et al.*, 1955). The resistant cultivars inoculated with both pathogens developed wilt symptoms similar to those plants which were wounded mechanically and inoculated with the bacterium. In the case of resistant cultivar the only pathway of entry of bacterial cells was through root injury.

Napiere (1980) and Napiere and Quinio (1980) found that wilt disease development occurred earlier and with a higher mortality rate in both wilt-resistant and susceptible tomato cultivars grown in *Pseudomonas solanacearum* and *M. incognita*-infested soil. The susceptibility of young fresh plum trees to *P. syringae* increased when they were infected with *Macroposthonia xenoplax*. Bacterial wilt of alfalfa caused by *Clavibacter michiganense* subsp. *insidiosum* increased in the presence of *Ditylenchus dipsaci* and the resistant cultivars to the wilt bacterium became diseased (Hawn, 1963, 1965, 1971). On the other hand, cultivars with nematode resistance remained relatively free from wilt when exposed to both pathogens (Hawn and Hanna, 1966).

To obtain maximum benefit from wilt-resistant cultivars it has been suggested to control the nematodes responsible for breaking the resistance to the bacterial pathogens. Resistance to vascular wilts caused by fungi and bacteria lies in the endodermis, which is normally avoided by fungi. The endodermis contains large quantities of antifungal substances like phenols, naphthols and anthrols which in later stages give rise to benzo-quinones, naphthoquinones and anthroquinones, some of which are highly antifungal and antibacterial (Goodman *et al.*, 1967). Invasion of root-knot nematodes results in pericycle origin of syncytia and galls, and the endo-dermis is not properly differentiated in the affected areas. The absence of endodermis removes the barrier against invasion by vascular wilt bacteria and fungi.

11.2.4 Nematode acts as vector

The nematode may act as a vector by carrying the bacterial inoculum externally. In this type of relationship the role of the nematode is mainly in carrying the pathogen from soil to plant or from the outer plant tissue to meristematic tissue. Pitcher and Crosse (1956, 1958) found that *Rhodo-coccus fascians* (*Corynebacterium fascians*) was unable to reach the meristem of the strawberry plant without the aid of *Aphelenchoides fragariae* or *A. ritzemabosi*. Without the active participation of the invading nematodes the bacteria, however, cannot reach the meristem and the plants remain healthy. Crosse and Pitcher (1952), Pitcher and Crosse (1958) and Pitcher (1963) demonstrated that *A. fragariae* or *A. ritzemabosi* and *R. fascians* are necessary to produce 'cauliflower disease'.

Infection of the rubber-bearing plants *Taraxacum koksaghys* and *Scorzonera tau* by *Tylenchus* and *Aphelenchus* was reported by Kalinenko (1936). His study showed that several bacterial pathogens were carried externally and internally by the nematodes and that they could in turn infect the plants. The bacterial pathogens *Erwinia carotovora* and *Xanthomonas phaseoli* are carried with the nematodes. The progress of these two pathogens in the host following penetration by the nematodes was followed and the extent of tissue disorganization attributable to each of the nematodes was recorded. The impact of the nematodes upon the host is succinctly described as traumatic and mechanically destructive whereas the bacteria were regarded as being enzymatically degradative.

The yellow ear-rot or 'tundu' disease of wheat is another example of this type of obligatory relationship, requiring both the nematode *Anguina tritici* and the bacterium *Clavibacter tritici* (*Corynebacterium michiganense* pv. *tritici*) for expression of the complex syndrome (Gupta and Swarup, 1972; Bird, 1981; Pathak, 1982). Transmission of *Clavibacter michiganense* subsp. *insidiosum* by the stem and bulb nematode, *Ditylenchus dipsaci*, in alfalfa wilt was demonstrated by Hawn (1963, 1965, 1971). The role of nematodes as vectors of bacterial plant pathogens is discussed in detail in Chapter 12 of this book.

11.2.5 Nematode root infection increases foliar bacterial diseases

Root-knot and lesion nematode infections have a systemic effect which could lead to an increase in susceptibility of other plant parts not actually infested by the nematode (Bowman and Bloom, 1966; Faulkner and Skotland, 1965; Faulkner *et al.*, 1970). Nematode–bacterium interactions have also been reported in the case of foliar bacterial diseases. Root infection with *Meloidogyne javanica* greatly increased the severity of gladiolus scab caused by *Pseudomonas marginata* (El-Goorani *et al.*, 1974). Bacterial canker of tomato induced by *Clavibacter* (*Corynebacterium*) *michiganense* was also increased when the roots were infected with *Meloidogyne* spp. (De Moura *et al.*, 1975). These studies show the systemic nature of physiological changes brought about by root-knot nematodes in favouring infection of foliar bacterial plant pathogens.

11.2.6 Nematode infection changes the rhizosphere microflora

Root exudates exert a powerful influence on the size and composition of the bacterial population next to the roots during the entire life of the plant.

They affect the life processes of the plant and also the plant's resistance to soil-borne pathogens. With increasing proximity to plant roots there are increasing numbers of microorganisms, bacteria being the most responsive organisms of the microflora to root exudates (Bhuwaneswari *et al.*, 1965). Injury and rupture of host cells can change the pattern of root exudates and influence the rhizosphere microflora and fauna. This changed pattern could affect the soil-borne plant pathogens. Physiological changes induced within the root tissue by endoparasitic nematodes bring about qualitative changes in root exudates. Root exudates stimulate the dormant stages of the fungi in the rhizosphere. Similarly, stimulation, reproduction and colonization of several soil-borne fungi are governed by root exudates before root penetration.

Bergeson *et al.* (1970) and Golden and Van Gundy (1972) demonstrated that growth-promoting substances for plant-pathogenic fungi are found in root exudates of *Meloidogyne*-infected roots. Root exudates of plants infected with *M. incognita* contain greater amounts of carbohydrates, proteins and amino acids than the root exudates of nematode-free plants (Endo, 1971; Krusberg, 1963; Sitaramaiah, 1989). Golden and Van Gundy (1972) using the ^{14}C tracer technique found that *Meloidogyne*-infected tomato roots began to leak labelled electrolytes five days after infection, and maximum leakage was observed three to five weeks after nematode infection. The host root exudates were shown to favour germination and infection of dormant propagules of vascular wilt pathogens (Bergeson *et al.*, 1970). A similar mechanism of stimulation of bacterial pathogens might result in plants infested with nematodes in rhizosphere soil. However, evidence to support this mechanism of change of rhizosphere bacterial flora in the presence of nematode-infected plants needs experimentation.

11.3 BACTERIAL PATHOGENS AND THEIR EFFECTS ON NEMATODES

Effects of interactions of plant-parasitic nematodes with bacterial and fungal pathogens are either harmful or beneficial to them. Plant-parasitic nematodes facilitate stimulation, penetration, establishment, growth and multiplication of bacterial pathogens by causing favourable changes within the host plant. But most of these interactions are of disadvantage to the nematode itself. *Fusarium oxysporum* decreases invasion and development of *Heterodera schachtii* (Jorgenson, 1970) and *Helminthosporium gramineum* or *F. moniliforme* or *Heterodera avenae* (Gill and Swarup, 1977). Pitcher (1963) suggested that the bacteria modifies host tissue extensively, which does not favour nematode multiplication. Lucas *et al.* (1955) reported that infection of tobacco roots by the wilt bacterium *Pseudomonas solanacearum* caused a decrease of *Meloidogyne incognita* in roots. The content of giant cells degenerated following bacterial invasion, leaving virtually empty cells. This resulted in the death of the nematode. These adverse effects on nematodes are not unexpected as the obligate parasite and the fungal or bacterial pathogen are sharing and competing for the same host substrate. The unfavourable effects are due to the destruction of feeding sites in obligate sedentary endoparasitic nematodes and impaired nutrition due to harmful byproducts produced during bacterial colonization.

11.3.1 Production of toxins in bacteria–nematode association

Limited information is available on this aspect of interaction. The association between *Anguina funesta* (*Anguina agrostis*) and *Clavibacter* sp. (*Corynebacterium rathayi*) infesting *Lolium rigidum* is an example of this type. Galls induced by *A. funesta* in the annual ryegrass become toxic to animals when colonized by the bacterium (Lanigan *et al.*, 1976; Stynes *et al.*, 1979). Bird *et al.* (1980) concluded that toxin production is associated with an interaction between nematode-infected plant cells and the bacterium.

11.3.2 Nematode inhibits disease development

Suppression of carnation wilt caused by *Pseudomonas caryophylli* by *Ditylenchus* spp. was reported by Stewart and Schindler (1956). However, Lucas and Krusberg (1956) found that *Tylenchorhynchus claytoni* did not increase the severity of bacterial wilt caused by *Xanthomonas solanacearum* in wilt-resistant tobacco plants grown in soil infested with the bacterium and the nematode. They noticed reduced symptoms. The authors suggested that weakening of the tobacco roots by the nematode provides a less congenial environment for the bacterium. The bacterium, however, requires vigorously actively growing plants. In this particular case the nematode was unable to penetrate deep into the xylem tissues of tobacco root. The authors concluded that the bacterial infections are less likely to occur with this ectoparasitic nematode than with *Meloidogyne* spp., which can reach the xylem tissue for feeding. The possibility cannot be ruled out in free-living nematodes as they may reduce soil-borne bacterial diseases by lowering the primary bacterial inoculum through feeding. Further work on this aspect is essential.

11.4 PROSPECTS

Disease complexes involving nematodes and other microorganisms are common in nature. The exact role of nematodes in disease complexes involving bacterial pathogens is still not completely elucidated. A single-organism–plant relationship does not exist in nature. In the complex biotic environment of the soil, the pathogens are always influenced by associated microorganisms. There is an urgent need to integrate plant pathological–nematological work with interrelationships between pathogens in causing plant diseases. The complex biological interactions must receive more critical appraisal through carefully controlled experimental conditions.

Physiological changes in disease complexes are a subject for future study for a better understanding of the intricacies involved in these complexes. A team research approach with greater participation of nematologists, plant pathologists and breeders is essential for devising complementary methods of control of disease complexes, including the development of crop varieties resistant to several pathogens at a time.

REFERENCES

Bergeson, G.B. (1972) Concepts of nematode–fungus associations in plant disease complexes: A review. *Experimental Parasitology*, **32**, 301–14.

Bergeson, G.B., Van Gundy, S.D. and Thomason, I.J. (1970) Effect of *Meloidogyne javanica* on rhizosphere microflora and Fusarium wilt of tomato. *Phytopathology*, **60**, 1245–9.

Bhuwaneswari, K., Sadasivan, T.S. and Sulochana (1965) Rhizosphere microflora in relation to soil borne diseases of plants, in *Advances in Agricultural Sciences and their Applications. Madras Agricultural Journal*, pp. 538–48.

Bird, A.F. (1974) Plant response to root-knot nematode. *Annual Review of Phytopathology*, **12**, 69–85.

Bird, A.F. (1981) The *Anguina–Corynebacterium* association, in *Plant Parasitic Nematodes*, Vol. III (eds B.M. Zuckerman and R.A. Rohde), Academic Press, New York, pp. 303–23.

Bird, A.F., Stynes, B.A. and Thomson, W.W. (1980) A comparison of nematode and bacteria colonized gall induced by *Anguina agrostis* in *Lolium rigidum*. *Phytopathology*, **70**, 1104–9.

Bookbinder, M.G., Bloom, J.R. and Lukezic, F.L. (1982) Interactions among selected endoparasitic nematodes and three Pseudomonads on alfalfa. *Journal of Nematology*, **14**, 105–9.

Bowman, P. and Bloom, J.R. (1966) Breaking the resistance of tomato varieties to Fusarium wilt by *Meloidogyne incognita*. *Phytopathology*, **56**, 871.

Brzeski, M.W. (1970) The interrelationships of nematodes and other plant pathogens in plant diseases. *Journal of Parasitology*, Section II, Part 3, **56**, 509–13.

Buddenhagen, I.W. (1965) The relation of plant pathogenic bacteria to the soil, in *Ecology of Soil-Borne Plant Pathogens* (eds K.F. Baker and W.C. Snyder), University of California Press, Berkeley, pp. 269–82.

Buddenhagen, I.W. and Kelman, A. (1964) Biological and physiological aspects of bacterial wilt caused by *Pseudomonas solanacearum*. *Annual Review of Phytopathology*, **2**, 203–30.

Caubel, G. and Samson, R. (1984) Effect of the stem nematode *Ditylenchus dipsaci* on the development of 'Cafe au lait' bacteriosis in garlic (*Allium sativum*) caused by *Pseudomonas fluorescens*. *Agronomie*, **4**, 311–13.

Crosse, J.E. (1968) Plant pathogenic bacteria in soil, in *The Ecology of Soil Bacteria* (eds T.R.C. Gray and D. Parkinson), Liverpool University Press, Liverpool, pp. 522–72.

Crosse, J.E. and Pitcher, R.S. (1952) Studies in the relationship of eelworms and bacteria to certain plant diseases. 1. The etiology of strawberry cauliflower disease. *Annals of Applied Biology*, **39**, 475–84.

Davide, R.G. (1972) Influence of root-knot nematodes on the severity of bacterial wilt and Fusarium wilt of tomato. *Philippine Phytopathology*, **8**, 78–81.

De Moura, R.M., Echandi, E. and Powell, N.T. (1975) Interaction of *Corynebacterium michiganense* and *Meloidogyne incognita* on tomato. *Phytopathology*, **65**, 1332–5.

Devay, J.E., Lownsberry, B.F., English, W.H. and Lambright, H. (1967) Activity of soil fumigants in relation to increased growth response and control of decline and bacterial canker in trees of *Prunus persica*. *Phytopathology*, 57, 809.

Dimond, A.E. (1972) The origin of symptoms of vascular wilt diseases, in *Phytotoxins in Plant Diseases* (eds R.K.S. Wood, A. Ballio and A. Graniti), Academic Press, London, pp. 289–309.

Duarte, M. De L.R. and De Albuquerque, F.C. (1971) [Tomato diseases in the Amazon Region] in Spanish. *Meloidogyne* attacks favour infection by *Pseudomonas* and *Fusarium*. Instituto de Pesquisase. *Experimentacao Agropecarias do Norte, Fitotecnia*, 2, 9–34.

El-Goorani, M.A., Abo-El-Dahab, M.K. and Mehiar, F.F. (1974) Interaction between root-knot nematode and *Pseudomonas marginata* on gladiolus corms. *Phytopathology*, 64, 271–2.

Endo, B.Y. (1971) Nematode induced syncytia (giant cells). Host–parasite relationship of Heteroderidae, in *Plant Parasitic Nematodes*, Vol. II (eds B.M. Zuckerman, W.F. Mai and R.A. Rohde), Academic Press, New York, pp. 91–117.

Faulkner, L.R. and Skotland, C.B. (1965) Interactions of *Verticillium dahliae* and *Pratylenchus minyus* in Verticillium wilt of peppermint. *Phytopathology*, 55, 583–6.

Faulkner, L.R., Bolander, W.J. and Skotland, C.B. (1970) Interaction of *Verticillium dahliae* and *Pratylenchus minyus* in Verticillium wilt of peppermint: Influence of the nematode as determined by a double root technique. *Phytopathology*, 60, 100–3.

Feldmesser, J. and Goth, R.W. (1970) Association of a root-knot with bacterial wilt of potato. *Phytopathology*, 60, 1014.

Fukudome, N. and Sakasegawa, Y. (1972) Interaction between root-knot nematode and other diseases. II. Influence of the root-knot nematode (*Meloidogyne incognita*) on the occurrence of Granville wilt on tobacco. *Proceedings of the Association for Plant Protection of Kyushu*, 18, 100–2.

Gallegly, M.E., Jr and Walker, J.C. (1949) Plant nutrition in relation to disease development. V. Bacterial wilt of tomato. *American Journal of Botany*, 36, 613–23.

Gill, J.S. and Swarup, G. (1977) Effect of interaction between *Heterodera avenae* Woll., 1924, *Fusarium moniliforme* and *Helminthosporium gramineum* on barley plants and nematode reproduction. *Indian Journal of Nematology*, 7, 42–5.

Golden, J.K. and Van Gundy, S.D. (1972) Influence of *Meloidogyne incognita* on root rot development by *Rhizoctonia solani* and *Thielaviopsis basicola* in tomato. *Journal of Nematology*, 4, 225.

Goodman, R.N., Kiraly, Z. and Zaitlin, M. (1967) *The Biochemistry and Physiology of Infectious Plant Disease*, Van Nostrand, Princeton, NJ.

Gupta, P. and Swarup, G. (1972) Ear-cockle and yellow ear-rot disease of wheat. II. Nematode bacterial association. *Nematologica*, 18, 320–4.

Harrison, B.D. (1964) The transmission of plant viruses in soil, in *Plant Virology* (eds M.K. Corbett and H.D. Sisler), University of Florida Press, Gainesville, pp. 118–44.

Hawn, E.J. (1963) Transmission of bacterial wilt of alfalfa by *Ditylenchus dipsaci* (Kuhn). *Nematologica*, 8, 65–8.

Hawn, E.J. (1965) Influence of stem nematode infestation on the development of bacterial wilt in irrigated alfalfa. *Nematologica*, 11, 39.

Hawn, E.J. (1971) Mode of transmission of *Corynebacterium insidiosum* by *Ditylenchus dipsaci*. *Journal of Nematology*, 3, 420–1.

Hawn, E.J. and Hanna, M.R. (1966) Influence of stem nematode infestation on bacterial wilt reaction and forage yield of alfalfa varieties. *Canadian Journal of Plant Science*, 47, 203–8.

Hunger, F.W.T. (1901) Een bacterie-ziekte den tomaat. *Mededlingen Plantentuin, Batavia*, **48**, 4–57.

Hunt, D.T., Griffin, G.D., Murray, J.J., Pedersen, M.W. and Paeaden, R.N. (1971) The effects of root-knot nematodes on bacterial wilt in tobacco. *Phytopathology*, **61**, 256–9.

Husain, A. and Kelman, A. (1958) Relation of slime production to mechanism of wilting and pathogenicity of *Pseudomonas solanacearum*. *Phytopathology*, **48**, 155–65.

Jatala, J. and Martin, C. (1977a) Interactions of *Meloidogyne incognita acrita* and *Pseudomonas solanacearum* on *Solanum chacoense* and *S. sparsipilum*. *Proceedings of the American Phytopathological Society*, **4**, 178.

Jatala, J. and Martin, C. (1977b) Interaction of *Meloidogyne incognita acrita* and *Pseudomonas solanacearum* on field grown potatoes. *Proceedings of the American Phytopathological Society*, **4**, 177–8.

Jenkins, S.F., Jr (1972) Interaction of *Pseudomonas solanacearum* and *Meloidogyne incognita* on bacterial wilt resistant and susceptible cultivars of tomato. *Phytopathology*, **62**, 767.

Johnson, H.A. (1966) Studies on the Granville wilt–root-knot interaction on flue-cured tobacco. MS thesis, North Carolina State University, Raleigh.

Johnson, H.A. and Powell, N.T. (1969) Influence of root-knot nematodes on bacterial wilt development in flue-cured tobacco. *Phytopathology*, **59**, 486–91.

Jorgenson, E.C. (1970) Antagonistic interaction of *Heterodera schachtii* Schmidt and *Fusarium oxysporum* (Wool.) on sugarbeets. *Journal of Nematology*, **2**, 393–8.

Kalinenko, V.I. (1936) The inoculation of phytopathogenic microbes into rubber-bearing plants by nematodes. *Phytopathologische Zeitschrift*, **9**, 407–16.

Kelman, A. (1953) The bacterial wilt caused by *Pseudomonas solanacearum*. *North Carolina Agricultural Experimental Station, Technical Bulletin*, **99**, 194 pp.

Kessler, K.J., Jr (1966) Xylem sap as a growth medium for four tree wilt fungi. *Phytopathology*, **56**, 1165–9.

Khan, M.W. (1984) Ecological niche of plant nematodes in soil biotic community, in *Progress in Microbial Ecology* (eds K.G. Mukerjee, V.P. Agnihotri and R.P. Singh), Print House (India), Lucknow, pp. 165–77.

Krusberg, L.R. (1963) Host response to nematode infection. *Annual Review of Phytopathology*, **1**, 219–40.

Lanigan, G., Payne, A.L. and Frahn, J.L. (1976) Origin of toxicity in parasitized annual rye grass (*Lolium rigidum*). *Australian Veterinary Journal*, **52**, 244–6.

Lele, V.C., Durgapal, J.C., Agarwal, D.K. and Sethi, C.L. (1978) Crown and root gall of grape (*Vitis vinifera* L.) in Andhra Pradesh. *Current Science*, **47**, 280.

Libman, G., Leach, J.G. and Adams, R.E. (1964) Role of certain plant parasitic nematodes in infection of tomatoes by *Pseudomonas solanacearum*. *Phytopathology*, **54**, 151–3.

Lucas, G.B. and Krusberg, L.R. (1956) The relationship of the stunt nematode to Granville wilt resistance in tobacco. *Plant Disease Reporter*, **40**, 150–2.

Lucas, G.B., Sasser, J.N. and Kelman, A. (1955) The relationship of root-knot nematodes to Granville wilt resistance in tobacco. *Phytopathology*, **45**, 537–40.

Mace, M.E. and Solit, E. (1966) Interactions of 3-indole acetic acid and 3-hydroxy-tyramine in Fusarium wilt of banana. *Phytopathology*, **56**, 245–7.

Mayol, P.S. and Bergeson, G.B. (1969) The role of secondary invaders in premature breakdown of plant roots infected with *Meloidogyne incognita*. *Journal of Nematology*, **1**, 17.

Mountain, W.B. (1965) Pathogenesis by soil nematodes, in *Ecology of Soil-Borne Plant Pathogens* (eds K.F. Baker and W.C. Snyder), University of California Press, Berkeley, pp. 285–301.

Munnecke, D.E., Chandler, P.A. and Starr, M.P. (1963) Hairy root (*Agrobacterium rhizogenes*) of field roses. *Phytopathology*, 53, 788–99.

Napiere, C.M. (1980) Varying inoculum levels of bacteria, nematodes and the severity of tomato bacterial wilt. *Annals of Tropical Research*, 2, 129–34.

Napiere, C.M. and Quinio, A.J. (1980) Influence of root-knot nematode on bacterial wilt severity in tomato. *Annals of Tropical Research*, 2, 29–39.

Nigh, E.L. (1966) Incidence of crown gall infection in peach as affected by the Javanese root-knot nematode. *Phytopathology*, 56, 150.

Nirula, K.K. and Paharia, K.D. (1970) Role of root-knot nematodes in spread of brown rot in potatoes. *Indian Phytopathology*, 23, 158–9.

Pathak, K.N. (1982) Studies on the relationship between ear-cockle nematode *Anguina tritici* and *Corynebacterium tritici* in the tundu disease complex of wheat. PhD thesis, Indian Agricultural Research Institute, New Delhi, India.

Pegg, G.F. and Sequeira, L. (1968) Stimulation of aromatic biosynthesis in tobacco plants infected by *Pseudomonas solanacearum*. *Phytopathology*, 58, 476–83.

Pitcher, R.S. (1961) Nematodes as primary incitants of root rot disease, in *Recent Advances in Botany*, Vol. 1, Toronto University Press, Toronto, pp. 477–81.

Pitcher, R.S. (1963) The role of plant-parasitic nematodes in bacterial diseases. *Phytopathology*, 53, 35–9.

Pitcher, R.S. (1965) Interrelationships of nematodes and other pathogens of plants. *Helminthological Abstracts*, 34, 1–17.

Pitcher, R.S. (1978) Interactions of nematodes with other pathogens, in *Plant Nematology* 3rd edn (ed. J.F. Southey), Her Majesty's Stationery Office, London, pp. 63–77.

Pitcher, R.S. and J.E. Crosse (1956) An etiological association between nematodes (*Aphelenchoides* spp.) and *Corynebacterium facians* in strawberry disease. *Proceedings of the International Congress Zoology*, Copenhagen, August 1953, p. 376.

Pitcher, R.S. and Crosse, J.E. (1958) Studies in the relationship of eelworms and bacteria to certain plant diseases II. Further analysis of the strawberry cauliflower disease complex. *Nematologica*, 3, 244–56.

Powell, N.T. (1963) The role of plant parasitic nematodes in fungus diseases. *Phytopathology*, 53, 28–35.

Powell, N.T. (1971a) Interactions between nematodes and fungi in disease complexes. *Annual Review of Phytopathology*, 9, 253–74.

Powell, N.T. (1971b) Interactions of plant parasitic nematodes with other disease causing agents, in *Plant Parasitic Nematodes*, Vol. II (eds B.M. Zuckerman, W.F. Mai and R.A. Rohde), Academic Press, New York, pp. 119–36.

Powell, N.T. (1979) Internal synergisms among organisms inducing disease, in *Plant Disease*, Vol. IV (eds J.G. Horsfall and E.B. Cowling), Academic Press, New York, pp. 113–33.

Powell, N.T. and Nusbaum, C.J. (1960) The black shank-root-knot complex in flue-cured tobacco. *Phytopathology*, 50, 899–906.

Raski, D.J. and Hewitt, W.B. (1963) Plant parasitic nematodes as vectors of plant viruses. *Phytopathology*, 53, 39–47.

Reddy, P.P., Singh, D.B. and Ramkishun (1979) Effect of root-knot nematodes on the susceptibility of Pusa Purple Cluster brinjal to bacterial wilt. *Current Science*, 48, 915–16.

Schuster, M.L. (1959) Relation of root-knot nematodes and irrigation water to the incidence and dissemination of bacterial wilt of bean. *Plant Disease Reporter*, 43, 27–32.

Sellam, M.A., Rushdi, M.H. and Gendi, D.M. (1982) Interrelationship of *Meloidogyne incognita* Chitwood and *Pseudomonas solanacearum* on tomato. *Egyptian Journal of Phytopathology*, 12, 35–42.

Sequeira, L. (1964) Inhibition of indole acetic acid oxidase in tobacco plants infected by *Pseudomonas solanacearum*. *Phytopathology*, **54**, 1078–83.

Sequeira, L. (1965) Origin of indole acetic acid in tobacco plants infected by *Pseudomonas solanacearum*. *Phytopathology*, **55**, 1232–6.

Sequeira, L. and Kelman, A. (1962) The accumulation of growth substances in plants infected by *Pseudomonas solanacearum*. *Phytopathology*, **52**, 439–48.

Sequeira, L. and Williams, P.H. (1964) Synthesis of indole acetic acid by *Pseudomonas solanacearum*. *Phytopathology*, **54**, 1240–6.

Sitaramaiah, K. (1989) Host–parasite relationships of plant parasitic nematodes, in *Perspectives in Plant Pathology* (eds V.P. Agnihotri, N. Singh, H.S. Chaube, U.S. Singh and T.S. Dwivedi), Today and Tomorrow's Printers and Publishers, New Delhi, pp. 155–65.

Sitaramaiah, K. and Singh, R.S. (1990) Interactions of nematodes with fungal and bacterial plant pathogens, in *Progress in Plant Nematology* (eds S.K. Saxena, M.W. Khan, A. Rashid and R.M. Khan), CBS Publishers and Distributors, New Delhi, pp. 195–208.

Sitaramaiah, K. and Sinha, S.K. (1984a) Interaction between *Meloidogyne javanica* and *Pseudomonas solanacearum* on brinjal. *Indian Journal of Nematology*, **14**, 1–5.

Sitaramaiah, K. and Sinha, S.K. (1984b) Histological aspects of *Pseudomonas* and root-knot nematode wilt complex in brinjal. *Indian Journal of Nematology*, **14**, 175–8.

Stewart, R.N. and Schindler, A.F. (1956) The effect of some ectoparasitic and endoparasitic nematodes on the expression of bacterial wilt in carnations. *Phytopathology*, **46**, 219–22.

Stynes, B.A., Petterson, D.S., Lloyd, J., Payne, A.L. and Lanigan, G.W. (1979) The production of toxin in annual rye grass, *Lolium rigidum* infected with a nematode, *Anguina* sp. and *Corynebacterium rathayi*. *Australian Journal of Agriculture Research*, **30**, 201–9.

Taylor, C.E. (1972) Nematode transmission of plant viruses. *Pest Articles and News Summaries*, **18**, 269–82.

Temiz, K. (1968) *Pseudomonas solanacearum* un bazi domates cesitlerine enfeksiyonunda bitki paraziti nematodlarin rolu uzerine arastirmalar. *Yalova-Bache. Kult. Arast. Egit. markezi Derg*, **1**, 17–28.

Thakur, S.C. (1985) Interaction of nematode, *Meloidogyne incognita* and bacterium *Pseudomonas solanacearum* on tomato plants. MSc thesis, Rajendra Agricultural University, Pusa, India.

Walker, J.C. (1972) *Plant Pathology*, McGraw-Hill, New York.

Wallace, H.R. (1963) *The Biology of Plant Parasitic Nematodes*, Edward Arnold, London.

Weischer, B. (1968) Wechselwirkungen Zwischen Nematoden und anderen Schaderregern an Nutzpflanzen. *International Nematology Symposium, Antibes, 1965*, 91–107.

Zutra, D. and Orion, D. (1982) Crown gall bacteria (*Agrobacterium radiobacter* var. *tumefaciens* on cotton in Israel. *Plant Disease*, **66**, 1200–1.

12

Nematodes as vectors of bacterial and fungal plant pathogens

M. Wajid Khan and K.N. Pathak

It is presumptive that the first association between nematodes and bacteria and fungi was that of nematodes feeding on bacteria or fungi. The vector relationship existed as soon as nematodes carried and defaecated viable bacteria or fungal spores into the environment (Poinar and Hansen, 1986). Microbitrophic, free-living nematodes feeding on bacterial cells which grow on damaged plant tissue are reported to be vectors of such bacteria (Armstrong *et al.*, 1976). The literature shows several instances where plant-parasitic nematodes assist plant-pathogenic bacteria as carriers (Table 12.1). Some seed gall and foliar nematodes act as specific vectors of bacterial plant pathogens and an obligate aetiological relationship is established between the two, resulting in manifestation of a new disease syndrome in host plants (Taylor, 1990). The 'tundu' or yellow ear-rot disease of wheat, annual ryegrass toxicity and cauliflower disease of strawberry are such examples. Some aspects of the subject of this chapter have been reviewed by Pitcher (1963, 1965), Powell (1971), Bird (1981), Khan (1984), Poinar and Hansen (1986), Swarup and Gokte (1986), Paruthi and Bhatti (1990) and Taylor (1990).

The studies accomplished on this subject have focused primarily on those bacteria which establish an obligatory relationship with foliar nematodes, developing a new disease syndrome. Other relationships have received little study and provide inadequate information. A lack of sufficient information on all aspects of this chapter subject restrict us to consider mainly vector relationships between *Clavibacter–Anguina* in yellow ear-rot or 'tundu' disease of wheat and annual ryegrass toxicity (ARGT), *Rhodococcus–Aphelenchoides* in cauliflower disease of strawberry, *Clavibacter–Ditylenchus* in alfalfa wilt and *Dilophospora alopecuri–Anguina* in twist disease of cereals. Additionally some other little-known associations, in which vector relationships are not established, are mentioned.

Table 12.1 Nematodes implicated as vectors of bacterial or fungal plant pathogens

Nematode	Bacterial/fungal plant pathogen	Host	Disease	Reference
	Bacterial pathogens			
Anguina tritici	Clavibacter tritici	Wheat	Yellow ear-rot or 'tundu'	Carne (1926); Cheo (1946); Gupta and Swarup (1972); Pathak and Swarup (1984)
Anguina funesta	Clavibacter sp.	Annual ryegrass	Annual ryegrass toxicity	Bird and Stynes (1977); Stynes and Wise (1980); Schneider (1981); Riley and McKay (1990)
Aphelenchoides ritzemabosi or A. fragariae	Rhodococcus fascians	Strawberry	Cauliflower	Crosse and Pitcher (1952); Pitcher and Crosse (1958)
Ditylenchus dipsaci	Clavibacter michiganense subsp. insidiosum	Alfalfa	Wilt	Hawn (1963); Hawn (1971)
Ditylenchus dipsaci	Pseudomonas fluorescens	Garlic	'Café au lait' bacteriosis	Caubel and Samson (1984)
Ditylenchus dipsaci	Bacterium rhaponticum	Rhubarb	Crown rot	Metcalfe (1940)
Globodera pallida	Pseudomonas solanacearum	Potato	Wilt	Jensen (1978)

Free-living microbitrophic nematodes	Agrobacterium tumefaciens			Wasilewska and Webster (1975); Armstrong et al. (1976); Poinar and Hansen (1986)
	Erwinia amylovora			
	E. carotovora			
	Pseudomonas phaseolicola			
	P. syringae			
	P. tolasii	Mushroom	Blackspot	Haseman and Ezell (1934)
	Fungal plant pathogens			
Anguina tritici	Dilophospora alopecuri	Cereals	Twist	Atanasoff (1925); Munjal and Kaul (1961)
A. tritici	Ustilago tritici*	Wheat	Smut	Bedi et al. (1959)
A. tritici	Tilletia foetida*	Wheat	Bunt	Mathur and Misra (1961)
A. tritici	Neovossia indica*	Wheat	Karnal bunt	Paruthi and Bhatti (1980)
Pristionchus lheritieri	Fusarium oxysporum lycopersici*		Wilt	Jensen (1967)
	Verticillium dahliae*		Wilt	Jensen and Siemer (1971)

*Vector relationships of these fungi with nematodes are not established. Smut fungi were only recorded along with A. tritici. The wilt fungi were carried internally in the alimentary canal of the nematode.

12.1 CLAVIBACTER–ANGUINA

Two species of *Anguina, A. tritici* and *A. funesta,* are of considerable interest as vectors of *Clavibacter* spp. *A. tritici* vectors *C. tritici* and their association gives rise to a disease of wheat known as yellow ear-rot or 'tundu'. *A. funesta* acts as vector of *Clavibacter* sp. and they jointly cause annual ryegrass toxicity (Riley and McKay, 1990).

12.1.1 Yellow ear-rot or 'tundu' disease of wheat

The interaction of *Anguina tritici* and *Clavibacter tritici* causes a disease in wheat commonly known as 'tundu' in India. It is also known as yellow ear-rot, spike blight or yellow slime. This disease complex was first reported by Hutchinson (1917) from India as a bacterial disease and subsequently from Egypt (Fahmy and Mikhail, 1925; Sabet, 1953, 1954), Western Australia (Carne, 1926) and China (Cheo, 1946). Inoculations of plants with bacteria alone by Fahmy and Mikhail (1925) were not successful. Carne (1926) found that bacteria were carried into plants by nematodes. Chaudhuri (1935) at Lahore (now in Pakistan) claimed that bacteria alone could cause the disease and the presence of the nematode, *A. tritici,* was not necessary. This hypothesis was disproved later. The findings of Cheo (1946), Vasudeva and Hingorani (1952), Sabet (1953), Swarup and Singh (1962), Gupta and Swarup (1968), Midha (1969), Pathak (1982) and Ray (1987) strongly claimed that the presence of *A. tritici* was essential for the expression of the disease syndrome in wheat.

Ear-cockle nematode, Anguina tritici

Anguina tritici was the first nematode to be observed as a plant parasite (Needham, 1743). The name *Vibrio tritici* was given to this nematode by Steinbuch (1799). Chitwood (1935) revived the genus *Anguina* enacted by Scopoli (1777) and named the ear-cockle nematode of wheat as *A. tritici* (Steinbuch, 1799) Chitwood, 1935. Dovaine (1857) elucidated the life history of this nematode. For morphological details, the Commonwealth Institute of Helminthology (CIH) descriptions of plant-parasitic nematodes Set 1, No. 13, can be referred to. Paruthi and Bhatti (1990) have described the morphology, biology, host range and economic importance of this nematode in detail. The nematode is reported from most of the Western European countries, Russia, Israel, Syria, Pakistan, India, Egypt, China, Australia, New Zealand, Australia, the USA, Brazil, Iraq, Ethiopia, Romania and Yugoslavia. The disease is widely encountered in the wheat-growing belt of northern India. In addition to its main host – wheat – spelt, rye, emmer, barley and triticale are also reported as hosts of this nematode (Paruthi and Bhatti, 1990).

Infected wheat plants show basal enlargement of the stem, and curling, crinkling and twisting of leaf blades due to ectoparasitic feeding of the nematodes. Infected plants mature more slowly than non-infected ones. Culms become shorter and seed-heads smaller. All the flower primordia in a single head may be infected. A single plant may have healthy and diseased heads. Kernels are replaced by galls or cockles. Developing galls are greener and smoother than healthy kernels. Seed galls are smaller, shorter and darker in colour compared to healthy seeds. Mature galls are hard, with walls composed of collenchyma. The central cavity of the gall remains filled with a mass of second-stage juveniles. Frequently there is single cavity but galls may be compound, showing two or three cavities.

If the galls are present in soil or sown with seed, second-stage juveniles emerge in moist soil from the softened walls of the galls, the time varying considerably with soil temperature, moisture and thickness of the gall wall (Leukel, 1924, 1957; Goodey, 1932). As early as 1926 Carne from Australia had observed that the germination of wheat seed preceded release of juveniles from galls. Pathak and Swarup (1983) recorded earliest release of juveniles from galls, on the fourth day, while normally sprouting of wheat seed started within three days. Swarup and Gokte (1986) have reviewed the earlier work and have described that second-stage infective juveniles migrate to the growing point of the radicle through the ruptured coleorhiza (rupture is created through emergence of the radicle from the embryo of the sprouting wheat seed) and from there the juveniles move to the growing point of the shoot (Gupta, 1966; Midha, 1969; Pathak, 1982). Juveniles feed ectoparasitically on the growing point. With the initiation of flower primordia, they invade the floral tissues and further development of staminate as well as ovarial tissues is checked. The juveniles appear to prefer staminate tissue only. When ovary is also aborted, grain formation may not take place. After invasion of floral tissues, there is a succession of quick moults before they become adults. Amphimictic reproduction occurs and after egg laying the adults soon die. Eggs hatch into second-stage juveniles. Morphological changes associated with transition of freshly hatched juveniles into dauer juveniles have been described by Bird (1981). Paruthi and Bhatti (1990) have also given an account of anhydrobiosis of the nematode.

The 'tundu' bacterium, Clavibacter tritici

The 'tundu' bacterium was known as *Corynebacterium tritici* until 1977, when some changes were introduced in the nomenclature of plant-pathogenic coryneform bacteria. There is now general agreement amongst bacterial taxonomists that plant-pathogenic coryneform bacteria do not belong in the genus *Corynebacterium* (Davis, 1986). There has been, however, less agreement about their new taxonomic position. Some taxonomists preferred to retain them in the genus *Corynebacterium* until the generic classification of the bacteria could be more clearly resolved (Dye and Kemp, 1977; Carlson and Vidavar, 1982). On the basis of numerical phenetic studies, Dye and Kemp (1977) suggested that 11 of the 13 plant-pathogenic coryneform bacteria should be placed in one of the two species, *Corynebacterium michiganense* or *C. flaccumfaciens*. They further suggested that the pathogens within these two species be designated as pathovars and not subspecies because of lack of differential characters. The 'tundu' bacterium was, therefore named as *C. michiganense* pv. *tritici*. Carlson and Vidavar (1982), after studying profiles of cellular proteins in polyacrylamide gels, bacterial production of strains, as well as other characters of plant-pathogenic coryneform bacteria, suggested that the bacteria require classification as species and subspecies but not as pathovars. Formal proposals largely based on chemotaxonomic characters have been published that classify plant-pathogenic coryneform bacteria into the genera *Rhodococcus*, *Arthrobacter*, *Curtobacterium* and *Clavibacter* (Goodfellow, 1984; Collins *et al.*, 1981; Collins and Jones, 1983; Davis *et al.*, 1984). Therefore, *Clavibacter tritici* (Carlson and Vidaver) Davis *et al.* 1984 is the current name of 'tundu' bacterium (Riley and McKay, 1990).

Association between the ear-cockle nematode and 'tundu' bacterium

Association of *Anguina tritici* and *Clavibacter tritici* in wheat results in the development of 'tundu' syndrome under favourable temperature and moisture conditions. The disease syndrome has been characterized by the production of bright yellow slime or gum on the leaf surface of young plants and also on the abortive ears and leaves while they remain still in the boot leaf. Infected ears either do not emerge or emerging spikes look narrow and short. Bacterial mass replaces the grains partly or completely. Stalks bearing infected ears show various degrees of distortions (Hutchinson, 1917; Carne, 1926; Vasudeva and Hingorani, 1952; Sabet, 1954; Gupta, 1966; Gupta and Swarup, 1968). In addition to juvenile and bacterial concentrations, the development of the disease appears to be dependent on environmental conditions like temperature, humidity, age of the seedlings and the source of the galls. Midha (1969), under experimental conditions, observed that the minimum juvenile population to induce the disease was 10^4 juveniles. He further observed that a mixture of 10^{4-5} juveniles and 0.4 OD bacterial suspension resulted in production of severe disease syndrome. Relative humidity of 100% for 72 hours and an average temperature of 25°C are optimal for the fullest expression of 'tundu' symptoms in 25-day-old wheat seedlings (Swarup and Gokte, 1986). Ray (1987) observed that the role of temperature was more important than relative humidity for disease severity.

The exact mode of vectoring of the bacterium by the nematode has been a controversial issue. Evidence from the work of Cheo (1946), Vasudeva and Hingorani (1952), Sabet (1953) and Gupta and Swarup (1968), though, indicated that the nematode acts as a vector and carries the bacterium on to the plants, and that the presence of *Anguina tritici* is essential for the expression of 'tundu' and *Clavibacter tritici* is unable to cause the disease in the absence of the nematode, but the precise mechanism of transmission was not fully established for a considerably long time. Sabet (1954) claimed the presence of the bacterium as a contaminant in soil and its transport by the nematode to the growing points of wheat seedlings. Cheo (1946), however, suggested that the bacterium is carried within the nematode galls and showed that bacteria alone either from pure culture or from scrappings of diseased plants could not cause yellow ear-rot. Hypodermal injections of the bacterial suspension into young seedlings were also not successful. The disease developed only when galls contaminated with bacteria were used as inocula. Bacteria were claimed to be carried mainly within the nematode galls. Later work (Gupta, 1966; Gupta and Swarup, 1972; Pathak and Swarup, 1984; Gokte and Swarup, 1988) supported the view and showed that the bacterium was originally associated with the nematode juveniles inside the galls.

Gupta (1966) demonstrated that bacteria were present on the body surface of juveniles. Development of 'tundu' did not occur when surface-sterilized juveniles of *Anguina tritici* were used as inocula by Gupta and Swarup (1972). Wheat seedlings infected with surface-sterilized juveniles of *A. tritici* also did not show 'tundu' symptoms when the bacterium was inoculated at the growing point. The juveniles obtained from dry galls yielded the 'tundu' bacterium. Ray (1987) failed to induce either ear-cockle or 'tundu' when the floral tissues of wheat were inoculated with *A. tritici* juveniles contaminated with *Clavibacter tritici*. The transport of bacterial cells picked up by juveniles from soil to the growing point of the seedlings as claimed by Sabet (1954) was not substantiated. Recent studies of Pathak and Swarup (1984), Ray (1987) and Gokte and Swarup (1988) also support this view. Pathak and Swarup (1984) found 40–55% incidence of the bacterium in the wheat seed galls of *A. tritici* obtained from different locations in India. They further determined through soil sterilization that field soil was not a source of the bacterium and concluded that the nematode and the bacterium have an obligate relationship for cohabitation in the gall. Therefore, the juveniles carrying bacterial cells are released from galls produced from ears of wheat infested with ear-cockle nematode. Gokte and Swarup (1988) in a direct observation found a number of bacterial cells sticking to the body surface of juveniles. The distribution and concentration of bacterial cells on the body surface showed no definite pattern. Some juveniles were heavily loaded with bacterial cells, but on most the concentration was not high and the cells were scattered throughout the body length. Juveniles plated on an artificial medium (yeast, glucose, chalk, agar) yielded bright-yellow glistering colonies which belonged to *C. tritici*, *Bacillus subtilis*, *B. pumilus*, *B. cereus* and two isolates of *Pseudomonas* sp. designated 1 and 2 by the authors. They further demonstrated conclusively that bacterial spores were not present inside the nematode body. They examined the role of bacterial flora associated with the juveniles, which suggested that some of these bacteria could be larvicidal and inactivate the juveniles of *A. tritici*.

A mechanism for adhesion of coryneform plant-pathogenic bacteria to nematode cuticle has now been determined (Riley and McKay, 1990). Some details are presented later under annual ryegrass toxicity. This mechanism may be true for *Clavibacter tritici* and *Anguina tritici* association (Riley and McKay, 1990).

The nematode interferes considerably with protein, nitrogen and sugar metabolism of the host. The quantitative and qualitative changes in proteins during the course of development and advancement of disease is considered as reflecting efforts on the part of the infected host to contain the nematode. Apparently, all these biochemical changes are activated and triggered by *Anguina tritici* infecting wheat. Pathak *et al.* (1983) observed a considerable increase in protein as well as acidic amino acids during the 'tundu' phase of the disease. Accumulation of protein, due to its high water solubility, may lead to an increase of osmotically bound water, which in turn would permit conservation during the manifestation of disease development (Epstein and Cohn, 1971). Accumulation of amino acids in nematode-infected wheat tissue may serve as readily available storage compounds in moisture-sensitive infected tissue which may be utilized by bacteria during the 'tundu' phase. Feeding of the juveniles perhaps plays an obligatory role. Since transfer of the bacterium at the growing point of the seedlings or inoculation of floral tissue of wheat with *A. tritici* juveniles contaminated with the bacteria has not been successful for development of 'tundu', possibly some changes within the host are essential for the bacterium to show its pathogenicity. Ray (1987) suggested that the infection of wheat plants by nematodes at the seedling stage is necessary for 'tundu' or even ear-cockle development.

12.1.2 Annual ryegrass toxicity (ARGT)

Neurotoxicity in sheep caused by ingestion of grass seed galls was first observed during 1943–44 in Oregon (Haag, 1945). Similar symptoms in sheep and cattle consuming seed galls infected with the nematode and a yellow bacterium were discovered in 1967. A similar report of neurotoxicity in sheep grazing on annual ryegrass (*Lolium rigidum*) infected with the nematode and the bacterium was made in 1971 from Western Australia (Bird, 1981). Lanigan *et al.* (1976) located that seed galls colonized by the bacterium were the cause of the syndrome in grazing sheep. The clinical symptoms of annual ryegrass toxicity have been described. After feeding on the infected galls, cattle, sheep and horses suffer from nervous disorders characterized by falling, trembling of muscles and lack of coordination. Sheep show staggering and collapse. Sheep may regain their feet after some time or continue to have periods of violent convulsions until they die (Shaw and Muth, 1949; Bird, 1981).

The ARGT nematode, Anguina funesta

The seed-gall nematode on annual ryegrass (*Lolium rigidum*) was recognized as *Anguina agrostis* (Steinbuch, 1799), Filipjev, 1936 (Bird, 1981). The identity of *Anguina* found in *L. rigidum* in Australia which is associated with *Clavibacter* sp. causing toxicity to grazing animals and sheep in particular has been a subject of debate in recent years. This nematode was described as *A. funesta* Price, Fisher and Kerr, 1979 (Price *et al.*, 1979a, 1979b), but later synonymized with *A. agrostis* (Stynes and Bird, 1980) because both were found to be morphologically indistinguishable. Fisher *et al.* (1984) questioned the morphological basis of the synonymy and listed ecological differences between the two. Riley *et al.* (1988) resolved the taxonomic status of the seed-gall nematode involved in ARGT using allozyme electrophoresis. They concluded that the nematode associated with ARGT is undoubtedly *A. funesta* and is distinct from *A. agrostis* infesting *Agrostis capillaris*. The host range of *A. funesta* includes *Lolium* and *Festuca* spp. Only cases of toxicity to livestock occur in association with *Anguina* and bacteria on *L. rigidum* (Price *et al.*, 1979a) and *Festuca rubra* (Galloway, 1961).

The ARGT bacterium, Clavibacter sp.

The bacterium *Corynebacterium rathayi* together with the nematode *Anguina agrostis* was claimed to cause annual ryegrass toxicity (Bradbury, 1973a, 1973b). Following Dye and Kemp (1977), this bacterium was later designated as *C. michiganense* pv. *rathayi* (Bird, 1981). The latest changes in the nomenclature of coryneform bacteria recognize it as *Clavibacter rathayi* (Smith) Davis *et al.*, 1984 (Davis, 1986). Some doubts have been expressed about the involvement of *C. rathayi* in annual ryegrass toxicity. *Clavibacter* sp. is now considered to be involved in ARGT (Riley, 1987) whereas *C. rathayi* is recognized on *Dactytis glomerata* (Riley and McKay, 1990).

Bird (1981) gave some characteristics of the ARGT bacterium. The bacterium is made up of short rods with bluntly rounded ends, occurring singly or in pairs joined end to end. The rods remain surrounded by a capsule which gives them a rounded appearance. The capsules may enhance adherence to nematodes (Bird and Stynes, 1977).

Association between the seed-gall nematode and ARGT bacterium

Infection of ryegrass with *Clavibacter* sp. does not occur in the absence of *Anguina funesta*. Annual ryegrass infected with ARGT bacterium appears to be similar to wheat infected with 'tundu' bacterium. Spikelets become filled and glumes covered with a mass of bacteria. The seed galls colonized by the bacterium look bright yellow in colour (Bird *et al.*, 1980). The bacterial colonization in *A. funesta*-induced galls make them toxic to animals, as the yellow bacteria-colonized galls contain the toxic agent (Lanigan *et al.*, 1976). The toxicity is reported to be present in the walls of the yellow galls. The plant component of the gall is 20–30 times more toxic than the bacterial component (Stynes *et al.*, 1979). A comparison of non-toxic nematode gall with bacteria colonized yellow gall with the aid of high-resolution technology revealed 25–30 nm diameter particles associated with the bacterium and the gall wall (Bird *et al.*, 1980). However, the identity status and the relation of these particles with toxicity are not well known. A group of glycolipid toxins has been isolated from seed heads and annual ryegrass infected by *Clavibacter* sp. (Vogel *et al.*, 1981; Frahn *et al.*, 1984).

The nematode juveniles act as a vector and carry the bacterium within the plant (Price, 1973). It acquires the bacterium as a surface contaminant. The mode of contamination and transmission of the bacterium by juveniles in ARGT seems to be similar to those observed in 'tundu'. Bird *et al.* (1980) showed that bacteria-colonized galls contained bacteria closely packed in a regular array, both along the walls of the cells making up the gall walls and within the galls themselves. They considered that close packing of infective juveniles and bacteria in an anhydrous condition was important for their survival under extreme environmental conditions.

The galls colonized by bacteria liberate the juveniles contaminated with the bacterium in winter and are carried to the growing point of the seedlings where they feed, grow and mature into adults. The associated bacteria undergo rapid development in the nematode-induced galls and galls may be fully packed with bacteria.

Using light microscopy, and both scanning and transmission electron microscopy, it has been demonstrated that the bacterial cells remain adhered to the cuticle. The adhesion involves an attractive force of considerable magnitude and a specific interaction exists between the nematode cuticle and bacterium capsule (Bird and Stynes, 1977; Bird, 1981). A model for the mechanism of adhesion of bacteria to the surface of animal cells has been proposed which involves specific interaction of two substances – adhesins on the surface of bacteria, and receptors on the animal cells (Jones, 1977). The nature of adhesion of the microorganisms to *Anguina* juveniles has been examined (Bird, 1985; Bird and McKay, 1987). Microbial adhesion to animal cells follows excretion of insoluble polysaccharides to facilitate attachment, and such a process was observed by Bird (1985) in the association of *Clavibacter* and *Anguina*. In a recent study, Riley and McKay (1990) determined the specificity of the interaction of *Clavibacter* spp. and *Dilophospora alopecuri* adhesins for the receptors of the nematodes with which they are associated. A high degree of specificity and different levels of adhesion were found. The bacterial strains and nematode populations were important determinants of the level of adhesion. This level of adhesion is significant: strong adhesion would favour the vector nematodes to introduce the microbes into the plant hosts.

Rhodococcus fascians (= *Corynebacterium fascians*) and *Aphelenchoides fragariae* or *A. ritzemabosi* together cause a disease of strawberry called cauliflower.

12.2.1 Cauliflower disease of strawberry

This disease has been reported to occur in the UK and other European countries as well as in North America. Ritzema-Bos (1891) proposed the term 'cauliflower' to describe a malformation of the inflorescence and shortening and thickening of the aerial parts of strawberry plants. He considered *Aphelenchoides fragariae*, an ectoparasitic nematode, as the causal agent of the disease since this nematode, which was described by him as a new species, was consistently found in the diseased plants. Aetiological investigations carried out later showed that *A. fragariae* was not the sole cause of the disease because apparently healthy plants also yielded moderate populations of the nematode (Crosse and Pitcher, 1952). Studies by Lacey (1936, 1942) provided evidence that a bacterium, now called *Rhodococcus fascians*, was also associated with the disease. Her researches remained inconclusive in establishing the obligatory association of the bacterium (*R. fascians*) and the nematode (*A. fragariae*) for development of the cauliflower syndrome. Research by Crosse and Pitcher (1952) and Pitcher and Crosse (1958) on the aetiology of the cauliflower disease complex provided ample evidence of the obligatory association of the two pathogens for full cauliflower syndrome. According to Pitcher (1965) cauliflower disease is a complex requiring the nematode and certain strains of the bacterium and both organisms actively participate in the development of the full disease syndrome.

The nematodes, Aphelenchoides *spp.*

Two species of *Aphelenchoides*, *A. ritzemabosi* and *A. fragariae*, act as vectors of *R. fascians* and play an obligatory role in the cauliflower disease complex on strawberry. The morphological details and some other aspects of *A. ritzemabosi* are given in detail in CIH descriptions of plant-parasitic nematodes, Set 3, No. 32, and of *A. fragariae* in Set 5, No. 74.

Aphelenchoides ritzemabosi (Schwartz, 1911) Steiner and Buhrer, 1932 is an obligate parasite and feeds endoparasitically on mesophyll cells and ectoparasitically on buds and growing points. A large number of plant species across the world form its host index. Strawberry is an important host on which it is usually found with *A. fragariae* (Siddiqi, 1974).

Aphelenchoides fragariae (Ritzema Bos 1891) Christie 1932 is an obligate parasite of above-ground parts of plants. It was originally described by Ritzema-Bos (1890) from strawberry plants suffering from cauliflower disease in Kent, England. Its extensive host range includes a large number of angiosperms and ferns. The nematode is reported from a number of European countries, Japan and the USA. The nematode feeds ectoparasitically or endoparasitically. It causes crimp or spring dwarf in strawberry and feeds ectoparasitically, living within folded crown and runner buds (Franklin, 1950). Twisting and puckering of leaves, discoloured areas with a rough surface, dwarfed leaves with crinkled margins, reddening of petioles, short internodes of runners, reduced flower trusses and death of the crown buds are shown by the infected plants of strawberry. The nematode is successfully transferred from ferns to strawberry and from strawberry to ferns.

The cauliflower bacterium, Rhodococcus fascians

The cauliflower bacterium, *Rhodococcus fascians* (Tilford) Goodfellow as it is currently known, is widespread and remains in healthy strawberries as saprophyte, becoming pathogenic when the plants are invaded by nematodes. There is also evidence of its independent saprophytic existence in the field (Pitcher and Crosse, 1958). According to Pitcher (1963), *R. fascians* is present universally as a harmless epiphyte on many plants, including strawberry. Only a few strains are capable of inducing cauliflower symptoms, and they need to be carried to the meristem of a crown bud by the nematode for the development of cauliflower disease.

Association between Aphelenchoides spp. and Rhodococcus fascians

Strawberry plants infected with both *Aphelenchoides ritzemabosi* and certain strains of *Rhodococcus fascians* resemble small cauliflowers. Cauliflower is a manifestation of fundamental disturbances in the growth regulatory system of the plant (Pitcher and Crosse, 1958). Various grades of deformation are found. In extreme cases, plants are stunted and flowers are deformed. The petals fail to develop or become small and greenish. The sepals become very enlarged. Stamens and receptacle are malformed, so that fruits do not develop. Axillary buds are continually produced in the crowns. Strawberry plants attacked by nematodes alone show alaminate leaves, which have reduced lamina with a very small petiole (Crosse and Pitcher, 1952).

Although Lacey (1936) originally implicated *Rhodococcus fascians* in cauliflower disease of strawberry, the aetiological investigations by Crosse and Pitcher (1952) established that typical cauliflower symptoms were produced only by a combination of *Aphelenchoides ritzemabosi* and strains of *R. fascians* isolated from cauliflower strawberries. They showed that cauliflower crowns developed in two ways, either following alaminate leaves or growing direct from a secondary crown bud. The bacterium without eelworm did not induce any abnormality. The association of the bacterium and nematode was essential to the production of symptoms but only certain strains of *R. fascians* induced cauliflower. The authors also claimed that alaminate leaves did not differ fundamentally from the basic cauliflower symptoms and considered that it was unlikely that alaminate leaves resulted entirely from the ectoparasitic feeding of the nematode. From results of their subsequent study (Pitcher and Crosse, 1958) on the cauliflower disease complex, they modified some earlier conclusions. They showed that alaminate leaves are purely nematode-induced symptoms, nematodes acting as primary plant parasites, and continue to produce alaminate leaves in the absence of bacterium. *R. fascians* played no direct role in the aetiology of this symptom. Under experimental conditions, *R. fascians* produced leaf gall symptoms on aseptic seedlings in tubes, but vegetative plants or meristems exposed to the bacterium did not cause any symptom. It was concluded that strawberry cauliflower is a primarily hyperplastic bacterial symptom, modified by the activities of the nematode, which is also essential as a vector of the bacterium. Apart from vectoring, the nematodes by feeding provide a useful metabolite or modify the host substrate to favour the bacterium (Pitcher, 1963). Therefore, both organisms actively contribute to the cauliflower syndrome. The exact mechanisms of contamination and transport of the bacterium by the nematodes have not been fully elucidated. It is believed that the bacterium attaches to the cuticle of the nematodes and is transported to the crown tissue of strawberry plants (Poinar and Hansen, 1986; Taylor, 1990).

12.3 *CLAVIBACTER–DITYLENCHUS*

Clavibacter michiganense subsp. *insidiosum* (McCulloch) Davis *et al.* 1984 (= *Corynebacterium insidiosum*), which causes bacterial wilt in alfalfa, is transmitted by the nematode *Ditylenchus dipsaci* Kuhn into alfalfa crown buds resulting in increased incidence of bacterial wilt. Studies of Hawn (1963, 1971) provided evidence of transmission of the bacterial cells by *D. dipsaci*. Hawn (1963) found frequent association of *C. michiganense* subsp. *insidiosum* and *D. dipsaci* in the same alfalfa stands in southern Alberta, Canada. Well-developed and typical symptoms of the respective diseases caused by these pathogens were often found in the same plants. *C. michiganense* subsp. *insidiosum* was isolated from distilled water-washed and homogenized bodies of the nematodes obtained from buds of wilt-infected alfalfa plants. A glasshouse study on the effect of *D. dipsaci* on the initiation and development of bacterial wilt in alfalfa showed that *D. dipsaci* acted as transmitting agent and carried *C. michiganense* subsp. *insidiosum* into the crown bud and placed in the conducive infection court (Hawn, 1963). It was believed that *C. michiganense* subsp. *insidiosum* was carried on the body of the nematodes. The possibility of internal transmission within the body of the nematodes was discounted because the lumen diameter of adult males and females was too narrow to allow ingestion of mature bacterial cells and surface-disinfected nematodes in artificial culture remained free of bacterial growth (Hawn, 1963). Hawn (1971) demonstrated that numerous bacterial cells adhered to the cuticle of nematodes exposed to the wilt bacteria suspension. Therefore, it was confirmed that *C. michiganense* subsp. *insidiosum* is transmitted from wilt-infected to healthy alfalfa plants, externally on the body surface of *D. dipsaci*.

In addition to vectoring of *Clavibacter michiganense* subsp. *insidiosum* by *Ditylenchus dipsaci*, there are reports of similar vectoring by the same or other nematodes. These associations have not been studied in detail, however. The causal agent of 'cafe au lait' bacteriosis, *Pseudomonas fluorescens*, is reported to be carried in its host, garlic, by *D. dipsaci* (Caubel and Samson, 1984) whereas the causal agent of rhubarb crown rot, *Bacterium rhaponticum*, finds entry into its host through the carrier *D. dipsaci* (Metcalfe, 1940). *Globodera pallida* is reported to transmit *Pseudomonas solanacearum* in potatoes (Jensen, 1978).

A report by Kalinenko (1936) claimed that stylet-bearing nematodes have the ability to inoculate bacteria from their intestine into intact plant cells, but the possibility of this mode of transport is unlikely as the opening of the stylet in most tylenchids ranges from 0.2 to 1.0 μm in diameter, which cannot give passage to most plant-pathogenic bacteria.

12.4 SAPROZOIC NEMATODE VECTORS OF PATHOGENIC BACTERIA

Injured or damaged plant tissue provides a favourable site for bacterial growth, which in turn attracts saprozoic or bacterivorous nematodes. The role of such nematodes in protecting and transporting the pathogenic bacteria cannot be precluded. However, free-living nematodes are seldom considered as accomplices in plant disease complexes. Microbitrophic nematodes have been reported to ingest and excrete bacteria which were found to be viable. Females of *Pristionchus lheritieri* ingested more bacteria (*Agrobacterium tumefaciens*) and retained them for up to 27 hours in viable form (Wasilewska and Webster, 1975). A bibliography on this topic has appeared in the literature (Armstrong *et al.*, 1976). Other plant-pathogenic bacteria which are carried by free-living nematodes include *Erwinia amylovora*, *E. carotovora*, *Pseudomonas phaseolicola* and *P. syringae*. The causal agent of blackspot on mushroom, *P. tolasii*, is carried through *Cruznema tripartirum* (Linslow) (= *Rhabditis lamdiensis* Maupas) (Steiner, 1933; Haseman and Ezell, 1934). These type of association have been discussed by Poinar and Hansen (1986).

12.5 NEMATODES AS VECTORS OF FUNGAL PLANT PATHOGENS

Theoretically, propagules or spores of fungal plant pathogens can be vectored by nematodes externally on the body surface or internally in the alimentary canal. A well-known example is *Dilophospora alopecuri* (Fr.) Fr.

which causes twist disease on cereals and grasses. The conidia of the fungus are transmitted by *Anguina tritici*. The symptomatology of the disease has been described by several workers (Atanasoff, 1925; Johnson and Leukel, 1946; Munjal and Kaul, 1961; Bamadadian, 1973). Atanasoff (1925) in his extensive studies addressed himself to two questions: first, whether *D. alopecuri* is a pathogen capable of causing the disease on some cereals on its own as observed in nature; and second, whether the nematode, *A. tritici* (= *Tylenchus tritici*) favours the parasitism of the fungus on the same plant. Atanasoff carried out several experiments to answer these two questions. The answer was negative for the first and positive for the second question. The fungus was pathogenic but it could establish itself successfully only on the plants infected with the nematode. Atanasoff (1925) referred to it as co-parasitism, where one pathogen depended absolutely upon the other for its successful parasitism.

12.5.1 Twist disease

Twist disease, also called plumed spore or *Dilophospora* disease, is widely distributed in Europe, the USA, India, Australia and Iran on wheat, barley, oats and several grasses (Paruthi and Bhatti, 1990). Sprauge (1950) lists it as a well-known disease on an extensive range of Poaceae. The symptoms of the disease are distinctive. Initially, young leaves show spiralling and rolling. Small light-coloured round or oblong spots or blotches first appear on the upper leaves of plants. Black dots consisting of one or more pycnidia appear in the centre of the spots. Several pycnidia soon develop around the original ones and the black dots enlarge to cover the whole blotch. The blotches are seen on both leaf surfaces but are more pronounced on the lower surface of the leaves. The infected leaves become yellow and die. Symptoms of the disease also appear on leaf sheaths. The fungus spreads in the concavity of the sheaths and enclosed young leaves. The upper central portion of the plant is thus firmly bound. Affected plants remain very small, form no ears and die. On the plants, where the symptoms appear shortly before ear formation, the ears remain attached to the upper leaf sheath by the fungus growth. The ears become abnormal and twisted. Some plants show free ears but are deformed and destroyed by the fungus and look charred. The ear deformity may be partial or total.

12.5.2 The pathogen *Dilophospora alopecuri*

Atanasoff (1925) gave a description of *Dilophospora alopecuri*. Walker and Sutton (1974) re-described it. The pycnidia are solitary in young lesions immersed in the mesophyll, being globose, brown, glabrous, unilocular or very rarely multilocular. The ostiole is central, often papillate. In older leaf lesions and leaf sheaths, pycnidia are aggregated into rows between the veins. Conidiogenous cells are formed directly from the inner cells of the entire pycnidial wall, enteroblastic, phialidic, separate or aggregated into short conidiophores. The individual cells of conidiophores are conidiogenous, and produce conidia in succession. Conidia are cylindrical or tapered towards the base, hyaline, zero to tri-septate and smooth, with several simple to dichotomously or irregularly branched appendages. Septa and appendages seem to be formed when conidia attain their full size (Walker and Sutton, 1974).

12.5.3 The mode of vectoring

The pathogenicity of *Dilophospora alopecuri* on cereals is fully dependent on the seed-gall nematode, *Anguina tritici*. The nematode acts as carrier of fungal conidia. The conidia are transported to the growing point of the plant. Atanasoff (1925) observed that the nematode juveniles, while creeping and crawling between the leaf sheaths on their way to the growing point and later to the young ears of the plant, carry fungal conidia and deposit them on the growing point of the plant. The appendages (bristles) of the conidia help them to adhere to the nematode body. Transport of the conidia to the growing point of the plant cannot be accomplished without the nematodes. Examination of the nematodes has shown the presence of conidia on their body. It has also been suggested that the role of the nematode is not merely mechanical in transporting the conidia but the nematodes, by sucking or injuring or affecting the plant cells of the tender leaves, facilitate the penetration and establishment of the fungus (Atanasoff, 1925). The nematodes cannot form seed galls on the ears attacked by the fungus; if they form any gall, the galls are subsequently destroyed by the fungus. Nematodes die in the case of a severe attack of twist disease.

In recent years, adhesion of conidia of *Dilophospora alopecuri* to the cuticle of *Anguina agrostis* (now *A. funesta*), the vector in ARGT, was examined by Bird and McKay (1987). Riley and McKay (1990) showed that conidia of *D. alopecuri* from two hosts adhered to varying degrees to all seven populations of seed-gall nematode, *Anguina* spp. The adhesin-receptor mechanism was operative in their adhesion. The adhesion indicated that the nematodes were important as vectors in the disease caused by *D. alopecuri*.

The simultaneous occurrence or association of *Anguina tritici* with smut fungi like *Tilletia foetida* (Mathur and Misra, 1961), *Ustilago tritici* (Bedi et al., 1959; Paruthi and Bhatti, 1982) and *Neovossia indica* (Paruthi and Bhatti, 1980) in wheat has been recorded. Their exact relationship with the nematode and the significance of these observations are not known.

Transmission of plant-pathogenic fungi internally by the nematodes and its significance in plant disease epidemiology have received little attention. Jensen (1967) demonstrated that spores of *Fusarium* and *Verticillium* survived passage through the alimentary canal of a microbitrophic nematode. Jensen and Siemer (1971) examined the protection of ingested microconidia of *F. oxysporum* f. sp. *lycopersici* and conidia of *V. dahliae* by *Pristionchus lheritieri* from certain biocides and indicated that the nematode can shield ingested spores from fungicidal treatments.

12.6 CONCLUSIONS

The role of nematodes as vectors in transmitting secondary plant pathogens (bacteria or fungi) which are not potent enough to enter and establish without their aid is very interesting and of great significance, particularly in cases where new types of disease syndromes are produced that are not possible by either pathogen alone. Mechanisms of contamination and transport have been investigated substantially but some gaps in our knowledge still persist. In ear-cockle 'tundu' complex, the ear-cockle phase is important for both nematode and bacteria. Seed galls are produced in this phase only. The second-stage juveniles from such galls carry 'tundu' bacteria on their body. How bacteria pass into the next generation of juveniles is not known. Probably a certain degree of bacterial multiplication might be occurring within floral tissue at the endophytic stage of juveniles, so that bacteria adhere to the cuticle of infective second-generation juveniles. The nature and significance of physiological and histopathological transformation caused by feeding of the nematodes in the development of the complex are little known. Specific physical and biochemical adhesion of coryneform bacteria on the body surface of *Anguina* juvenile reveals a highly advanced association between the two. These types of associations appear to favour the secondary pathogen more than the nematodes. Nematodes are vulnerable to deleterious effects of associated secondary pathogens. 'Tundu'-affected wheat tillers do not produce nematode galls showing adverse effect of the bacteria on the nematode. All these indicate much more complex interactions than a simple vector relationship.

There are a good number of instances where a simple vector relationship between nematode and other bacterial or fungal plant pathogens has been demonstrated. However, it appears more likely that even the simple transmission of bacterial cells or fungal spores by the nematode is associated with biochemical alterations of the host (in favour of transmitted microorganism) created by salivary secretions of the vector nematode.

Many species of microbitrophic free-living nematodes are found associated with degenerative disease conditions or with organic decay. Such nematodes ingest bacteria and fungal spores and excrete viable spores or bacterial cells. In these types of associations free-living saprozoic nematodes direct and mobilize microorganisms towards the vulnerable area of the plant root system. Pathogenic fungi or bacteria ingested in one site may be transported and deposited in another site of the rhizosphere. The role of free-living nematodes as vectors of pathogenic microorganisms (bacteria or fungi) in protecting the pathogens from adverse conditions, including fungicides and antibiotics in the soil, is an interesting area of research, and modern tools may be of help in determining its exact significance in the soil ecosystem.

REFERENCES

Armstrong, J.M., Jatala, P. and Jensen, H.J. (1976) Bibliography of nematode interactions with other organisms in plant disease complexes. *Station Bulletin, Agricultural Experiment Station, Oregon State University, Corvallis*, No. 623, 152 pp.

Atanasoff, D. (1925) The Diplophospora disease of cereals. *Phytopathology*, **15**, 11–40.

Bamdadian, A. (1973) The importance and situation of wheat disease in Iran, in *CENTO Panel of Pests and Diseases of Wheat*, Tehran University, College of Agriculture, Karanj, Iran, pp. 57–63.

Bedi, K.S., Chohan, J.S. and Chahal, D.S. (1959) Simultaneous occurrence of *Ustilago tritici* (Pers.) Rost. and *Anguina tritici* (S.) G.Ben. in a single ear of wheat. *Indian Phytopathology*, **12**, 187.

Bird, A.F. (1981) The *Anguina-Corynebacterium* association, in *Plant Parasitic Nematodes*, Vol. III (eds B.M. Zuckerman and R.A. Rohde), Academic Press, London, pp. 303–23.

Bird, A.F. (1985) The nature of the adhesion of *Corynebacterium rathayi* to the cuticle of the infective larva of *Anguina agrostis*. *International Journal for Parasitology*, **15**, 301–8.

Bird, A.F. and McKay, A.C. (1987) Adhesion of conidia of the fungus *Dilophospora alopecuri* to the cuticle of nematode *Anguina agrostis*, the vector in annual ryegrass toxicity. *International Journal for Parasitology*, **17**, 1239–47.

Bird, A.F. and Stynes, B.A. (1977) The morphology of a *Corynebacterium* sp. parasitic on annual ryegrass. *Phytopathology*, **67**, 828–30.

Bird, A.F., Stynes, B.A. and Thomson, W.W. (1980) A comparison of nematode and bacteria-colonized galls induced by *Anguina agrostis* in *Lolium rigidum*. *Phytopathology*, **70**, 1104–9.

Bradbury, J.F. (1973a) *Corynebacterium rathayi. Commonwealth Mycological Institute Descriptions of Pathogenic Fungi and Bacteria*, No. 376, 2 pp, Commonwealth Agricultural Bureau.

Bradbury, J.F. (1973b) *Corynebacterium tritici. Commonwealth Mycological Institute Description of Pathogenic Fungi and Bacteria*, No. 377, 2 pp, Commonwealth Agricultural Bureau.

Carlson, R.R. and Vidaver, A.K. (1982) Taxonomy of *Corynebacterium* plant pathogens, including a new pathogen of wheat, based on polyacrylamide gel electrophoresis of cellular proteins. *International Journal of Systematic Bacteriology*, **32**, 315–26.

Carne, W.M. (1926) Ear-cockle (*Tylenchus tritici*) and a bacterial disease (*Pseudomonas tritici*) of wheat. *Journal of Department of Agriculture, West Australia*, **31**, 508–12.

Caubel, G. and Samson, R. (1984) Effect of the stem nematode *Ditylenchus dipsaci* on the development of 'Cafe au lait' bacteriosis in garlic (*Allium sativum*) caused by *Pseudomonas fluorescens. Agronomie*, **4**, 311–13.

Chaudhuri, H. (1935) A bacterial disease of wheat in the Punjab. *Proceedings of the Indian Academy of Science*, **2**, 579.

Cheo, C.C. (1946) A note on the relation of nematode (*Tylenchulus tritici*) to the development of bacterial disease of wheat caused by *Bacterium tritici* Hutchinson. *Annals of Applied Biology*, **33**, 446–9.

Chitwood, B.G. (1935) Nomenclatorial notes. *Proceedings of the Helminthological Society of Washington*, **2**, 51–4.

Collins, M.D. and Jones, D. (1983) Reclassification of *Corynebacterium flaccumfaciens*, *Corynebacterium betae*, *Corynebacterium oortii* and *Corynebacterium poinsettiae* in the genus *Curtobacterium* as *Curtobacterium flaccumfaciens* Comb. nov. *Journal of General Microbiology*, **129**, 3545–8.

Collins, M.D., Jones, D. and Kroppenstedt, R.M. (1981) Reclassification of *Corynebacterium ilicis* (Mandel, Guba and Litsky) in the genus *Arthrobacter* as *Arthrobacter ilicis* comb. nov. *Zentrabi Bakteriologie Hyg Abstract 1 Orig. Reihec*, **2**, 318–23.

Crosse, J.E. and Pitcher, R.S. (1952) Studies in the relationship of eelworms and bacteria to certain plant diseases. 1. The etiology of strawberry cauliflower disease. *Annals of Applied Biology*, **39**, 475–84.

Davis, M.J. (1986) Taxonomy of plant-pathogenic corneform bacteria. *Annual Review of Phytopathology*, **24**, 115–40.

Davis, M.J., Gillaspie, A.G., Jr, Vidaver, A.K. and Harris, R.W. (1984) *Clavibacter*: A new genus containing some phytopathogenic coryneform bacteria including *Clavibacter xyli* subsp. *xyli* sp. nov.; subsp. nov. and *Clavibacter xyli* subsp. *cynodontis* subsp. nov.; pathogens that cause ratoon stunting disease of sugarcane and Bermudagrass stunting disease. *International Journal of Systematic Bacteriology*, **34**, 107–17.

Dovaine, C. (1857) Recherches sur l'anguillule duble niell consideree au point de vue de l'histoire naturelle et de l'agriculture, Paris, 88 pp.

Dye, D.W. and Kemp, W.J. (1977) A taxonomic study of plant pathogenic *Corynebacterium* species. *New Zealand Journal of Agricultural Research*, **20**, 563–82.

Epstein, E. and Cohn, E. (1971) Biochemical changes in terminal galls caused by an ectoparasitic nematode, *Longidorus africanus*: Amino acids. *Journal of Nematology*, **4**, 334–40.

Fahmy, T. and Mikhail, T. (1925) The bacterial disease of wheat caused by *Pseudomonas tritici* Hutchinson. *Agricultural Journal of Egypt*, **1**, 64–72.

Fisher, J.M., McKay, A.C. and Dube, A.J. (1984) Observations on growth of adult of *Anguina funesta* (Nematoda: Anguinidae). *Nematologica*, **30**, 463–9.

Frahn, J.L., Edgar, J.A., Jones, A.J., Cockrum, P.A., Anderton, N. and Culvenor, C.C.J. (1984) Structure of the corynetoxins, metabolites of *Corynebacterium rathayi* responsible for toxicity of annual ryegrass (*Lolium rigidum*) pastures. *Australian Journal of Chemistry*, **37**, 165–82.

Franklin, M.T. (1950) Two species of *Aphelenchoides* associated with strawberry bud disease in Britain. *Annals of Applied Biology*, **37**, 1–10.

Galloway, J.H. (1961) Grass seed poisoning in livestock. *Journal of the American Veterinary and Medical Association*, **139**, 1212–14.

Gokte, N. and Swarup, G. (1988) On the association of bacteria with larvae and galls of *Anguina tritici*. *Indian Journal of Nematology*, **18**, 313–18.

Goodey, T. (1932) The genus *Anguillulina* Gerv & V.Ben., 1859, vel *Tylenchus* Bastian, 1865. *Journal of Helminthology*, **10**, 75–180.

Goodfellow, M. (1984) Reclassification of *Corynebacterium fascians* (Tilford) Dowson in the genus *Rhodococcus* as *Rhodococcus fascians* comb. nov. *Systematic Applied Microbiology*, **5**, 225–9.

Gupta, P. (1966) Studies on ear-cockle and 'tundu' disease of wheat. PhD thesis, Indian Agricultural Research Institute, New Delhi.

Gupta, P. and Swarup, G. (1968) On the ear-cockle and yellow ear-rot disease of wheat. I. Symptoms and histopathology. *Indian Phytopathology*, **22**, 318–22.

Gupta, P. and Swarup, G. (1972) Ear-cockle and yellow ear-rot disease of wheat. II. Nematode bacterial association. *Nematologica*, **18**, 320–4.

Haag, J.R. (1945) Toxicity of nematode infesting Chewings's fescue seed. *Science,* **102**, 406-7.

Haseman, L. and Ezell, H.O. (1934) The eelworm *Rhabditis lambiensis,* a new pest of mushrooms. *Journal of Economic Entomology,* **27**, 1189-91.

Hawn, E.J. (1963) Transmission of bacterial wilt of alfalfa by *Ditylenchus dipsaci* (Kuhn). *Nematologica,* **9**, 65-8.

Hawn, E.J. (1971) Mode of transmission of *Corynebacterium insidiosum* by *Ditylenchus dipsaci. Journal of Nematology,* **3**, 420-1.

Hutchinson, O.M. (1917) A bacterial disease of wheat in the Punjab. *Memoires of the Department of Agriculture in India, Bacteriological Series 1,* **17**, 169-75.

Jensen, H.J. (1967) Do saprozoic nematodes have a significant role in epidemiology of plant diseases? *Plant Disease Reporter,* **51**, 98-102.

Jensen, H.J. (1978) Interrelations of nematodes and other organisms in disease complexes. *International Potato Centre Report of the Second Planning Conference on the Developments in the Control of Nematode Pests of Potatoes,* CIP, Lima, Peru.

Jensen, H.J. and Siemer, S.R. (1971) Protection of *Fusarium* and *Verticillium* propagules from selected biocides following ingestion by *Pristionchus lheritieri. Journal of Nematology,* **3**, 23-7.

Johnson, A.G. and Leukel, R.W. (1946) *Dilophospora alopecuri* on wheat in South Carolina. *Plant Disease Reporter,* **30**, 327.

Jones, G.W. (1977) The attachment of bacteria to the surface of animal cells, in *Microbial Interactions* (ed. J.L. Reissing), Chapman and Hall, London, pp. 139-76.

Kalinenko, V.O. (1936) The inoculation of phytopathogenic microbes into rubber bearing plants by nematodes. *Phytopathologische Zeitschrift,* **9**, 407-16.

Khan, M.W. (1984) Ecological niche of plant nematodes in soil biotic community, in *Progress in Microbial Ecology* (eds K.G. Mukerji, V.P. Agnihotri and R.P. Singh), Print House (India), Lucknow, pp. 165-77.

Lacey, M.S. (1936) Studies in bacteriosis. XXII. I. The isolation of a bacterium associated with fasciation of sweet peas, cauliflower of strawberry plants and 'leafy gall' of various plants. *Annals of Applied Biology,* **23**, 302-10.

Lacey, M.S. (1942) Studies in bacteriosis. XXV. Studies on a bacterium associated with leafy galls, fasciations and cauliflower disease of various plants. Part IV. The inoculation of strawberry plants with *Bacterium fascians* Tilford. *Annals of Applied Biology,* **29**, 11-15.

Lanigan, G., Payne, A.L. and Frahn, J.L. (1976) Origin of toxicity in parasitized annual ryegrass *(Lolium rigidum). Australian Veterinary Journal,* **52**, 244-6.

Leukel, R.W. (1924) Investigations on the nematode disease of cereals caused by *Tylenchus tritici. Journal of Agriculture Research,* **27**, 925-56.

Leukel, R.W. (1957) Nematode disease of wheat and rye. *Farmers Bulletin USDA,* No. 1607, 1-16.

Mathur, R.S. and Misra, M.P. (1961) Simultaneous occurrence of *Tilletia foetida* (Wallr.) Liro and *Anguillulina tritici* (S.) G. Ben. in the same ear and grains of wheat in Pauri Garhwal, Uttar Pradesh. *Current Science,* **30**, 307.

Metcalfe, G. (1940) *Bacterium rhaponticum* (Millard) Dowson, a cause of crown-rot disease of rhubarb. *Annals of Applied Biology,* **27**, 502-8.

Midha, S.K. (1969) Studies on the ear-cockle nematode and the development of ear cockle and 'tundu' disease of wheat, PhD thesis, Indian Agricultural Research Institute, New Delhi, India.

Munjal, R.L. and Kaul, T.N. (1961) *Dilophospora* leaf spot of wheat in India. *Indian Phytopathology,* **14**, 13-16.

Needham, T. (1743) A letter concerning certain chalky tubulous concretions called malm, with some microscopical observations on the farina of the red lily, and of worms discovered in smutty corn. *Philosophical Transactions of the Royal Society*, 42, 634–41.

Paruthi, J.P. and Bhatti, D.S. (1980) Concomitant occurrence of *Anguina tritici* and Karnal bunt fungus (*Neovossia indica*) in wheat seed galls. *Indian Journal of Nematology*, 10, 256–7.

Paruthi, I.J. and Bhatti, D.S. (1982) Simultaneous occurrence of tundu and loose smut of wheat. *Indian Journal of Nematology*, 12, 166–7.

Paruthi, I.J. and Bhatti, D.S. (1990) Wheat seed gall nematode *Anguina tritici*: A review. *Current Nematology*, 1, 107–36.

Pathak, K.N. (1982) Studies on the relationship between ear-cockle nematode *Anguina tritici* and *Corynebacterium tritici* in the 'tundu' disease complex of wheat, PhD thesis, Indian Agricultural Research Institute, New Delhi, India.

Pathak, K.N. and Swarup, G. (1983) Emergence of juveniles from *Anguina tritici* galls, numbers reaching growing point of wheat seedlings and relationship between inoculum levels and number and size of galls produced. *Indian Journal of Nematology*, 13, 155–60.

Pathak, K.N. and Swarup, G. (1984) Incidence of *Corynebacterium michiganense* pv. *tritici* in the ear-cockle nematode (*Anguina tritici*) galls and pathogenicity. *Indian Phytopathology*, 37, 267–70.

Pathak, K.N., Dasgupta, D.R. and Swarup, G. (1983) Biochemical changes in *Anguina tritici* infected wheat tissue. *Tropical Plant Science Research*, 1, 235–8.

Pitcher, R.S. (1963) Role of plant parasitic nematodes in bacterial diseases. *Phytopathology*, 53, 35–9.

Pitcher, R.S. (1965) Interrelationships of nematodes and other pathogens of plants. *Helminthological Abstracts* (Series B), 34, 1–17.

Pitcher, R.S. and Crosse, J.E. (1958) Studies in the relationship of eelworms and bacteria to certain plant diseases. II. Further analysis of the strawberry cauliflower disease complex. *Nematologica*, 31, 244–56.

Poinar, G.O., Jr, and Hansen, E.C. (1986) Associations between nematodes and bacteria. *Helminthological Abstracts* (Series B), 55, 62–91.

Powell, N.T. (1971) Interaction of plant parasitic nematodes with other disease-causing agents, in *Plant Parasitic Nematodes*, Vol. II (eds B.M. Zuckerman, W.F. Mai and R.A. Rohde), Academic Press, New York, pp. 114–36.

Price, P.C. (1973) Investigation of a nematode–bacterium disease complex affecting Wimmera ryegrass, PhD thesis, University of Adelaide, Australia.

Price, P.C., Fisher, J.M. and Kerr, A. (1979a) Annual ryegrass toxicity: Parasitism of *Lolium rigidum* by a seed-gall forming nematode (*Anguina* sp.). *Annals of Applied Biology*, 91, 359–69.

Price, P.C., Fisher, J.M. and Kerr, A. (1979b) On *Anguina funesta* n. sp. and its association with *Corynebacterium* sp. in infecting *Lolium rigidum*. *Nematologica*, 25, 76–85.

Ray, S.N. (1987) Studies on some aspects of *Anguina tritici* nematode infesting wheat, MSc thesis, Rajendra Agricultural University, Pusa, India.

Riley, I.T. (1987) Serological relationships between strains of coryneform bacteria responsible for annual ryegrass toxicity and other plant pathogenic coryne-bacteria. *International Journal of Systematic Bacteriology*, 37, 153–9.

Riley, I.T. and McKay, A.C. (1990) Specificity of the adhesion of some plant pathogenic microorganisms to the cuticle of nematodes in the genus *Anguina* (Nematoda: Anguinidae). *Nematologica*, 35, 90–103.

Riley, I.T., Reardon, T.B. and McKay, A.C. (1988) Electrophoretic resolution of species boundaries in seed-gall nematodes, *Anguina* spp. (Nematoda: Anguinidae) from some graminaceous hosts in Australia and New Zealand. *Nematologica*, 34, 401–11.

Ritzema-Bos, J. (1890) De Bloemkoolziekte der aardbeien, veroorzakt door

Aphelenchus fragarae nov. spec. Maanblad voor Natuurwefenschlappen, **16**, 107–17.

Ritzema-Bos, J. (1891) Zwei neu Nematodenkrankheiten der Erdbeerpflanze. *Zeitschrift Pflanzenkrankheiten*, **1**, 1–16.

Sabet, K.A. (1953) On the source and mode of infection with the yellow slime disease of wheat. *Bulletin, Faculty of Agriculture, Cairo University*, **421**, 1–15.

Sabet, K.A. (1954) Pathological relationship between host and parasite in the yellow slime disease of wheat. *Bulletin, Faculty of Agriculture, Cairo University*, **43**, 1–10.

Schneider, D.J. (1981) First report of annual ryegrass toxicity in the republic of South Africa. *Onderstepoort Journal of Veterinary Research*, **48**, 251–5.

Scopoli, J.A. (1777) Introductio ad historiam naturalem sistens genera lapidum, plantarum, et animalium Prague (N.S.).

Shaw, H.M. and Muth, O.H. (1949) Some types of forage poisoning in Oregon cattle and sheep. *Journal of the American Medical Association*, **114**, 315–17.

Sprauge, R. (1950) *Diseases of Cereals and Grasses in North America (Fungi Except Smuts and Rusts)*, Ronald Press, New York.

Steinbuch, J.G. (1799) Das Grasalchen *Vibrio agrostis*. *Naturoforschung Halle*, **28**, 233–50.

Steiner, G. (1933) *Rhabditis lambdiensis*, a nematode possibly acting as a disease agent in mushroom beds. *Journal of Agricultural Research*, **46**, 427–35.

Stynes, B.A. and Bird, A.F. (1980) *Anguina agrostis*, the vector of annual ryegrass toxicity in Australia. *Nematologica*, **26**, 475–90.

Stynes, B.A. and Wise, J.L. (1980) The distribution and importance of annual ryegrass toxicity in Western Australia and its occurrence in relation to cropping rotations and cultural practices. *Australian Journal of Agricultural Research*, **31**, 557–69.

Stynes, B.A., Petterson, D.S., Lloyd, J., Payne, A.L. and Lanigan, G.W. (1979) The production of toxin in annual ryegrass, *Lolium rigidum*, infected with a nematode, *Anguina* sp. and *Corynebacterium rathayi*. *Australian Journal of Agricultural Research*, **30**, 201–9.

Swarup, G. and Gokte, N. (1986) Nematode diseases in wheat, in *Plant Parasitic Nematodes of India: Problems and Progress* (eds G. Swarup and D.R. Dasgupta), Indian Agricultural Research Institute, New Delhi, pp. 300–11.

Swarup, G. and Singh, N.J. (1962) A note on the nematode bacterium complex in tundu disease of wheat. *Indian Phytopatholgy*, **15**, 294–5.

Taylor, C.E. (1990) Nematode interactions with other pathogens. *Annals of Applied Biology*, **116**, 405–16.

Vasudeva, R.S. and Hingorani, M.K. (1952) Bacterial disease of wheat caused by *Corynebacterium tritici* (Hutchinson) Bergey et al. *Phytopathology*, **42**, 291–3.

Vogel, P., Petterson, D.S., Berry, P.H., Frahn, J.L., Anderton, N., Cockrum, P.A., Edgar, J.A., Jago, M.V., Lanigan, G.W., Payne, A.L. and Culvenor, C.C.J. (1981) Isolation of a group of glycolipid toxins from seed heads of annual ryegrass (*Lolium rigidum* Gaud.) infected by *Corynebacterium rathayi*. *Australian Journal of Experimental Biology and Medical Science*, **59**, 455–67.

Walker, J. and Sutton, B.C. (1974) *Dilophia* Sacc. and *Dilophospora* Desm. *Transactions of the British Mycological Society*, **62**, 231–41.

Wasilewska, L. and Webster, J.M. (1975) Free-living nematodes as disease factors of men and his crops. *International Journal of Environmental Studies*, **7**, 201–4.

13

The role of fungi in fungus–nematode interactions

A. Hasan

A fungus is an essential component of the interacting system of a fungus–nematode complex disease and plays an important role in the disease aetiology. Since the publication of Atkinson's report (1892) that *Fusarium* wilt of cotton was more severe in the presence of root-knot nematode (*Meloidogyne* spp.) than in its absence, a large volume of data has been accumulated to date which firmly establishes the involvement of plant-parasitic nematodes in interactions with fungal plant pathogens on various crop plants (Powell, 1971, 1979; Bergeson, 1972; Taylor, 1979; Mai and Abawi, 1987; Riedel, 1988; Taylor, 1990). Fungus–nematode interactions have been classified in a number of ways and roles played by nematodes in such interactions have been examined thoroughly and are well documented (Pitcher, 1965, 1978; Powell, 1971, 1979; Bergeson, 1972).

The role of fungi in their interactions with nematodes has not been delineated and defined in most of the studies. A synergistic relationship is the common feature of fungus–nematode interaction and diseases caused by the fungal pathogens become more pronounced and may appear earlier when plants are infected with nematodes three to four weeks prior to fungus infection (Evans, 1987; Mai and Abawi, 1987; Francl *et al.*, 1988; Hasan, 1988, 1989; Gray *et al.*, 1990). But sometimes fungal diseases are suppressed by the nematode (Orion and Netzer, 1981; Grey *et al.*, 1990). Some soil fungi which are normally known to be non-parasitic on plants become parasitic on roots infected with nematodes (Powell, 1971; Powell *et al.*, 1971). In general, in synergistic interactions nematodes provide an opportunity for fungal pathogens to show their greater pathogenic capabilities. The behaviour of fungi in nematode-infected roots, in general, becomes more aggressive. There is evidence to show that in some interactions the capabilities of the nematodes are also increased in plant roots infected with fungi. Some possible roles of fungi in their interactions can be identified, although not all are equally evident or discrete. The role of fungi, in fact, is related to the assumed behaviour of fungal pathogens in a dormant or active state under changed conditions created through infection of the host by plant-parasitic nematodes. Additionally, in a given interaction the involved fungus may play more than one role, or the roles may be interlinked.

Attempts in this chapter are made to identify and examine the role of fungi in their interactions with nematodes. This chapter, by necessity, may duplicate some of the aspects related to fungus–nematode interactions discussed elsewhere in this book.

13.1 ACTIVATION OF FUNGAL PATHOGENS

Dormant fungal spores or other propagules are activated by root exudates in the rhizosphere. Actively growing roots secrete a variety of chemicals such as carbohydrates, sugars, amino acids, growth regulators, organic acids, alkaloids, terpenoids, phenolic compounds and minerals. These exudates may influence soil fungistasis, serve as a nutrient base in the soil, influence fungal orientation, provide an instant energy source to the invading pathogen, break the dormancy in resting propagules or suppress germination and growth of spores, and thus influence the microbial activities of the rhizosphere and rhizoplane microflora (Rovira, 1965; Walter, 1965). Changes in the pattern of root exudation may influence the mode of parasitism of soil-inhabiting microorganisms. Such changes may be brought about by several factors, including plant-parasitic nematodes (Hale *et al.*, 1971; Golden and Van Gundy, 1972; Wang and Bergeson, 1974; Wang *et al.*, 1975; Van Gundy *et al.*, 1977; Melakeberhan *et al.*, 1985). The maximum leakage of electrolytes from *Meloidogyne incognita*-infected tomato plants occurred three to five weeks after nematode inoculation (Wang and Bergeson, 1974).

Activated fungal spores germinate and reach the host root, where they penetrate and eventually may become involved in a synergistic interaction with the nematode. For this activation and growth of the fungus the energy demand is met from root exudates.

13.2 AGGRESSIVENESS OF FUNGAL PATHOGENS

Histophysiological changes caused by nematodes are the basis of increased aggressiveness of fungal pathogenic damage to the host. Of several nematodes of economic importance, root-knot nematodes, *Meloidogyne* spp., are most thoroughly studied and are commonly found involved in synergistic interactions with wilt-causing or root-rot fungi. Chemical secretions of root-knot nematodes (Bird, 1974; Hussey, 1989) induce a chain of reactions within the plant tissue that ultimately result in the development of giant cells which function as a nutrient sink to the developing juveniles, adults and egg-laying females. Similar features are shown by some other endoparasitic nematodes (Dropkin, 1969; Bird, 1974; Bird and Loveys, 1975).

Apart from these structural features, some of the biochemical changes occurring during the development of syncytia/giant cells have been characterized and quantified. Cellulose and lignin have been found to decrease by 31% and 36% and carbohydrates and pectins by 36%, while amino acids, DNA, hemicelluloses, lipids, minerals, nucleotides, organic acids, proteins and RNA increased by 304%, 70%, 36%, 154%, 4%, 29%, 67%, 80% and 87%, respectively, in tomato roots infected with root-knot nematode-infected plants (Owens and Specht, 1966). Thus, nutritionally enriched medium in nematode-infected plants is the cause for rapid growth of the invading fungus, which results in synergistic interactions. The *Meloidogyne*-induced giant cells are reported to contain maximum concentrations of DNA, RNA, sugars and photosynthates and to have an intense metabolic activity in terms of protein synthesis and active transport of amino acids at about three to four weeks after nematode infection (Bird, 1974; Wang and Bergeson, 1974; Webster, 1975; Bird and Loveys, 1975). Moreover, this is the period in which the nematodes have been shown to exert their greatest predisposing effects on their host plants to fungus infection and subsequent disease development (Porter and Powell, 1967; Powell, 1971, 1979).

A number of studies underline the significance of a nematode-infected tissue from which the fungal pathogen derives aggressiveness and becomes pronounced. Melendez and Powell (1967) observed that hyphae of *Fusarium oxysporum* f. *nicotianae* invaded *Meloidogyne incognita*-induced galls on a susceptible cultivar of flue-cured tobacco. The fungal hyphae were highly vigorous in giant cells. Nearby vessel elements and some female nematodes, and the gelatinous matrix of the egg masses, were also invaded. This aggressive colonization by the fungus caused collapse of the giant cells. Khan and Müller (1982) using a weak pathogenic isolate of *Rhizoctonia solani* and root-knot nematode, *M. hapla*, on radish observed that fungus growth on galls was much more luxuriant, with numerous sclerotia, in comparison to the non-galled region where mycelial growth was poor, with no sclerotial formation. Necrosis of roots was not evident even after three weeks of inoculation in the non-galled region. The mycelium traversed the cortical portion of the galls rapidly and colonized giant cells. Root tissues were obliterated at several places and there was no evidence of the female nematodes and giant cells. This differential colonization of nematode-affected and unaffected tissues, as was evidenced by histopathological studies of roots infected with both the organisms combined, was implicated by the fact that galled roots provided nutrients which were deficient in non-galled roots (Powell and Nusbaum, 1960; Melendez and Powell, 1967; Polychronopoulos *et al.*, 1969; Batten and Powell, 1971; Powell, 1971). The in vitro test also indicated that more rapid and profuse growth of the fungus mycelium occurred in the culture medium supplemented with root-gall extract than in the medium supplemented with root extract of nematode-uninfected plants (Melendez and Powell, 1970; Powell, 1971; Ribeiro and Ferraz, 1983).

In the case of vascular wilt fungi, fungus colonization was confined not only to the root galls of *Meloidogyne*-infected plants, but spread vigorously into the vascular system of non-galled areas of root, stem and leaves (Noguera, 1983; Hillocks, 1985; Jeffers *et al.*, 1985; Mousa and Hague, 1987; Storey and Evans, 1987). Histopathological studies of celery roots infected with *Pythium polymorphon* plus *M. hapla*, and with *P. polymorphon* alone, revealed that mycelium development in non-galled roots at 48 and 72 hours after inoculation was similar to that in galled roots at 24 and 48 hours, respectively. However, the percentage of galled samples colonized was greater than the percentage of *M. hapla*-free samples colonized. But autoclaved/double filter-sterilized aqueous extracts of galled and non-galled roots supplemented with 1% glucose equally supported the growth of *P. polymorphon*. Thus, the results were at variance with the earlier views that galled roots provided nutrients for fungus colonization.

Sporulation, on the other hand, was greater in medium supplemented with root-gall extracts than in the medium with extracts from non-galled roots in vitro (Starr and Aist, 1977). Brodie and Cooper (1964) also supported the observation with regard to sporulation, however, in *Pythium debaryanum*–root-knot complexes on cotton. Differences in the intensity of sporulation but not in the mycelial growth on medium supplemented with galled and non-galled extracts suggests either that the nutritional requirement of growth and sporulation are dissimilar, or that the findings warrant further experimentation. Staining of roots in acid fuchsin and lactophenol showed perithecia production in strawberry inoculated with *Gnomonia comari* and *Pratylenchus penetrans* but not in plants inoculated the fungus alone. Where growing evidence implicated nematode infections with rapid fungus colonization of the host, speculations were also made that prior nematode inoculation might suppress the establishment of fungus, as was noticed in case of *Fusarium oxysporum* and *Heterodera daverti* interaction on subterranean clover, *Trifolium subterraneum* (Nordmeyer and Sikora, 1983).

Preinoculation of strawberry, eggplant or potato with *Pratylenchus penetrans*, which restricts itself to the cortex only and causes extensive damage to it, helped the microsclerotia of *Verticillium dahliae* to germinate and colonize various plants vigorously (Mountain and McKeen, 1965; McKinley and Talboys, 1979; Nicot and Rouse, 1987; Storey and Evans, 1987). This indicates that the nature of growth stimulatory factor(s) emanating from nematode infection might be similar in both *Pratylenchus* and *Globodera/Meloidogyne*, although their mode of parasitsm is altogether different – the former produces lesions in the cortex, a negative host-parasite relationship, while the latter induces the development of syncytia/giant cells in vasculer parenchyma, a highly balanced positive host–parasite relationship.

The syncytial/giant cell cytoplasm was reported to be rapidly disintegrated as the fungus colonized, and this appears to be a characteristic feature of cyst/root-knot–fungus disease complexes (Melendez and Powell, 1967; Golden and Van Gundy, 1975; Starr and Mai, 1976; Starr and Aist, 1977; Fattah and Webster, 1983). Temporal studies pertaining to the ultrastructural response of giant cell development in tomato infected with *Fusarium oxysporum* f. sp. *lycopersici* and *Meloidogyne javanica* revealed that during the first week of fungus–nematode invasion there was little evidence of wilt or of the presence of fungus mycelium, and giant cells developed normally. In the second week, hyphae spread throughout the xylem vessels adjacent to giant cells. The giant cell cytoplasm became less electron dense, nuclei became spherical, the nuclear membrane swelled and ruptured and nuclear chromatin contained scattered electron-dense spheres surrounded by an electron-light halo – in other words, nuclear atrophy occurred. Within the third week, hyphae and spores ramified throughout the xylem elements, accompanied by necrosis of giant cells. The cell-free culture filtrate of *F. oxysporum* f. sp. *lycopersici*, when applied to *M. javanica*-infected tomato roots, brought about similar ultrastructural changes in the giant cells as were observed in roots infected with the same fungus (Fattah and Webster, 1981, 1983; Webster, 1985; Fattah and Webster, 1989). It was further noted that structural changes in giant cells started taking place before hyphae invaded them. This implies that disintegration of syncytia or giant cells is influenced more by fungal secretions of a translocatable nature than the nutrition depletion brought about by fungus colonization (Webster, 1985; Fattah and Webster, 1989).

The occurrence of root colonization by various fungi to their greatest extent within three to four weeks after nematode infection (Powell and Nusbaum, 1960; Powell, 1971; Bergeson, 1975; Fattah and Webster, 1983; Webster, 1985; Overstreet and McGawley, 1988), the development of synergistic interactions in plants where nematodes precede by three to four weeks the fungus infection, and the development of parasitic second-stage juveniles of root-knot or cyst nematodes to adult stage during this period appear to be highly synchronized and provide additional support to the physiological/nutritional basis of disease complexes.

The nutrient status of nematode-infected roots thus paves the way for development of disease complexes. The fungal pathogen benefits from the nutritionally enhanced status of the nematode-infected host. It shows greater aggressiveness and causes greater pathogenic damage to the host. In other words, this enhanced aggressive role is conferred to the fungal pathogen by the nematode through histophysiological and biochemical changes induced by it in the host. The nematode is finally depleted due to overdominance of the fungal pathogen.

13.3 EFFECT OF FUNGI ON NEMATODES

Fungi in general in synergistic interactions with nematodes dominate in the complex and suppress the development and reduce the resulting population density of the nematode. Some pathogenic fungi produce metabolites that suppress hatching of nematode juveniles (Vaishnav *et al.*, 1985; Mani *et al.*, 1986; Ciancio *et al.*, 1988).

The population density of nematodes in their interactions with fungi is affected by the interactive effects of the pathogens. Most of the reports of interactions between the migratory endoparasites such as *Pratylenchus* spp. and pathogenic fungi show that the population of the nematode generally increases (Mountain and McKeen, 1962, 1965; Faulkner and Skotland, 1965; Powell, 1971; Siti, 1979; Vrain, 1987; Hasan, 1988). In contrast, the sedentary endoparasites such as *Meloidogyne*, *Heterodera* and *Globodera* show reduced population density in the presence of wilt fungi (*Fusarium*, *Verticillium*) or in a few instances in the presence of root-rot fungi (*Pythium*, *Rhizoctonia* and *Phytophthora*) (Powell, 1971; Salem, 1980; Hasan, 1984; Nordmeyer and Sikora, 1983; Ribeiro and Ferraz, 1983; Al-Hazmi, 1985; Griffin and Thyr, 1986; Starr and Veech, 1986; Griffin *et al.*, 1988; Hasan, 1989; Starr *et al.*, 1989; Gray *et al.*, 1990).

The repression of sedentary endoparasites has been implicated with the impairment of nutrient supply to the developing and egg-laying adult nematodes available to them through syncytia or giant cells induced by the invading juveniles. This is borne out by the fact that disintegration of syncytia/giant cells occurs in fungus-infected plants. However, the role of a translocatable fungus metabolite of allelopathic nature (repressive to the nematode) cannot be discounted in the light of observations made by Fattah and Webster (1989) that the ultrastructural modifications of giant cells in *Fusarium* wilt-resistant and susceptible tomato cultivars treated with *F. oxysporum* f. sp. *lycopersici* culture filtrates were indistinguishable from those modifications in giant cells of *Fusarium*-infected plants. It is further substantiated from the findings that *Meloidogyne incognita* was suppressed on tomato even though it was physically isolated from *F. oxysporum* f. sp. *lycopersici*, i.e. nematode on one half and fungus on the other half of the split-root system (A. Hasan, unpublished data). Changes in *Pratylenchus minyus* population on peppermint infected with *Verticillium dahliae* when the two organisms were kept isolated on the double root systems further support the above observation (Faulkner *et al.*, 1970). A shift in sex ratio in favour of males in *Meloidogyne* spp. when crowding occurred within the roots (Tyler, 1933; Triantaphyllou, 1960) or in *Heterodera* (= *Globodera*) *rostochiensis* on tomato plants inoculated with *Rhizoctonia solani*, *V. albo-atrum* or grey sterile fungus (Ketudat, 1969) could also account for decrease in population due to the fact that these species are predominantly partheno-genetic and population increase is not the function of males but of females in any population. Although these results apparently indicated the role of altered sex ratio on the outcome of the nematode population, they ultimately related it to the nutrition stress-mediated change. But certain reports claiming an increase in population of these endoparasites including *Rotylenchulus reniformis*, a semi-endoparasite, neither support nor discount the nutritional or allelopathic effects on nematode multiplication (Ross, 1965; Hazarika and Roy, 1974; Tchatchoua and Sikora, 1978; Nordmeyer and Sikora, 1983; Overstreet and McGawley, 1988; MacGuidwin and Rouse, 1990). A significant increase in numbers of *Heterodera daverti* cysts on subterranean clover was reported with inoculation of 10^6 macroconidia of *F. oxysporum* plus 500 or 2000 juveniles of the nematode or with 10^4 macroconidia plus 1000 juveniles. Conversely, the cyst production decreased significantly when 500 juveniles were combined with 10^2 or 10^4 macroconidia (Nordmeyer and Sikora, 1983).

Little headway has been made to offer an explanation for the increase in population density of *Pratylenchus* spp. However, based on our limited knowledge, increase in population of this nematode on various plants infected with *Verticillium* wilt fungi could be ascribed to their vagrant nature (moving to fresh feeding sites when toxicity and necrotic lesions develop on the tissues fed by them) within the root tissues as well as to growth-promoting substances produced in response to fungus–host–nematode interactions (Faulkner and Skotland, 1965; Faulkner and Bolander, 1969). The latter assumption appears to be highly subjective due to lack of sufficient data to support it. And if that was the case, the multiplication of sedentary nematodes should have also been favoured. Also, in certain instances, the population of *Pratylenchus* spp. decreased on various host plants infected with wilt fungi (Mountain and McKeen, 1962; Bergeson, 1963; Conroy *et al.*, 1972; Burpee and Bloom, 1978; Mauza and Webster, 1982; Martin *et al.*, 1982; Riedel *et al.*, 1985; Rowe *et al.*, 1985). A positive correlation between migratory or sedentary endoparasitic nematode populations and root growth has also been suggested – however, with little experimental evidence – which indirectly supports the nutritional basis of nematode multiplication (Nordmeyer and Sikora, 1983; Vrain, 1987). But this view could also be subject to criticism in light of the fact that root systems are excessively reduced where synergistic interaction occurs irrespective of the nature of the host–parasite relationship of the nematodes.

Most of the disease complexes are of polycyclic nature and unless the role of first, second or third-generation nematodes on disease complexes is worked out intensively, the correlation of disease complexes with the final nematode population will appear to be of little consequence, since synergistic interactions had been found to be independent of nematode population changes (increase or decrease).

13.4 PATHOGENIC BEHAVIOUR OF SAPROPHYTIC FUNGI

Most of the soil fungi are inherently saprophytic and obtain their nutrition through decomposition of organic materials. This decomposition is an essential biological transformation that ensures recycling of the material in an ecosystem. Some reports indicate that some common soil fungi become parasitic on roots infected with root-knot nematodes. Melendez and Powell (1969) were first to recognize this ability in a saprophytic fungus. While studying the influence of root-knot infection on root decay of tobacco by *Pythium ultimum* and *Trichoderma harzianum*, they observed that *T. harzianum* – a saprophyte – was able to damage the roots. *P. ultimum* also caused extensive root necrosis, though this fungus is known to attack only seedlings, causing damping-off disease, and is not parasitic on plants of advanced age. In a later study, Powell *et al.* (1971) demonstrated that a number of soil fungi which are common soil inhabitants and are non-parasitic on tobacco, caused necrosis of roots infected with root-knot nematode, *Meloidogyne incognita*. These fungi included *Curvularia trifolii, Botrytis cinerea, Aspergillus ochraceus, Penicillium martensii, T. harzianum* and *P. ultimum*. Earlier research (Porter and Powell, 1967; Powell and Nusbaum, 1960) on disease complexes in tobacco had shown that *M. incognita* usually involves those fungi which are capable of causing the respective disease by themselves without the aid of the nematode. Interactions of *M. incognita* and the fungi involved were recognized by increased disease incidence and rapid disease development. Therefore, the parasitic behaviour of saprophytic fungi under the influence of nematode-infected roots is a unique feature and underlines that this transformation may become significant in soil in relation to plant damage.

13.5 RESISTANCE BREAKING BY FUNGI

If nematodes can break resistance in crop cultivars to fungus infection, it is logical to presume that fungal pathogens might also be involved in reducing resistance of cultivars/genotypes to nematode infection, especially of the endoparasitic ones. Out of 31 cultivars/lines of chilli pepper (*Capsicum annuum*) tested in glasshouse for their resistance to *Meloidogyne incognita*, one (G-31-1-6) gave a highly resistant response: five were resistant and 16 were moderately resistant to infection of this nematode. Under field conditions, G-31-1-6 remained unaffected but of the resistant cultivars/lines, Jowala and C-3 partially lost their resistance, and of 16 moderately resistant cultivars/lines eight showed a susceptible response to infection of this nematode. Soil and root examination revealed the association of *Rhizoctonia solani* and *Pythium aphanidermatum*. Under controlled conditions, when Jowala (resistant) and Longthin Faizabadi (moderately resistant) cultivars were inoculated with *M. incognita* and *R. solani* or *P. aphanidermatum*, they were found to be losing their resistance, i.e. Jowala gave a moderately resistant response while Longthin Faizabadi gave a susceptible response to infection of this nematode. Thus, the fungi were implicated in the breaking down of resistance in these cultivars to *M. incognita* infection (Hasan, 1985).

Similarly, in a greenhouse study on the effect of *Rhizoctonia solani, Sclerotium rolfsii* and *Verticillium dahliae* on expression of resistance to *Meloidogyne incognita* in tomato cultivars F-24-C8 (immune), Heinz-1409 (resistant) and T-4 (moderately resistant), it was shown that *V. dahliae* reduced the resistance of all three cultivars, *R. solani* of T-4 only and *S. rolfsii* of Heinz-1409 only (Hasan and Khan, 1985). Khan and Husain (1989) also noted that in presence of *R. solani*, resistance of cowpea cultivars to nematodes, *M. incognita* and *Rotylenchulus reniformis* was broken. If future studies substantiate these observations, a fresh look at the strategy of developing nematode-resistant cultivars will be warranted as these fungi are of wide occurrence in various soils.

13.6 SELECTIVE INTERACTIONS OF SOME FUNGAL PATHOGENS

The tomato wilt fungus *Fusarium oxysporum* f. sp. *lycopersici* was reported to be non-pathogenic to cucumber, whereas if plants were simultaneously attacked by *Meloidogyne incognita* and this fungus, wilt symptoms developed. But the reverse was untrue, i.e. when cucumber wilt pathogen, *F. oxysporum* f. sp. *cucumerinum*, non-pathogenic to tomato plants, attacked them simultaneously with *M. incognita* no effect was found (Hirano and Kawamura, 1971; Pelcz *et al.*, 1983). *Verticillium albo-atrum* interacted with lesion nematodes *Pratylenchus penetrans* and *P. vulnus* on balsam (*Impatiens balsamina*) but interaction did not occur with the fungus in the presence of *P. fallax*, *P. thornei* or *P. crenatus* (Müller, 1977). Similarly, the degree of interaction of *V. dahliae* on potato cultivar Superior varied with species: *P. penetrans*, *P. crenatus* and *P. scribneri* (Riedel *et al.*, 1985). *Fusarium* wilt in alfalfa caused by *F. oxysporum* f. sp. *medicaginis* increased synergistically from combined inoculations of *M. hapla* on cultivars Ranger and Nevada Synthetic XX, while less fungal infection and vascular discoloration occurred on cultivar Moapa 69. Ranger alfalfa showed synergism to this fungus in the presence of *Ditylenchus dipsaci*, while cultivars Lahontan and Moapa 69 exhibited merely an additive effect (Griffin and Thyr, 1988; Griffin, 1990).

The causation of wilt disease in cucumber by tomato wilt pathogen in the presence of *Meloidogyne incognita* and failure of cucumber wilt pathogen in the presence of the same nematode, differential interaction of wilt fungus of different species of the same nematode on the same host or differential interaction of the same wilt fungus and the lesion nematode but on different host cultivars show that fungus–nematode interactions are of a highly specific nature and might be determined at the molecular level by specific gene combinations which differ from one cultivar–pathogen system to another.

13.7 CONCLUSIONS

Disease complexes resulting from synergistic interactions between patho-genic fungi and plant-parasitic nematodes present an interesting area of study. In this system, the role of the nematode – one of the two interacting pathogens – has been well documented, but the role of the fungal component has attracted inadequate attention. The roles described in this chapter may be well or ill defined, precise or imprecise or interlinked, but the fungal component has a significant bearing over the course of development of the disease complex and the subsequent impact on the host in relation to damage and on the nematode in relation to population growth. The dominance of fungi in later stages of disease complexes and suppression of the nematode population density may be significant in population dynamics of the nematode and disease epidemiology. There is hardly any objective study or data to draw any specific conclusion. This aspect needs to be examined in some detail in studies dealing with diseases of complex aetiology involving fungi and nematodes.

The influence of the fungal pathogen on resistance of a crop cultivar to the nematode is an important aspect and offers a new area of study. Resistance breaking by the nematode has been examined in a number of studies but very few attempts have been made otherwise. This phenomenon when fully established may complicate the situation for breeders in their efforts to manage nematodes through host resistance. This study area may create a new awareness and generate information to elucidate the mechanism of fungus–nematode interactions.

The parasitic behaviour of saprophytic fungi on nematode-infected roots is fascinating and may be very significant in the soil ecosystem. Unfortunately, no serious effort seems to have been made to examine the magnitude and importance of this acquired behaviour of saprophytic fungi in the given circumstances.

REFERENCES

Al-Hazmi, A.S. (1985) Interaction of *Meloidogyne incognita* and *Macrophomina phaseolina* in root-rot disease complex of French bean. *Phytopathologische Zeitschrift*, 113, 311–16.

Atkinson, G.F. (1892) Some diseases of cotton. *Bulletin of the Alabama Agricultural Experiment Station*, 41, 61–5.

Batten, C.K. and Powell, N.T. (1971) The *Rhizoctonia–Meloidogyne* disease complex in flue-cured tobacco. *Journal of Nematology*, 3, 164–9.

Bergeson, G.B. (1963) Influence of *Pratylenchus penetrans* alone and in combinations with *Verticillium albo-atrum* on peppermint. *Phytopathology*, 53, 1164–6.

Bergeson, G.B. (1972) Concepts of nematode–fungus associations in plant disease complexes: A review. *Experimental Parasitology*, 32, 301–14.

Bergeson, G.B. (1975) The effect of *Meloidogyne incognita* on the resistance of four musk-melon varieties to Fusarium wilt. *Plant Disease Reporter*, 56, 1022–6.

Bird, A.F. (1974) Plant response to root-knot nematode. *Annual Review of Phytopathology*, 12, 69–85.

Bird, A.F. and Loveys, B.R. (1975) The incorporation of photosynthates by *Meloidogyne javanica*. *Journal of Nematology*, 7, 111–13.

Brodie, B.B. and Cooper, W.E. (1964) Relation of parasitic nematodes to post-emergence damping-off of cotton. *Phytopathology*, 54, 1023–7.

Burpee, L.L. and Bloom, J.R. (1978) Influence of *Pratylenchus penetrans* on the incidence and severity of Verticillium wilt of potato. *Journal of Nematology*, **10**, 95–9.

Ciancio, A., Logrieco, A., Lamberti, F. and Bottalico, A. (1988) Nematicidal effects of some *Fusarium* toxins. *Nematologia Mediterranea*, **16**, 137–8.

Conroy, J.J., Green, R.J., Jr and Ferris, J.M. (1972) Interaction of *Verticillium albo-atrum* and root-lesion nematode *Pratylenchus penetrans* in tomato root at controlled inoculum densities. *Phytopathology*, **62**, 362–6.

Dropkin, V.H. (1969) Cellular responses of plants to nematode infections. *Annual Review of Phytopathology*, **7**, 101–22.

Evans, K. (1987) The interaction of potato cyst nematodes and *Verticillium dahliae* on early and main crop potato cultivars. *Annals of Applied Biology*, **110**, 329–39.

Fattah, F.A. and Webster, J.M. (1981) Effect of culture filtrate of *Fusarium oxysporum* f. sp. *lycopersici* on the ultrastructure of giant cells induced by *Meloidogyne javanica* in tomato. *Journal of Nematology*, **13**, 457–8.

Fattah, F.A. and Webster, J.M. (1983) Ultrastructural changes caused by *Fusarium oxysporum* f. sp. *lycopersici* in *Meloidogyne javanica* induced giant cells in Fusarium resistant and susceptible tomato cultivars. *Journal of Nematology*, **15**, 128–35.

Fattah, F.A. and Webster, J.M. (1989) Ultrastructural modifications of *Meloidogyne javanica* induced giant cells caused by fungal culture filtrates. *Revue de Nematologie*, **12**, 197–210.

Faulkner, L.R. and Bolander, W.J. (1969) Interaction of *Verticillium dahliae* and *Pratylenchus minyus* in Verticillium wilt of peppermint. *Phytopathology*, **59**, 868–70.

Faulkner, L.R. and Skotland, C.J. (1965) Interaction of *Verticillium dahliae* and *Pratylenchus minyus* in Verticillium wilt of peppermint. *Phytopathology*, **55**, 583–6.

Faulkner, L.R., Bolander, W.J. and Skotland, C.B. (1970) Interaction of *Verticillium dahliae* and *Pratylenchus minyus* in Verticillium wilt of peppermint: Influence of nematode as determined by double root technique. *Phytopathology*, **60**, 100–3.

Francl, L.J., Rowe, R.C., Riedel, R.M. and Maddan, L.V. (1988) Effects of three soil types on potato early dying disease and associated yield reduction. *Phytopathology*, **78**, 159–66.

Golden, J.K. and Van Gundy, S.D. (1972) Influence of *Meloidogyne incognita* on root development by *Rhizoctonia solani* and *Thielaviopsis basicola* in tomato. *Journal of Nematology*, **4**, 225.

Golden, J.K. and Van Gundy, S.D. (1975) A disease complex of okra and tomato involving the nematode *Meloidogyne incognita* and the soil inhabiting fungus, *Rhizoctonia solani*. *Phytopathology*, **65**, 265–73.

Gray, F.A., Griffin, G.D., Johnson, D.A, Eckert, J.W. and Kazimir, J.E. (1990) Inter-relationships between *Meloidogyne hapla* and *Phytophthora megasperma* f. sp. *medicaginis* in seedling damping-off and root infection of alfalfa. *Phytopathology*, **80**, 228–32.

Griffin, G.D. (1990) Pathological relationship of *Ditylenchus dipsaci* and *Fusarium oxysporum* f. sp. *medicaginis* on alfalfa. *Journal of Nematology*, **22**, 333–6.

Griffin, G.D. and Thyr, B.D. (1986) The importance of nematode resistance on the interaction of *Meloidogyne hapla* and *Fusarium oxysporum* on alfalfa. *Phytopathology*, **76**, 843–4.

Griffin, G.D. and Thyr, B.D. (1988) Interaction of *Meloidogyne hapla* and *Fusarium oxysporum* f. sp. *medicaginis* on alfalfa. *Phytopathology*, **78**, 421–5.

Griffin, G.D., Gray, F.A. and Johnson, D.A. (1988) Effect of *Meloidogyne hapla* on resistance and susceptibility of alfalfa to *Phytophthora megasperma* f. sp. *medicaginis*, in *Report of 31st North American Alfalfa Improvement Center at Beltsville, Maryland, Utah State University, London, USA*, p. 23.

Hale, M.G., Foy, C.L. and Shay, F.J. (1971) Factors affecting root exudation. *Advances in Agronomy*, **23**, 89–109.

Hasan, A. (1984) Synergism between *Heterodera cajani* and *Fusarium udum* attacking *Cajanus cajan*. *Nematologia Mediterranea*, **12**, 159–62.

Hasan, A. (1985) Breaking resistance in chilli to root-knot nematode by fungal pathogens. *Nematologica*, **31**, 210–17.

Hasan, A. (1988) Interaction of *Pratylenchus coffeae* and *Pythium aphanidermatum* and/or *Rhizoctonia solani* on chrysanthemum. *Phytopathologische Zeitschrift*, **123**, 227–32.

Hasan, A. (1989) Efficacy of certain nonfumigant nematicides on the control of pigeonpea wilt involving *Heterodera cajani* and *Fusarium udum*. *Phytopathologische Zeitschrift*, **126**, 335–42.

Hasan, A. and Khan, M.N. (1985) The effect of *Rhizoctonia solani*, *Sclerotium rolfsii* and *Verticillium dahliae* on the resistance of *Meloidogyne incognita*. *Nematologia Mediterranea*, **13**, 133–6.

Hazarika, B.P. and Roy, A.K. (1974) Effect of *Rhizoctonia solani* on the reproduction of *Meloidogyne incognita* on eggplant. *Indian Journal of Nematology*, **4**, 246.

Hillocks, R.J. (1985) The effect of root-knot nematode on vascular resistance to *Fusarium oxysporum* f. sp. *vasinfectum* in the stems of cotton plants. *Annals of Applied Biology*, **107**, 213–18.

Hirano, K. and Kawamura, T. (1971) Changes in parasitism of *Fusarium* spp. in plants exposed to root-knot nematode and Fusarium complex. *Technical Bulletin, Faculty of Horticulture, Chiba University*, **19**, 29–33.

Hussey, R.S. (1989) Disease inducing secretions of plant-parasitic nematodes. *Annual Review of Phytopathology*, **27**, 123–41.

Jeffers, D.P., Garber, R.H. and Roberts, P.A. (1985) Factors affecting Fusarium wilt development in four cotton selections. *Phytopathology*, **75**, 1347.

Ketudat, U. (1969) The effects of some soil-borne fungi on the sex ratio of *Heterodera rostochiensis* on tomato. *Nematologica*, **15**, 229–33.

Khan, M.W. and Müller, J. (1982) Interaction between *Rhizoctonia solani* and *Meloidogyne hapla* on radish in gnotobiotic culture. *Libyan Journal of Agriculture*, **11**, 133–40.

Khan, T.A. and Husain, S.I. (1989) Relative resistance of six cowpea cultivars as affected by the concomitance of two nematodes and a fungus. *Nematologia Mediterranea*, **17**, 39–41.

MacGuidwin, A.E. and Rouse, D.I. (1990) Effect of *Meloidogyne hapla* alone and in combination with subthreshold populations of *Verticillium dahliae*, on disease symptomology and yield of potato. *Phytopathology*, **80**, 482–6.

Mai, W.F. and Abawi, G.S. (1987) Interaction among root-knot nematodes and Fusarium wilt fungi on host plants. *Annual Review of Phytopathology*, **25**, 317–38.

Mani, A., Sethi, C.L. and Devkumar (1986) Isolation and identification of nematoxins produced by *Fusarium solani* (Mart) Sacc. *Indian Journal of Nematology*, **16**, 247–51.

Martin, M.J., Riedel, R.M. and Rowe, R.C. (1982) *Verticillium dahliae* and *Pratylenchus penetrans*: Interactions in the early dying complex of potato in Ohio. *Phytopathology*, **72**, 640–7.

Mauza, B.E. and Webster, J.M. (1982) Suppression of alfalfa growth by concomitant populations of *Pratylenchus penetrans* and *Fusarium* species. *Journal of Nematology*, **14**, 364–7.

McKinley, R.T. and Talboys, P.W. (1979) Effect of *Pratylenchus penetrans* on development of strawberry wilt caused by *Verticillium dahliae*. *Annals of Applied Biology*, **92**, 347–57.

Melakeberhan, H., Brooke, R.C., Websters, J.M. and D'Auria, J.M. (1985) The influence of *Meloidogyne incognita* on the growth, physiology and nutrient content of *Phaseolus vulgaris*. *Physiological Plant Pathology*, **26**, 259–68.

Melendez, P.L. and Powell, N.T. (1967) Histological aspects of the Fusarium wilt–root-knot complex in flue-cured tobacco. *Phytopathology*, **57**, 286–92.

Melendez, P.L. and Powell, N.T. (1969) The influence of *Meloidogyne* on root decay in tobacco caused by *Pythium* and *Trichoderma*. *Phytopathology*, **59**, 1348.

Melendez, P.L. and Powell, N.T. (1970) The *Pythium* root-knot nematode complex in flue-cured tobacco. *Phytopathology*, **60**, 1303.

Mountain, W.B. and McKeen, C.D. (1962) Effect of *Verticillium dahliae* on the population of *Pratylenchus penetrans*. *Nematologica*, **7**, 261–6.

Mountain, W.B. and McKeen, C.D. (1965) Effects of transplant injury and nematodes on incidence of Verticillium wilt of eggplant. *Canadian Journal of Botany*, **43**, 619–24.

Mousa, E.M. and Hague, N.G.M. (1987) The influence of root-knot nematode, *Meloidogyne incognita*, on the development of *Fusarium oxysporum* f. sp. *glycines* in wilt resistant and wilt susceptible soybean cultivars. *Mededlingen van de Faculteit Landbouwwetenschappen Rijksunviersiteit, Gent*, **52**, 571–5.

Müller, J. (1977) Interactions between 5 species of *Pratylenchus* and *Verticillium albo-atrum*. *Zeitschrift für Pflanzenkrankheiten und Pflanzenschutz*, **84**, 215–20.

Nicot, P.C. and Rouse, D.I. (1987) Relationship between soil inoculum density of *Verticillium dahliae* and systemic colonization of potato stems in commercial fields over time. *Phytopathology*, **77**, 1346–55.

Noguera, R. (1983) Influence of *Meloidogyne incognita* on the colonization of *Fusarium oxysporum* f. sp. *lycopersici* in tomatoes. *Agronomia Tropical*, **33**, 103–10.

Nordmeyer, D. and Sikora, R.A. (1983) Studies on the interaction between *Heterodera daverti*, *Fusarium avenaceum* and *F. oxysporum* on *Trifolium subterraneum*. *Revue de Nematologie*, **6**, 193–8.

Orion, D. and Netzer, D. (1981) Suppressive effects of the root-knot nematode on Fusarium wilt of muskmelons. *Revue de Nematologie*, **4**, 65–70.

Overstreet, C. and McGawley, E.C. (1988) Influence of *Calonectria crotalariae* on reproduction of *Heterodera glycines* on soybean. *Journal of Nematology*, **20**, 457–67.

Owens, R.G. and Specht, H.N. (1966) Biochemical alterations induced in host tissues by root-knot nematodes. *Boyce Thompson Institute Contribution*, **23**, 181–98.

Pelcz, J., Skadow, K. and Fritzsche, R. (1983) Influence of *Meloidogyne incognita* on the host susceptibility of cucumber to *Fusarium oxysporum* f. sp. *lycopersici* and of tomato to *F. oxysporum* f. sp. *cucumerinum*. *Nematologica*, **29**, 443–53.

Pitcher, R.S. (1965) Interrelationships of nematodes and other pathogens of plants. *Helminthological Abstract*, **34**, 1–17.

Pitcher, R.S. (1978) Interactions of nematodes with other pathogens, in *Plant Nematology* (ed. J.F. Southey), Her Majesty's Stationery Office, London, pp. 63–77.

Polychronopoulos, A.G., Houston, B.R. and Lownsbery, B.F. (1969) Penetration and development of *Rhizoctonia solani* in sugar beet seedlings infected with *Heterodera schachtii*. *Phytopathology*, **59**, 482–5.

Porter, D.M. and Powell, N.T. (1967) Influence of certain *Meloidogyne* species on Fusarium wilt development in flue-cured tobacco. *Phytopathology*, **57**, 282–5.

Powell, N.T. (1971) Interactions between nematodes and fungi in disease complexes. *Annual Review of Phytopathology*, **9**, 253–74.

Powell, N.T. (1979) Internal synergisms among organisms inducing disease, in *Plant Disease*, Vol. IV (eds J.G. Horsfall and E.B. Cowling), Academic Press, New York, pp. 113–33.

Powell, N.T. and Nusbaum, C.J. (1960) The black shank–root-knot complex in flue-cured tobacco. *Phytopathology*, **50**, 899–906.

Powell, N.T., Melendez, P.L. and Batten, C.K. (1971) Disease complexes in tobacco involving *Meloidogyne incognita* and certain soil-borne fungi. *Phytopathology*, 61, 1332–7.

Ribeiro, C.A. and Ferraz, S. (1983) Studies on the interaction between *Meloidogyne javanica* and *Fusarium oxysporum* f. sp. *phaseoli* on bean (*Phaseolus vulgaris*). *Phytopathologia Brasileira*, 8, 439–46.

Riedel, R.M. (1988) Interactions of plant-parasitic nematodes with plant pathogens. *Agriculture Ecosystem and Environment*, 24, 281–92.

Riedel, R.M., Rowe, R.C. and Martin, M.J. (1985) Differential interaction of *Pratylenchus crenatus*, *P. penetrans* and *P. scribneri* in potato early dying disease. *Phytopathology*, 75, 419–22.

Ross, J.P. (1965) Predisposition of soybeans to Fusarium wilt by *Heterodera glycines* and *Meloidogyne incognita*. *Phytopathology*, 55, 361–6.

Rovira, A.D. (1965) Plant root exudates and their influence upon soil microorganisms, in *Ecology of Soil-Borne Plant Pathogens* (eds K.F. Baker and W.C. Snyder), University of California Press, Berkeley, pp. 170–86.

Rowe, R.C., Riedel, R.M. and Martin, M.J. (1985) Synergistic interactions between *Verticillium dahliae* and *Pratylenchus penetrans* in potato early disease. *Phytopathology*, 75, 412–18.

Salem, A.A.M. (1980) Effect of *Meloidogyne incognita* and *Fusarium oxysporum* f. sp. *vasinfectum* on cotton, *Gossypium barbadense*. *Egyptian Journal of Phytopathology*, 12, 27–30.

Siti, E. (1979) The interrelationships between *Pratylenchus thornei* and *Verticillium dahliae* and their effect on potatoes, PhD thesis, Volcani Center, Bet Dagaon, Israel.

Starr, J.L. and Aist, J.R. (1977) Early development of *Phythium polymorphon* on celery roots infected by *Meloidogyne hapla*. *Phytopathology*, 67, 497–501.

Starr, J.L. and Mai, W.F. (1976) Effect of soil microflora on the interaction of three plant-parasitic nematodes with celery. *Phytopathology*, 66, 1224–8.

Starr, J.L. and Veech, J.A. (1986) Susceptibility of root-knot nematodes in cotton lines resistant to the Fusarium wilt/root-knot complex. *Crop Science*, 26, 543–6.

Starr, J.L., Jeger, M.J., Martyn, R.D. and Schilling, K. (1989) Effects of *Meloidogyne incognita* and *Fusarium oxysporum* f. sp. *vasinfectum* on plant mortality and yield of cotton. *Phytopathology*, 79, 640–6.

Storey, G.W. and Evans, K. (1987) Interactions between *Globodera pallida* juveniles, *Verticillium dahliae* and three potato cultivars, with descriptions of associated histopathologies. *Plant Pathology*, 36, 192–200.

Taylor, C.E. (1979) *Meloidogyne* interrelationships with microorganisms, in *Root-Knot Nematodes (Meloidogyne Species): Systematics, Biology and Control* (eds F. Lamberti and C.E. Taylor), Academic Press, London, pp. 375–98.

Taylor, C.E. (1990) Nematode interactions with other pathogens. *Annals of Applied Biology*, 116, 405–16.

Tchatchoua, J. and Sikora, R.A. (1978) Untersuchungen uber die wechselbeziehungen zwischen *Rotylenchulus reniformis* Linford & Oliveira 1940 und *Verticillium dahliae* an baumwollpflanzen. *Mededlingen van de Faculteit Landbouwweteuschappen Rijksuniversiteit, Gent*, 43/2, 757–64.

Triantaphyllou, A.C. (1960) Sex determination in *Meloidogyne incognita* Chitwood, 1949 and intersexuality in *M. javanica* (Treub, 1885) Chitwood, 1949. *Annals of Institute of Phytopathology*, 3, 12–31.

Tyler, J. (1933) Reproduction without males in aseptic root cultures of the root-knot nematode. *Hilgardia*, 7, 391–415.

Vaishnav, M.U., Patel, H.R. and Dhruj, I.U. (1985) Effect of culture filtrates of *Aspergillus* spp. on *Meloidogyne arenaria*. *Indian Journal of Nematology*, 15, 116–17.

Van Gundy, S.D., Kirkpatrick, J.D. and Golden, J. (1977) The nature and role of metabolic leakage from root-knot nematode galls and infection by *Rhizoctonia solani*. *Journal of Nematology*, **9**, 113–21.

Vrain, T.C. (1987) Effect of *Ditylenchus dipsaci* and *Pratylenchus penetrans* on Verticillium wilt of alfalfa. *Journal of Nematology*, **19**, 379–83.

Walter, J.C. (1965) Host resistance as it relates to root pathogens and soil microorganisms, in *Ecology of Soil-Borne Plant Pathogens* (eds K.F. Baker and W.C. Snyder), University of California Press, Berkeley, pp. 314–20.

Wang, E.L.H. and Bergeson, G.B. (1974) Biochemical changes in root exudates and xylem sap of tomato plants infected with *Meloidogyne incognita*. *Journal of Nematology*, **6**, 194–202.

Wang, E.L.H., Hodges, T.K. and Bergeson, G.B. (1975) *Meloidogyne incognita* induced changes in cell permeability of galled roots. *Journal of Nematology*, **7**, 256–60.

Webster, J.M. (1975) Aspects of the host–parasite relationship of plant-parasitic nematodes. *Advances in Parasitology*, **13**, 225–50.

Webster, J.M. (1985) Interaction of *Meloidogyne* with fungi on crop plants, in *An Advanced Treatise on Meloidogyne, Vol. I: Biology and Control* (eds J.N. Sasser and C.C. Carter), North Carolina State University Graphics, Raleigh, pp. 183–92.

14

Biochemical and genetic basis of fungus–nematode interactions

Jean-Claude Prot

Plant-parasitic nematodes are primary plant pathogens. Independently, they can cause important plant diseases. However, Powell's (1971) conclusion of a review paper on interaction between nematodes and other disease-causing agents was that nematodes are major components of disease complexes and that interactions with other pathogens may be their principal economic hazard. All root-parasitic nematodes cause mechanical injuries as they penetrate within or feed on root tissues, providing ready avenues of entry for other pathogens. For many years, the role of nematodes in fungal disease interaction was limited to wounding agents. However, mechanical wounding does not always promote fungal penetration within the root tissues. Hart and Endo (1981) reported that wounding roots of celery prior to inoculation of *Fusarium oxysporum* f. sp. *apii* had no effect on the development of *Fusarium* yellows. Sumner and Minton (1987) reported that wounding roots with a knife did not increase wilt caused by *Fusarium oxysporum* f. sp. *tracheiphilum* race 1 in soya bean cv. Cobb, while the disease was more severe in the presence of *Belonolaimus longicaudatus* and *Pratylenchus brachyurus*. There seems to be something special about the role of some nematode wounds in fungal disease interaction (Pitcher, 1965).

Statements such as 'Nematodes predispose plants to fungal diseases' or 'Nematode infestations enhance fungal development and fungal disease symptoms in plants' are based on many research papers on interactions involving nematodes and fungi. Other researchers have indicated that nematodes may affect fungal resistance in plants and that they may render plants susceptible to fungi which are innocuous in the absence of nematodes (Batten and Powell, 1971; Powell, 1971). These results and statements imply that nematodes induce or produce factor(s) which can alter the susceptibility of their hosts to fungal diseases or enhance fungus ability to penetrate and develop within the plant tissues.

Most would agree that plant-parasitic nematodes induce physiological, biochemical and structural changes in their hosts. On the other hand, many abiotic and biotic factors can predispose plants to diseases that would otherwise occur to a lesser extent (Lockwood, 1988). In addition, almost any stress or stimulus seems to influence fungal disease development in plants. For example, Shawish and Baker (1982) reported that a gentle shaking of stems and leaves for 1 minute each day increases *Fusarium* wilt symptom expression in flax, pea and tomato. It will be purely speculative to discuss all changes induced by nematodes in their hosts that could potentially influence fungal disease developments. Because the importance and complexity of interactions between biotic and abiotic disease determinants have already been described (Wallace, 1978), this chapter will not discuss all interactions between nematodes and other biotic and abiotic factors. It will be limited to an interpretative evaluation of relevant literature which directly or indirectly supports or controverts biochemical and genetic bases of nematode–fungus interactions.

14.1 INOCULUM LEVEL

Observations indicate that plant predisposition to fungal diseases by plant-parasitic nematodes requires a minimum level of nematode infestation. Addition of 1000–2000 *Meloidogyne incognita* juveniles per plant significantly increased the number of plants of cotton cultivars Deltapine Smooth Leaf and Pima S-2 infected with *Verticillium albo-atrum*. However, addition of 250 and 500 juveniles per plant did not give a similar result (Khoury and Alcorn, 1973). A minimum initial population density of *M. incognita* was required to increase *Rhizoctonia solani* disease on cotton seedlings (Carter, 1975). A significant interaction occurred between *M. incognita* and *Fusarium oxysporum* f. sp. *vasinfectum* on cotton at a high nematode population, while no interaction was observed at a low nematode population level (Starr *et al.*, 1989).

Incubation period, incidence and symptom expression of some fungal diseases may also depend on the level of nematode infestation. The incubation period of *Cylindrocladium* black rot of peanut was shortened to three weeks when peanut plants were inoculated with 10^4 *Meloidogyne hapla* eggs compared to four weeks when 10^3 eggs per plant were used. Moreover, the highest incidence occurred with the highest inoculum level (Diomande and Beute, 1981). Wilt symptoms on tomato cultivar Matsudo-Ponderosa increased with the increasing number of *M. incognita* used to inoculate the plants (Kawamura and Hirano, 1967). Sumner and Johnson (1973) reported that symptom intensity of *Fusarium* wilt of water melon was significantly correlated with initial population of *M. incognita* juveniles. In controlled climate chamber experiments, the incidence of infection in tomato cultivar Bonny Best by *Verticillium albo-atrum* increased with increases in initial population density of *Pratylenchus penetrans* (Conroy *et al.*, 1972) and *Trichodorus christiei* (Conroy and Green, 1974).

An increase in incidence of a fungal disease related to an increase in nematode number which infects the plant may reflect the role of nematodes as wounding agents allowing the penetration of the fungus. The chances of infection by the fungus increase with increasing number of wounds. The shortening of the incubation period and an increase in severity of a fungal disease when the number of nematodes infecting the plant increase may also be related to a wounding action of the nematodes; however, they also suggest that the nematodes have to induce a certain level of physiological changes to modify the reaction of their host to the fungus.

14.2 SPECIFICITY OF NEMATODE–FUNGUS INTERACTIONS

Interactions between host plant, nematodes and fungi seem to be very specific and depend on the right combination between nematode species, plant species and(or) cultivar, and fungus.

Nematode species with different biology and feeding habits present differences in ability to predispose the same host to *Fusarium* wilt. Difference in ability to promote *Fusarium* wilt in cotton was observed between *Hoplolaimus galeatus* and two populations of *Belonolaimus longicaudatus* (Yang et al., 1976). *B. longicaudatus* promoted wilt development while *H. galeatus* did not. This difference was explained by the dissimilarity of root woundings caused by these two nematodes. *B. longicaudatus*, an ectoparasite with very long stylet, may provide access to the fungus to the vascular tissues of the plant. This cannot be achieved by *H. galeatus*, a migratory endoparasite which feeds primarily on cortical tissue. *Ditylenchus dipsaci* was unable to predispose alfalfa cultivar Moapa 90 to the infection of *F. oxysporum* f. sp. *medicaginis* (Griffin, 1990), while *Meloidogyne hapla* did (Griffin and Thyr, 1988). Griffin (1990) attributed this difference to variations in physiological effects of the two nematode species on alfalfa plant tissues.

Nematode species with closely related biology and feeding habits may present differences in ability to predispose a plant to infection by the same fungus. Wilt development was more rapid when *Fusarium oxysporum* f. sp. *vasinfectum* was associated with a population of *Belonolaimus longicaudatus* originally collected from cotton than with a population obtained from millet (Yang et al., 1976). Interaction between *Verticillium* spp. and lesion nematodes depends on the species of nematode involved. Müller (1977), studying interactions between five species of *Pratylenchus* and *V. albo-atrum* on balsam, reported that wilt was obtained when the fungus was combined with *Pratylenchus penetrans* and *P. vulnus* but wilting did not occur when combined with *P. crenatus*, *P. thornei* and *P. fallax*. *P. penetrans* interacted with *V. dahliae* in potato early dying disease on cultivar Superior while *P. crenatus* did not. *P. scribneri* also did not interact with *V. dahliae* except when high temperature stress occurred during tuberization (Riedel et al., 1985). *Fusarium* wilt symptoms appeared earlier and were more severe in chrysanthemum *Fusarium*-susceptible cultivar

Yellow Delaware when infected with *Meloidogyne javanica* than with *M. hapla* and *M. incognita* (Johnson and Littrel, 1969). These results suggest that nematodes which produce similar structural changes or wounds in the same host may induce different physiological or biochemical modifications resulting in a different degree of promotion of the infection and development of the same fungus.

A nematode species may interact differently with two fungi on the same plant. In a greenhouse experiment, *Verticillium* wilt was more severe in tomato cultivar Bonny Best infected with *Heterodera tabacum* than in uninfected plants, while *Fusarium* wilt was less severe in the presence of the nematode than in its absence (Miller, 1975). The alterations induced by one species of nematode may predispose its host to infection by one fungus and not another.

14.3 DELAY IN FUNGAL DISEASE DEVELOPMENT

There are many indications that the predisposition effect on plants to fungal diseases by root-knot nematodes reaches its maximum two to four weeks after nematode infection.

Meloidogyne incognita maximally predisposed tobacco plants to *Fusarium* wilt (Porter and Powell, 1967) and to root decay caused by *Pythium ultimum* (Meléndez and Powell, 1970) when the nematode was inoculated four weeks prior to fungus inoculations. When *M. incognita* inoculation preceded *Rhizochtonia solani* inoculation by at least ten days, the *M. incognita*-susceptible tobacco cultivars Dixie Bright 101 and Coker 316 exhibited more severe root-rot than when nematode and fungus were inoculated simultaneously (Batten and Powell, 1971). *R. solani* penetration and development in roots of radish was increased with prior *M. hapla* root infection (Khan and Müller, 1982). Griffin and Thyr (1986, 1988) reported that severity of *Fusarium* wilt of alfalfa increased when *M. hapla* inoculations preceded those of *F. oxysporum*. *Aspergillus ochraceus*, *Botrytis cinerea*, *Curvularia trifolli*, *Penicillium martensii*, *Pythium ultimum* and *Trichoderma harzianum*, which are considered non-pathogenic to tobacco, caused severe root necrosis in tobacco cultivar C136 when inoculated four weeks after *M. incognita* (Powell *et al.*, 1971). The incubation period of *Cylindrocladium* black rot of peanut was shortened in the presence of *M. hapla* and was the same whether the fungus was inoculated simultaneously with the nematode or two weeks later (Diomande and Beute, 1981). Van Gundy *et al.* (1977) reported that the development of infection on tomato plants by *R. solani* was delayed by three to four weeks when the fungus was inoculated simultaneously with *M. incognita*.

The delay in predisposition of plants to fungal diseases by root-knot nematodes suggests that these nematodes are not just wounding agents facilitating the penetration of the fungi within the roots. If their role in disease complexes with fungi was limited to wounding agents, the fungal invasion of the plant tissues would begin soon after nematode infection. The nematodes have to induce drastic structural, biochemical and/or physiological changes in the roots to render them suitable to fungal penetration and development. In the case of root-rot caused by *Rhizoctonia solani* on tomato, Van Gundy *et al.* (1977) attributed this delay to a modification of the root exudates which occurs three to four weeks after nematode infection, when a high concentration of nitrogenous compounds in nematode-infected root leachates is favourable for maximum virulence of the fungi. Wang and Bergeson (1974) suggested that changes in total sugar concentration in the xylem sap which reach maximum concentration four weeks after nematode inoculation may contribute to the enhancement of *Fusarium* wilt in tomato infected with *Meloidogyne incognita*.

14.4 BREAKING OF *FUSARIUM* RESISTANCE BY ROOT-KNOT NEMATODES

Young (1939) indicated that root-knot nematodes greatly decreased the resistance of tomato cultivars to *Fusarium* wilt. Since this early report, many authors have concluded that monogenic resistance (*I* gene) of tomato to *F. oxysporum* f. sp. *lycopersici* was rendered ineffective by infection with root-knot nematodes. Jenkins and Coursen (1957) reported that both *Meloidogyne hapla* and *M. incognita* induced wilting in *Fusarium* wilt-resistant tomato cultivar Chesapeake. Similar results were reported with cultivars Chesapeake (Bowman and Bloom, 1966) and Bradley (Goode and McGuire, 1967). Sidhu and Webster (1977) indicated that monogenic resistance to *F. oxysporum*. f. sp. *lycopersici* race 1 in tomato cultivar Chico III, possessing resistance gene *I-2*, was ineffective in the presence of *M. incognita*. Studying interaction between *M. hapla* and *F. oxysporum* f. sp. *medicaginis* on alfalfa, Griffin and Thyr (1988) reported that in cultivar Synthetic XX, resistant to both organisms, the nematode promoted the fungal infestation only at 30°C, where resistance to *M. hapla* was lost. Noguera (1982) suggested that the loss of *Fusarium* resistance in tomato cultivar Craigella GCR 161 when infected with *M. incognita* was associated with the lack of rishitin, an antifungal substance present in healthy plants but absent in nematode-infested plants.

In contrast, many other researchers have concluded that root-knot nematodes do not affect *Fusarium* wilt resistance in tomato. Hybrid Bohn-Tucker 463 remained immune to *Fusarium* wilt despite abundant root-knot infestation (Harrison and Young, 1941). McClellan and Christie (1949) indicated that root-knot nematodes had very little effect on the wilt susceptibility of tomato polygenic tolerant cultivar Marglobe. *Fusarium* wilt resistance of cultivar Toyonishiki was not affected by the presence of *Meloidogyne incognita* (Hirano *et al.*, 1979). Jones *et al.* (1976) reported that the monogenic resistance to *F. oxysporum*. f. sp. *lycopersici* race 1 and race 2 of cultivars Florida MH-1 and Manapal were not altered by either simultaneous or prior inoculation with *M. incognita*. Similar results were obtained by Abawi and Barker (1984) with tomato cultivars Nematex, Manapal and Floradel all with the *Fusarium* resistance gene *I*-1 and Florida MH-1 with the *Fusarium* resistance gene *I*-2. In addition, they observed that *Fusarium* resistance of Nematex was not reduced by the loss of *M. incognita* resistance at 35°C.

Inconsistencies between results may be due to differences in experimental conditions, cultivar nematode species associations, environmental and especially edaphic factors, and population levels of pathogens. For example, Jenkins and Coursen (1957), conducting experiments on the *Fusarium*-resistant tomato cultivar Chesapeake, reported that *Meloidogyne incognita* promoted wilt in 100% of the plants, whereas only 60% of the plants expressed wilt symptoms in the presence of *M. hapla*. Despite the discrepancies in results obtained under different experimental conditions, it is evident that under favourable experimental and environmental conditions root-knot nematode can alter the *Fusarium* resistance of their host. This suggests that root-knot nematodes may induce slight changes in host physiology resulting in predisposition of the host to *Fusarium* wilt.

A genetic basis of the physiological predisposition of tomato plants to *Fusarium oxysporum* f. sp. *lycopersici* by *Meloidogyne incognita* was given by Sidhu and Webster (1974). They tested the resistance to the nematode and the fungus inoculated separately and together of the F_2 progeny obtained by crossing tomato cultivars Small Fry (resistant to both parasites) and Wonder Boy (susceptible to both parasites). When the F_2 plants were inoculated with the nematode or the fungus alone, the ratio of plants falling into the four reaction classes was 9/16 resistant to both parasite, 3/16 resistant to the nematode and susceptible to the fungus, 3/16 susceptible to the nematode and resistant to the fungus, and 1/16 susceptible to both parasites. This ratio $9:3:3:1$ suggested the presence of two dominant genes segregating independently, one effective against *M. incognita* and the other against the fungus. When plants were inoculated with both parasites the $9:3:3:1$ ratio was modified to a $9:3:4$ ratio. This $9:3:4$ ratio characterized a recessive epistasis and indicated that plants which were genetically resistant to the fungus but not to the nematode showed a susceptibility to the fungus when infected by the nematode.

14.5 MODIFICATION OF THE RHIZOSPHERIC ENVIRONMENT AND FUNGAL DEVELOPMENT ON THE GALL SURFACE

Germinated chlamydospores of *Fusarium solani* f. sp. *phaseoli* were able to use root leachates as a substrate for growth and reproduction (Schroth and Hendrix, 1962). Wang and Bergeson (1974) suggested that changes in total sugars and amino acids of *Meloidogyne incognita*-infected plant root leachates contribute to the predisposition of tomato plants to *Fusarium* wilt. Exudates from galled roots may provide stimuli for germination of chlamydospores, accounting for the increase in numbers of propagules (Bergeson *et al.*, 1970) and colonies (Noguera and Smits, 1982) of *F. oxysporum* f. sp. *lycopersici* in the rhizosphere of tomato plants infected with *M. incognita* and *M. javanica* compared to those observed around non-galled roots.

Golden and Van Gundy (1975) reported that on tomato and okra infected with *Meloidogyne incognita*, *Rhizoctonia solani* sclerotia were formed only on gall tissues. Van Gundy *et al.* (1977) observed that 28 days after *M. incognita* inoculation of tomato roots, galled surface segments were abundantly colonized by *R. solani* sclerotia, which began to germinate. A similar observation was made by Khan and Müller (1982), who reported that following *M. hapla* infection mycelial growth of *R. solani* was more abundant on galled regions of radish roots than on non-galled areas.

Rhizoctonia solani was specifically attracted to *Meloidogyne incognita* gall tissues responding to stimuli which originated from the galled roots and passed through semi-permeable cellophane membranes (Golden and Van Gundy, 1975). Starr and Aist (1977) observed that *Pythium polymorphon* colonized preferentially and earlier galled areas of celery roots infected with *M. hapla* than non-galled root segments, suggesting that factors attractive to the fungus originated from galls caused by the root-knot nematode.

Van Gundy *et al.* (1977) demonstrated that severe root-rot caused by *Rhizoctonia solani* on Pixie Hybrid tomato may be associated with nutrient mobilization in root leachates induced by *Meloidogyne incognita*. When root leachates of plants inoculated simultaneously with the nematode and the fungus were permanently removed, no root-rot occurred. In contrast, when root leachates were not removed a severe root-rot developed. Moreover, when root leachates produced by *M. incognita*-infected plants were applied to roots of plants inoculated with *R. solani* alone, severe root-rot developed, whereas roots inoculated with *R. solani* receiving root leachates from control plants were free of decay. During the first 14 days after nematode infection, when carbohydrates were abundant and C/N ratio was high in *M. incognita*-infected root leachates, *R. solani* growth was stimulated in the rhizosphere and the fungus was attracted to the roots. Between 14 and 28 days after nematode infection, the C/N ratio was decreasing and at 28 days a low C/N ratio indicated a high level of nitrogen, favourable for parasitic development of *R. solani* (Weinhold *et al.*, 1972).

Biochemical modifications of root leachates induced by root-knot nematodes appear to enhance the colonization of the rhizosphere by pathogenic fungi, to attract them to gall tissues, and favour their growth at the gall surface. Moreover, they also appear to lower the numbers of actinomycetes, antagonistic to *Fusarium oxysporum* f. sp. *lycopersici*, in the rhizosphere. Bergeson *et al.* (1970) observed a highly significant reduction in the number of actinomycetes and a significant increase in number of *Fusarium* propagules in the rhizospheric soil surrounding roots inoculated simultaneously with *Meloidogyne javanica* and *F. oxysporum* f. sp. *lycoper-*

sici compared to those observed when the fungus was inoculated alone. A similar observation was made by Noguera and Smits (1982), who suggested that the reduction in number of actinomycetes antagonistic to *Fusarium* in the rhizosphere of *M. incognita*-infected plants may be partly responsible for the enhancement of the pathogenic effect of the fungus.

14.6 PENETRATION AND EARLY COLONIZATION OF ROOT-KNOT-INFECTED ROOTS

Tu and Cheng (1971) suggested that liberation of root leachates through wounds produced by *Meloidogyne javanica* in kenaf may stimulate *Macrophomina phaseoli* hyphal penetration within the roots. Powell and Nusbaum (1960), studying the invasion of *M. incognita*-infected tobacco roots by *Phytophthora parasitica* var. *nicotianae*, observed that fungal invasion and colonization was more extensive in galled than in non-galled tissue. The mycelium present in galled tissue was more vigorous and hyphae larger than in non-galled areas. Golden and Van Gundy (1975) observed that *Rhizoctonia solani* started to penetrate root-knot-induced galls four weeks after nematode inoculation. The fungus penetrated the galled tissue either directly or through openings created by the mature female nematodes to lay their eggs. After penetration the fungus showed a marked trophic intercellular growth towards the giant cells and physiological changes appeared in these cells prior to physical contact by the fungus. Fattah and Webster (1983) showed that considerable ultrastructural changes occurred in *Meloidogyne incognita*-induced giant cells prior to *Fusarium oxysporum* f. sp. *lycopersici* invasion. This suggests that giant cells are sensitive to a translocatable physiological factor produced by the fungus. The same authors confirmed this hypothesis by obtaining similar ultrastructural modifications of the giant cells by incubating root-knot nematode infected roots in cell-free culture filtrate of the fungus (Fattah and Webster, 1989).

Five to six weeks following root-knot nematode invasion, *Rhizoctonia solani* had extensively colonized the nematode-induced giant cells which degenerated within two or three days (Golden and Van Gundy, 1975). Similar observations on giant cell invasion and destruction by fungi were reported and described by Powell and Nusbaum (1960), Meléndez and Powell (1967), and Moussa and Hague (1988). In addition, Fattah and Webster (1983) reported that fungal hyphae were visible within the giant cells three weeks after fungal inoculation in *Fusarium*-susceptible tomato cultivar Pearson A-1 Improved and four weeks after fungal inoculation in *Fusarium*-resistant cultivar Pearson Improved. Usually, root-knot nematodes express their greatest capability for predisposing plants to fungal disease three to four weeks after they have infected the roots, when the giant cells they induce are metabolically most active (Bird, 1972; Bird and Loveys, 1975). Giant cells may serve as reservoirs of metabolites and an important source of food for the fungi and may enhance their development within the root tissues. After colonizing the giant cells, the fungi move into xylem tissues (Golden and Van Gundy, 1975; Van Gundy *et al.*, 1977).

14.7 TRANSLOCATION OF THE PREDISPOSITION EFFECT INDUCED BY NEMATODES

The predisposition of plant tissues to fungal infection by nematodes is not limited to the tissues they invade and structurally modify. Bowman and Bloom (1966) studied the breaking of *Fusarium* resistance in tomato cultivars Rutgers and Homestead by *Meloidogyne incognita* in a split-root experiment. They observed that wilt incidence was increased when the nematode and *F. oxysporum* f. sp. *lycopersici* were inoculated on opposite halves of the root system. Similar results were reported for the interaction between *M. incognita* and *F. oxysporum* f. sp. *lycopersici* by El-Sherif and Elwakil (1991) with the tomato cultivar Tropic, and by Faulkner *et al.* (1970) on peppermint with *Pratylenchus minyus* and *Verticillium dahliae*. Carter (1981) reported additive combined effects of *M. incognita* inoculation on the roots and hypocotyl wounding, facilitating the penetration of *Rhizoctonia solani* on the severity of seedling disease of cotton. Because the two organisms infected spatially separate tissues, this additive effect indicated a systemic effect of the nematode.

Sidhu and Webster (1977) studied the translocation of the nematode's predisposing effect by bending over four times the stems of *Fusarium*-resistant tomato plants (cultivar Chico III) to produce four adventitious root systems in addition to the primary root system. They inoculated the nematodes (*Meloidogyne incognita*) on the primary root system and the fungus (*F. oxysporum* f. sp. *lycopersici*) on one of the root systems (primary or one of the four adventitious). When the fungus was inoculated on the primary root system, wilt symptoms were observed on the entire plant. When the fungus was inoculated on one of the adventitious root systems, wilt developed at the site of fungal inoculation and on the portion of the plant between the site of fungal inoculation and the apex of the stem, whereas, wilt symptoms were minimal between the fungal inoculation site and the base of the plant. These results indicate that a predisposition factor produced or induced by the nematode can be transmitted at considerable distances from the nematode infection site to the upper foliage. Similar results were obtained by Hillocks (1986) and Nicholson *et al.* (1985). They observed that *Fusarium* wilt symptoms in cotton and anthracnose leaf blight in corn were worsened by the infection of the root system by *M. incognita* and *Pratylenchus hexincisus* when *F. oxysporum* f. sp. *vasinfectum* and *Colletotrichum graminicola* were inoculated on the stems, indicating a systemic effect of the nematodes on the host resistance to fungi.

Contradictory results were obtained in split-root experiments when the nematode and the fungi were inoculated on opposite parts of the root system, by Hillocks (1986) and Moorman *at al.* (1980). They did not observe any translocatable influence of *Meloidogyne incognita* on the development of *Fusarium oxysporum* f. sp. *vasinfectum* and *F. oxysporum* f. sp. *nicotianae* on cotton and tobacco, respectively. However, they observed an enhancement of the fungal infection when nematode and fungus were inoculated on the same half-root system. These apparent contradictory results seem to indicate that the nematodes may have two effects favouring fungal infection of their hosts: a localized effect, where the fungus penetration and initial development in the host is enhanced by the modifications induced by the nematodes at their feeding sites; and a systemic effect where inhibition of host resistance mechanisms occurs, resulting in a stimulation of the fungal development in tissues not infected by the nematodes.

14.8 CONCLUSION

There are numerous findings which have concluded that either there was no interaction between nematodes and fungi in disease complexes, or that nematodes did not predispose their hosts to fungal pathogens. It has also been reported that nematodes may suppress the infection of their host by pathogenic fungi (Orion and Netzer, 1981; Nordmeyer and Sikora, 1983). However, there are strong indications that nematodes, especially root-knot nematodes, may induce physiological and/or biochemical changes in their hosts which enhance the development of pathogenic fungi and/or predispose their host to fungal pathogens. Discrepancies between observations may probably result from differences in experimental conditions such as environmental and edaphic factors, nematode and fungus inoculum levels, or from differences in the considered nematode–fungus–plant (cultivar) combination.

Nematodes seem to favour all stages of fungal infection and development. By modifying the composition of the root leachates they can promote the growth of fungi in the rhizosphere and favour their pathogenic development. Moreover, these modifications of the rhizospheric environment may limit the development of organisms antagonistic to the pathogenic fungi. Their feeding sites and the cells they modify, especially the giant cells induced by root-knot nematodes, may serve as a favourable substrate which helps the fungi to establish within the plant and promote their development. Nematode-induced or produced factors appear to be translocated from the nematode feeding sites to other parts of their host, especially in the above-ground parts. These factors seem to modify the resistance of the host tissues to the fungi and/or directly stimulate fungal growth.

Results obtained by Van Gundy *et al.* (1977) and Golden and Van Gundy (1975) are direct indications that biochemical changes induced by nematodes in root leachates are responsible for the improvement of rhizosphere colonization by *Rhizoctonia solani* and its attraction by root-knot-induced galls, respectively. Observations made by Powell and Nusbaum (1960), Meléndez and Powell (1967), Golden and Van Gundy (1975) and Fattah and

Webster (1983) on the invasion of root-knot-induced giant cells by fungi provide direct evidence of their role in the enhancement of fungal development in the root tissues. Split-root system experiments (Bowman and Bloom, 1966; Faulkner *et al.*, 1970; El-Sherif and Elwakil, 1991), a bridging experiment (Sidhu and Webster, 1977) and experiments on inoculation of the two parasites (nematode and fungus) in spatially separated plant tissues (Nicholson *et al.*, 1985; Hillocks, 1986) provide strong direct evidence of a systemic effect of nematodes on fungal disease developments. However, the precise physiological and/or biochemical changes induced or produced by nematodes which predispose their hosts to fungal pathogens or directly enhance the invasion and development of pathogenic fungi in the host tissues are not known. Sidhu and Webster (1974) provided a genetic basis to the physiological predisposition of the tomato plant to *Fusarium* wilt by showing that a cultivar genetically resistant to the fungus has its resistance broken when infected by root-knot nematodes. The observations of Sidhu and Webster (1974) and similar observations of other workers raise questions such as 'What are the genes which make one nematode species capable of predisposing a plant to a fungal disease while another closely related nematode species is not?' or 'Do genes which control interactions between nematodes and fungi exist?' Answers to these questions and the understanding of nematode-induced physiological and biochemical changes induce in their hosts that are responsible for the predisposition of the host plants to fungal pathogens could be the necessary bases to develop control strategies against these parasites using resistant cultivars. More multidisciplinary research between biochemists, geneticists, pathologists and nematologists will be necessary to understand the interrelationships between nematodes, fungi and plants.

REFERENCES

Abawi, G.S. and Barker, K.R. (1984) Effects of cultivar, soil temperature, and population levels of *Meloidogyne incognita* on root necrosis and Fusarium wilt of tomatoes. *Phytopathology*, **74**, 433–8.

Batten, C.K. and Powell, N.T. (1971) The *Rhizoctonia–Meloidogyne* disease complex in flue-cured tobacco. *Journal of Nematology*, **3**, 164–9.

Bergeson, G.B., Van Gundy, S.D. and Thomason, I.J. (1970) Effect of *Meloidogyne javanica* on rhizoshere microflora and Fusarium wilt of tomato. *Phytopathology*, **60**, 1245–9.

Bird, A.F. (1972) Quantitative studies on the growth of syncytia induced in plants by root-knot nematodes. *International Journal of Parasitology*, **2**, 157–70.

Bird, A.F. and Loveys, B.R. (1975) The incorporation of photosynthates by *Meloidogyne javanica*. *Journal of Nematology*, **7**, 111–13.

Bowman, P. and Bloom, J.L. (1966) Breaking the resistance of tomato varieties to Fusarium wilt by *Meloidogyne incognita*. *Phytopathology*, **56**, 871 (Abstract).

Carter, W.W. (1975) Effects of soil temperatures and inoculum levels of *Meloidogyne incognita* and *Rhizoctonia solani* on seedling disease of cotton. *Journal of Nematology*, **7**, 229–33.

Carter, W.W. (1981) The effect of *Meloidogyne incognita* and tissue wounding on severity of seedling disease of cotton caused by *Rhizoctonia solani. Journal of Nematology*, **13**, 374–6.

Conroy, J.J. and Green, R.J., Jr (1974) Interactions of the root-knot nematode *Meloidogyne incognita* and the stubby root nematode *Trichodorus christiei* with *Verticillium albo-atrum* on tomato at controlled inoculum densities. *Phytopathology*, **64**, 1118–21.

Conroy, J.J., Green, R.J., Jr and Ferris, J.M. (1972) Interaction of *Verticillium albo-atrum* and the root lesion nematode, *Pratylenchus penetrans*, in tomato at controlled inoculum densities. *Phytopathology*, **62**, 362–6.

Diomande, M. and Beute, M.K. (1981) Effects of *Meloidogyne hapla* and *Macroposthonia ornata* on Cylindrocladium black rot of peanut. *Phytopathology*, **71**, 491–6.

El-Sherif, A.G. and Elwakil, M.A. (1991) Interaction between *Meloidogyne incognita* and *Agrobacterium tumefaciens* or *Fusarium oxysporum* f. sp. *lycopersici* on tomato. *Journal of Nematology*, **23**, 239–42.

Fattah, F. and Webster, J.M. (1983) Ultrastructural changes caused by *Fusarium oxysporum* f. sp. *lycopersici* in *Meloidogyne javanica* induced giant cells in *Fusarium* resistant and susceptible tomato cultivars. *Journal of Nematology*, **15**, 128–35.

Fattah, F.A. and Webster, J.M. (1989) Ultrastructural modifications of *Meloidogyne javanica* induced giant cells caused by fungal culture filtrates. *Revue de Nématologie*, **12**, 197–210.

Faulkner, L.R., Bolander, W.J. and Skotland, C.B. (1970) Interaction of *Verticillium dahliae* and *Pratylenchus minyus* in Verticillium wilt of peppermint: Influence of the nematode as determined by a double root technique. *Phytopathology*, **60**, 100–3.

Golden, J.K. and Van Gundy, S.D. (1975) A disease complex of okra and tomato involving the nematode *Meloidogyne incognita*, and the soil inhabiting fungus *Rhizoctonia solani. Phytopathology*, **65**, 265–73.

Goode, M.J. and McGuire, J.M. (1967) Relationship of root-knot nematode to pathogenic variability in *Fusarium oxysporum* f. sp. *lycopersici. Phytopathology*, **57**, 812 (Abstract).

Griffin, G.D. (1990) Pathological relationship of *Ditylenchus dipsaci* and *Fusarium oxysporum* f. sp. *medicaginis* on alfalfa. *Journal of Nematology*, **22**, 333–6.

Griffin, G.D. and Thyr, B.D. (1986) The importance of nematode resistance on the interaction of *Meloidogyne hapla* and *Fusarium oxysporum* on alfalfa. *Phytopathology*, **76**, 843–4.

Griffin, G.D. and Thyr, B.D. (1988) Interaction of *Meloidogyne hapla* and *Fusarium oxysporum* f. sp. *medicaginis* on alfalfa. *Phytopathology*, **78**, 421–5.

Harrison, A.L. and Young, P.A. (1941) Effect of root-knot nematode on tomato wilt. *Phytopathology*, **31**, 749–52.

Hart, L.P. and Endo, R.M. (1981) The effect of time exposure to inoculum, plant age, root development and root wounding on *Fusarium* yellows of celery. *Phytopathology*, **71**, 77–9.

Hillocks, R.J. (1986) Localised and systemic effects of root-knot nematode on the incidence and severity of Fusarium wilt in cotton. *Nematologica*, **32**, 202–8.

Hirano, K., Sugiyama, S. and Lida, W. (1979) Relation of the rhizosphere microflora to the occurrence of Fusarium wilt of tomato under presence of the root-knot nematode, *Meloidogyne incognita* (Kofoid & White) Chitwood. *Japanese Journal of Nematology*, **9**, 60–8.

Jenkins, W.R. and Coursen, B.W. (1957) The effect of the root-knot nematodes, *Meloidogyne incognita acrita* and *M. hapla* on Fusarium wilt of tomato. *Plant Disease Reporter*, **41**, 182–6.

Johnson, A.W. and Littrell, R.H. (1969) Effect of *Meloidogyne incognita*, *M. hapla* and M. *javanica* on the severity of Fusarium wilt of chrysanthemum. *Journal of Nematology*, 1, 122-5.

Jones, J.P., Overman, A.J. and Crill, P. (1976) Failure of root-knot nematode to affect Fusarium wilt resistance of tomato. *Phytopathology*, 66, 1339-41.

Kawamura, T. and Hirano, K. (1967) Studies on the complex disease caused by root-knot nematode and Fusarium wilt fungus in tomato seedlings. I. Development of wilt symptom and difference in disease incidence in tomato varieties. *Technical Bulletin, Faculty of Horticulture, Chiba University*, 15, 7-19.

Khan, M.W. and Müller, J. (1982) Interaction between *Rhizoctonia solani* and *Meloidogyne hapla* on radish in gnotobiotic culture. *Libyan Journal of Agriculture*, 11, 133-40.

Khoury, F.Y. and Alcorn, S.M. (1973) Effect of *Meloidogyne incognita acrita* on the susceptibility of cotton plants to *Verticillium albo-atrum*. *Phytopathology*, 63, 485-90.

Lockwood, J.L. (1988) Evolution of concepts associated with soilborne plant pathogens. *Annual Review of Phytopathology*, 26, 93-121.

McClellan, W.D. and Christie, J.R. (1949) Incidence of *Fusarium* infection as affected by root-knot nematodes. *Phytopathology*, 39, 568-71.

Meléndez, P.L. and Powell, N.T. (1967) Histological aspects of Fusarium wilt–root-knot complex in flue-cured tobacco. *Phytopathology*, 57, 286-92.

Meléndez, P.L. and Powell, N.T. (1970) The *Pythium*-root knot nematode complex in flue-cured tobacco. *Phytopathology*, 60, 1303 (Abstract).

Miller, P.M. (1975) Effect of the tobacco cyst nematode, *Heterodera tabacum*, on the severity of Verticillium and Fusarium wilts of tomato. *Phytopathology*, 65, 81-2.

Moorman, G.W., Huang, J.S. and Powell, N.T. (1980) Localized influence of *Meloidogyne incognita* on Fusarium wilt resistance of flue-cured tobacco. *Phytopathology*, 70, 969-70.

Moussa, E.M. and Hague, N.G.M. (1988) Influence of *Fusarium oxysporum* f. sp. *glycines* on the invasion and development of *Meloidogyne incognita* on soybean. *Revue de Nématologie*, 11, 437-9.

Müller, J. (1977) Interactions between five species of *Pratylenchus* and *Verticillium albo-atrum*. *Zeitschrift für Pflanzenkrankheiten und Pflanzenschutz*, 84, 215-20.

Nicholson, R.L., Bergeson, G.B., DeGennaro, F.P. and Viveiros, D.M. (1985) Single and combined effects of the lesion nematode and *Colletotrichum graminicola* on growth and anthracnose leaf blight of corn. *Phytopathology*, 75, 654-61.

Noguera, R. (1982) Atteraciones en la produccion de rishitia en racces y tilosas en tallos en la interaccion *Meloidogyne-Fusarium* en plantas de tomate. *Agronomia Tropical*, 32, 303-8.

Noguera, R. and Smits, B.G. (1982) Variaciones en la microflora de la rizosfera del tomate infectado con *Meloidogyne incognita*. *Agronomica Tropical*, 32, 147-54.

Nordmeyer, D. and Sikora, R.A. (1983) Studies on the interaction between *Heterodera daverti*, *Fusarium avenaceum* and *F. oxysporum* on *Trifolium subterraneum*. *Revue de Nématologie*, 6, 193-8.

Orion, D. and Netzer, D. (1981) Suppressive effect of the root-knot nematode on Fusarium wilt of muskmelons. *Revue de Nématologie*, 4, 65-70.

Pitcher, R.S. (1965) Interrelationships of nematodes and other pathogens of plants. *Helminthological Abstracts*, 34, 1-17.

Porter, D.M. and Powell, N.T. (1967) Influence of certain *Meloidogyne* species on Fusarium wilt development in flue-cured tobacco. *Phytopathology*, 57, 282-5.

Powell, N.T. (1971) Interaction of plant parasitic nematodes with other disease-causing agents, in *Plant Parasitic Nematodes*, Vol. II (eds B.M. Zuckerman, W.F. Mai and R.A. Rohde), Academic Press, New York, pp. 119-36.

Powell, N.T. and Nusbaum, C.J. (1960) The black shank–root-knot complex in flue-cured tobacco. *Phytopathology*, 50, 899-906.

Powell, N.T., Meléndez, P.L. and Batten, C.K. (1971) Disease complexes in tobacco involving *Meloidogyne incognita* and certain soil-borne fungi. *Phytopathology*, **61**, 1332-7.

Riedel, R.N., Rowe, R.C. and Martin, M.J. (1985) Differential interactions of *Pratylenchus crenatus, P. penetrans*, and *P. scribneri* with *Verticillium dahliae* in potato early dying disease. *Phytopathology*, **75**, 419-22.

Schroth, M.N. and Hendrix, F.F., Jr (1962) Influence of nonsusceptible plants on the survival of *Fusarium solani* f. *phaseoli* in soil. *Phytopathology*, **52**, 906-9.

Shawish, O. and Baker, R. (1982) Thigmomorphogenesis and predisposition of hosts to *Fusarium* wilt. *Phytopathology*, **72**, 63-8.

Sidhu, G. and Webster, J.W. (1974) Genetics of resistance in the tomato to root-knot nematode-wilt-fungus complex. *Journal of Heredity*, **65**, 153-6.

Sidhu, G. and Webster, J.W. (1977) Predisposition of tomato to the wilt fungus (*Fusarium oxysporum lycopersici*) by the root-knot nematode (*Meloidogyne incognita*). *Nematologica*, **23**, 436-42.

Starr, J.L. and Aist, J.R. (1977). Early development of *Pythium polymorphon* on celery roots infected by *Meloidogyne hapla. Phytopathology*, **67**, 497-501.

Starr, J.L., Jeger, M.J., Martyn, R.D. and Schilling, K. (1989) Effects of *Meloidogyne incognita* and *Fusarium oxysporum* f. sp. *vasinfectum* on plant mortality and yield of cotton. *Phytopathology*, **79**, 640-6.

Sumner, D.R. and Johnson, A.W. (1973) Effect of root-knot nematodes on Fusarium wilt of watermelon. *Phytopathology*, **63**, 857-61.

Sumner, D.R. and Minton, N.A. (1987) Interaction of Fusarium wilt and nematodes in Cobb soybean. *Plant Disease*, **71**, 20-3.

Tu, C.C. and Cheng, Y.H. (1971) Interaction of *Meloidogyne javanica* and *Macrophomina phaseoli* in kenaf root rot. *Journal of Nematology*, **3**, 39-42.

Van Gundy, S.D., Kirkpatrick, J.D. and Golden, J. (1977) The nature and role of metabolic leakage from root-knot nematode galls and infection by *Rhizoctonia solani. Journal of Nematology*, **9**, 113-21.

Wallace, H.R. (1978) The diagnosis of plant diseases of complex etiology. *Annual Review of Phytopathology*, **16**, 379-402.

Wang, E.L.H. and Bergeson, G.B. (1974) Biochemical changes in root exudates and xylem sap of tomato plants infected with *Meloidogyne incognita. Journal of Nematology*, **6**, 194-202.

Weinhold, A.R., Dodman, R.L. and Bowman, T. (1972) Influence of exogenous nutrition on virulence of *Rhizoctonia solani. Phytopathology*, **62**, 278-81.

Yang, H., Powell, N.T. and Barker, K.R. (1976) Interaction of concomitant species of nematodes and *Fusarium oxysporum* f. sp. *vasinfectum* on cotton. *Journal of Nematology*, **8**, 74-80.

Young, P.A. (1939) Tomato wilt resistance and its decrease by *Heterodera marioni. Phytopathology*, **29**, 81-9.

15

Interactions of nematodes with insects

Robin M. Giblin-Davis

There are over 3000 reported parasitic and phoretic associations between nematodes and insects, with less than ten involving phytoparasitic nematodes (Poinar, 1975). Nematodes are aquatic metazoans with very limited powers of dispersion. They are greatly benefited by a synchronized association with an insect host for increased mobility and protection during travel to an insect's breeding or feeding sites.

Two of the most important wilting diseases of trees in the world are caused by insect-transmitted aphelenchoidid nematodes. These are the pine wilt and the red ring diseases, which affect conifers and palms, respectively. Nematodes in such interactions can be phoretically associated with the insect and parasitic on the plant or parasitic on both hosts. In cases where the nematode is parasitic on both hosts the interaction can be dicyclic, where the nematode alternately completes its life cycle in the insect then the plant host, or phytocyclic/insect parasitic, where the nematode completes its life cycle on the plant host with one parasitic stage in the insect.

The purpose of this chapter is to introduce the reader to the growing body of knowledge concerning the taxonomy, biology and ecology of the diverse associations between phytoparasitic nematodes and insects. The first three associations to be discussed involve stylet-bearing nematodes from the family Aphelenchoididae (order: Aphelenchida). The last two associations to be discussed involve stylet-bearing nematodes from the superfamily Sphaerulariodea; families Fergusobiidae and Phaenopsitylenchidae (order: Tylenchida).

15.1 PINE WILT DISEASE

15.1.1 Overview and symptomatology

The pine wood nematodes *Bursaphelenchus xylophilus* and *B. mucronatus* are distributed throughout much of the temperate Northern Hemisphere. *B. xylophilus* causes the pine wilt disease, which has been responsible for epiphytotics to indigenous pines during the past few decades in southern Japan. It has been known since the early 1900s in Japan (Mamiya, 1984) and has been reported in southern China since the early 1980s (Rutherford *et al.*, 1990).

Pine wilt disease involves three organisms: a gymnosperm host, usually a conifer in the genus *Pinus*, the pine wood nematode, and an insect vector, almost always a longhorn beetle in the genus *Monochamus* (Cerambycidae) (Fig. 15.1). During primary transmission, dauer juveniles (J_{IV} stage) of *Bursaphelenchus xylophilus* are carried phoretically in the tracheae of their beetle host to young twigs of susceptible trees, where they enter through resin canals in wounds made during maturation feeding by the insect (Mamiya, 1984).

Tree host cellular responses progress after the invasion of the nematode, but this is dependent upon the nematode isolate, the conifer host and age, and the temperature. In a combination involving the most susceptible pine host and the most pathogenic nematode isolate under the most favourable temperature regime ($> 20°C$ mean daily summer temperatures), internal host responses can occur at the introduction site within one to three days post-inoculation (Mamiya, 1984). In that time, parenchymal cells begin to die, host respiration increases, water conductivity declines, stomatal closure and ethylene production increase, and phytotoxic monoterpenes and benzoic acid can be recovered (Mamiya, 1984). As soon as about six days post-inoculation, massive parenchymal cell death cascades through the tree host, preceding the cessation of oleoresin exudation.

Cessation of oleoresin exudation in artificial wounds in the trunk of the tree occurs between six and nine days with seedlings but can take as long as 20 days after inoculation in 25-year-old *Pinus densiflora* and is the first easily detectable internal symptom of pine wilt disease (Mamiya, 1984). This symptom concurs with an apparent increase in nematode density throughout the tree. At $\geqslant 20$ days, transpiration stops causing wilting and yellowing foliage concurrent with a nematode population bloom. At $\geqslant 30$ days the host tree dies, producing reddish pine needles. Mature susceptible trees, such as *P. densiflora*, and *P. thunbergii*, can have more than 1000 nematodes g^{-1} of wood (dry weight) 46–60 days post-inoculation (Mamiya, 1984). Natural infections of *Bursaphelenchus xylophilus* can cause host mortality in mature pines within 12 months in the warmer regions of Japan, where the pine wilt disease is considered the most devastating threat to native pines (Mamiya, 1984).

The dead or dying conifer is a suitable breeding host for the next generation of *Monochamus* spp. vectors. Nematodes brought into the conifer during oviposition of the beetle (secondary transmission) will moult from the dauer juvenile stage and enter the propagative phase to grow and reproduce on secondary fungi that are present as the pine host dies (Kobayashi, 1987). The cerambycid eggs hatch and develop through several larval instars while producing galleries, at first in the inner bark, cambium and outer

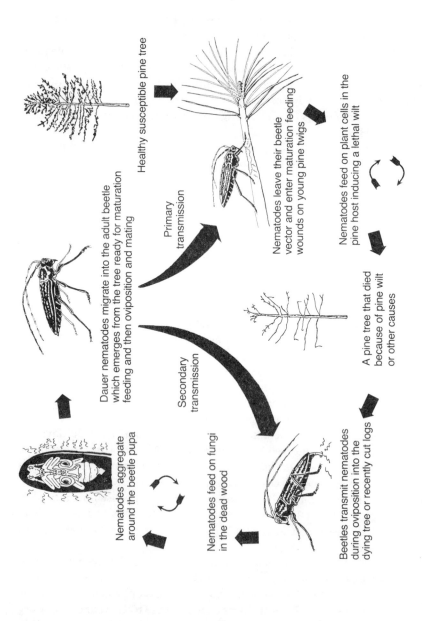

Healthy susceptible pine tree

Dauer nematodes migrate into the adult beetle which emerges from the tree ready for maturation feeding and then oviposition and mating

Primary transmission

Nematodes leave their beetle vector and enter maturation feeding wounds on young pine twigs

Nematodes feed on plant cells in the pine host inducing a lethal wilt

Secondary transmission

A pine tree that died because of pine wilt or other causes

Nematodes aggregate around the beetle pupa

Nematodes feed on fungi in the dead wood

Beetles transmit nematodes during oviposition into the dying tree or recently cut logs

Figure 15.1 Generalized representation of the association between the pine wood nematode, its *Monochamus* vector, and a conifer host.

sapwood and later in deeper woody tissue (Linit, 1988). The nematodes begin to produce a starvation-resistant predauer juvenile (J_{III}) as conditions in the log deteriorate, which occurs about the same time that the cerambycids pupate (Ishibashi and Kondo, 1977). The nematodes of this stage aggregate around the pupal chamber and moult to the dispersion-adapted dauer juvenile (J_{IV}), which migrates into the tracheae of the newly eclosed adult in the chamber (Mamiya, 1984).

15.1.2 Taxonomy of the pine wood nematodes

The pine wood nematode was originally described as the timber nematode, *Aphelenchoides xylophilus* (Steiner and Buhrer, 1934). It was recovered from dead wood of the longleaf pine, *Pinus palustris*, and was assumed to have no economic importance and to be associated with bark beetles in the southeastern USA. There was some confusion about the type specimens of *A. xylophilus* that resulted in a figure depicting the wrong spicules for this nematode in the publication where it was transferred into the genus *Bursaphelenchus* (Nickle, 1970). But clearly, the transfer of *A. xylophilus* into *Bursaphelenchus* was justified at that time.

In 1956, Rühm described the mycophagous *Bursaphelenchus fraudulentus* from Germany from the frass of *Cerambyx scopoli* (Cerambycidae) in *Prunus avium* (sweet cherry), or loosened bark of *Populus nigra* and *P. tremula* twigs where *Trypophloeus granulatus* (Scolytidae) was brooding. This nematode does not cause pathology in the plant hosts that have been tested, including *Pinus sylvestris* (Schauer-Blume, 1990). Mating studies suggest that *B. fraudulentus* is a valid species, regardless of its morphological similarities to *B. xylophilus* and to the subsequently described *B. mucronatus* (Schauer-Blume, 1990).

Mamiya and Kiyohara (1972) described *Bursaphelenchus lignicolus* from dead or dying Japanese red pine, *Pinus densiflora*, and Japanese black pine, *P. thunbergii*, in southwestern Japan where it was causing widespread disease. Mamiya and Enda (1979) described *B. mucronatus* from dead pine trees in Japan. *B. mucronatus*, morphologically very close to *B. lignicolus*, is distinguishable by the presence of a mucronate tail tip in females compared with a rounded tail in *B. lignicolus*. Moreover, *B. mucronatus* is less pathogenic to pines and has a wider distribution in Japan than *B. lignicolus* (Mamiya and Enda, 1979).

In 1981, studies showed that *Bursaphelenchus lignicolus* was a synonym of *B. xylophilus*, and that the closely related *B. mucronatus* was not reproductively compatible with *B. xylophilus*, thus justifying it as a valid species (Nickle *et al.*, 1981). The species status of *B. mucronatus* had been challenged by Baujard (1980) on the basis of variability in the morphology of the female tail terminus and the male bursal flap. Questions were raised because a French isolate of *Bursaphelenchus* discovered in the Landes Forest was morphologically and morphometrically similar to *B. xylophilus*,

B. mucronatus and *B. fraudulentus* (Baujard, 1980; Guiran *et al.*, 1985). The French isolate was also intermediate in pathogenicity to *Pinus pinaster* seedlings compared with *B. xylophilus* and *B. mucronatus* (Guiran *et al.*, 1985). Wingfield *et al.* (1983) reported an isolate of *B. xylophilus* from balsam fir, *Abies balsamea*, in Minnesota, USA, that resembles *B. mucronatus* in female tail morphology (mucronate form) but hybridized with *B. xylophilus* and not *B. mucronatus*.

Hybridization work (Guiran and Bruguier, 1989) confirmed that *Bursaphelenchus xylophilus* and *B. mucronatus* (both from Japanese red pine) are reproductively incompatible and that the *B. mucronatus* French isolate (from *Pinus pinaster* Landes Forest, Casteljaloux, France) can hybridize with both *B. xylophilus* and *B. mucronatus*. The *B. xylophilus* Minnesota isolate from balsam fir hybridizes with the *B. xylophilus* Japan isolate but not the *B. mucronatus* Japan isolate and can only produce viable offspring with the *B. mucronatus* French isolate when the parents are left with the F1 generation.

Other hybridization work with virulent and avirulent isolates of *Bursaphelenchus xylophilus* from the USA and Japan suggests incomplete compatibility and that virulence is dominant (Kiyohara and Bolla, 1990). Isolates of *B. xylophilus* from *Pinus thunbergii*, from Japan (one virulent and one avirulent isolate) produced viable F1 progeny when mated with North American isolates of *B. xylophilus* from Allepo pine, *P. halpensis*, from Arizona and Scots pine, *P. sylvestris*, from Missouri, but not with an isolate from Austrian pine, *P. nigra*, from New Jersey (Kiyohara and Bolla, 1990).

Guiran and Bruguier (1989) proposed that the Japanese and American isolates of pine wood nematodes originated from a mucronate tail form from Western Europe before the formation of the Atlantic Ocean during the Upper Cretaceous (80 million years ago). They describe the present-day scenario for the pine wood nematode as a 'circle of species' where species at the extremes of the distribution range are reproductively incompatible versus compatibility for species separated by less distance. The circle of pine wood nematode species was probably closed sometime around the late 1800s when *Bursaphelenchus xylophilus* from the southeastern USA could have been introduced into Japan. The first occurrence of the pine wilt disease in Japan was reported in 1905 by Yano in Nagasaki, Kyushu (Mamiya, 1984). The independent evolution that occurred during allopatry in both nematodes and pine hosts may have helped produce some of the pathogenic relationships observed when contact was re-established.

Guiran and Bruguier (1989) have proposed that all known species of 'pine wood' nematodes should be placed into a supraspecies. Accordingly, two species have been designated as follows: (1) *Bursaphelenchus* (suprasp. *xylophilus*) *xylophilus* for those isolates with females having a rounded tail terminus (r-form); and (2) *B.* (suprasp. *xylophilus*) *mucronatus* for those isolates with females having a mucronate tail terminus (m-form). Because this nomenclature is based upon a reasonable phylogenetic hypothesis, the supraspecies designation will be used for the remainder of the chapter, when possible.

The *Bursaphelenchus xylophilus* Minnesota isolate from balsam fir still complicates the classification of pine wood nematodes. The mating studies suggest that it probably belongs to *B.* (suprasp. *xylophilus*) *xylophilus*. The presence of the mucronate tail may indicate that it is an isolated population of the *B.* (suprasp. *xylophilus*) *xylophilus* stock that retained the primitive tail character or that it is an isolate of the rounded tail population of *B.* (suprasp. *xylophilus*) *xylophilus* with a character reversal for tail terminus shape. Other problems for the supraspecies designation come with the isolate of *B.* (suprasp. *xylophilus*) *xylophilus* from Austrian pine from New Jersey which is not reproductively compatible with Japanese isolates of *B.* (suprasp. *xylophilus*) *xylophilus* (Kiyohara and Bolla, 1990). In addition, there is a reportedly new species of *Bursaphelenchus* (morphologically similar to *B.* (suprasp. *xylophilus*) *xylophilus*) from *Pinus densiflora* from Kagoshima Prefecture (populations I and K) and *P. thunbergii* from Ooita Prefecture (population J) from Japan which is reproductively incompatible with other defined populations of *B.* (suprasp. *xylophilus*) *xylophilus* or *B.* (suprasp. *xylophilus*) *mucronatus* (Kiyohara and Bolla, 1990).

Kiyohara and Bolla (1990) showed that a single aspartate transaminase (GOT) isozyme band can be used to differentiate species of *Bursaphelenchus*, but not different pathotypes. Acid phosphatase, alcohol dehydrogenase and malate dehydrogenase are not useful in distinguishing *Bursaphelenchus* (suprasp. *xylophilus*) *xylophilus* and *B.* (suprasp. *xylophilus*) *mucronatus*. Zymograms produced from leucyl aminopeptidase and esterase are poor for resolution of *Bursaphelenchus* spp. because of equal within-species and between-species variation (Kiyohara and Bolla, 1990).

Webster *et al.* (1990) refer to the *Bursaphelenchus xylophilus* supraspecies as 'the pinewood nematode species complex' or 'PWNSC'. They prepared DNA probes from the non-transcribed spacer regions (NTS) between the 18 S and 28 S coding regions from ribosomal repeats from *B.* (suprasp. *xylophilus*) *xylophilus* (1.7 kb probe = pB × 6) and *B.* (suprasp. *xylophilus*) *mucronatus* (1.35 kb probe = pBm4). Using these probes and the *Caenorhabditis elegans* ribosomal lone probe (pCes370) as a DNA-positive control probe the authors produced a dot blot assay which required as few as two nematodes for a quantal response and which clearly segregated 16 isolates of the *B. xylophilus* supraspecies into two groups; *B.* (suprasp. *xylophilus*) *xylophilus* (14 geographical and host isolates: three from the USA, eight from Canada, two from Japan, and one from China) and *B.* (suprasp. *xylophilus*) *mucronatus* (four isolates: two from Japan, one from Norway, and one from France). Sequence divergence in the NTS region of the four isolates of *B.* (suprasp. *xylophilus*) *mucronatus* allowed for restriction analysis that demonstrated two subgroups ('Asian' and 'European') indicating where the isolates were collected. These data support the phylogenetic hypothesis of Guiran and Bruguier (1989).

Webster *et al.* (1990) suggest that the phoretic association of *Bursaphelenchus* (suprasp. *xylophilus*) with *Monochamus* spp., which often have overlapping ranges in North America, Europe and Asia, provides opportunities for some regional gene flow for the nematode. This might account for the relatively few hybridization failures observed in crosses of *B.* (suprasp. *xylophilus*) isolates. Bolla *et al.* (1988) suggest that pathotypes of *B.* (suprasp. *xylophilus*) which differ in genotype and conifer-host specificity could develop out of restrictions in gene flow caused by isolation of populations due to the host specificity and/or other aspects of the biology of the *Monochamus* vector. Thus, there may be enough gene flow in this system to prevent mating barriers from developing in *B.* (suprasp. *xylophilus*) except over long distances and enough regional isolation to allow for distinct pathotypes. Much more research is needed on the gene flow dynamics of the vector and the nematodes in this disease.

The males of the following five species within the genus *Bursaphelenchus* share a similarly distinctive spicule shape with a cucullus, and females of each species have a vulval flap: *B.* (suprasp. *xylophilus*) *xylophilus*, *B.* (suprasp. *xylophilus*) *mucronatus*, *B. fraudulentus*, *B. kolymensis* and *B. abruptus*. Only members of the *B.* (suprasp. *xylophilus*) have been shown to be pathogenic to plants (i.e. *Pinus*, *Abies*, etc.; Gymnosperms). The rest of the members of this group are apparently mycophagous.

15.1.3 Biology of the pine wood nematodes

Propagative phase

Bursaphelenchus (suprasp. *xylophilus*) *xylophilus* has a generation time of four to five days from egg to adult when cultured on the fungus *Botrytis cinerea* on potato dextrose agar (PDA) at 25°C (Mamiya, 1975). Generation time is temperature dependent and is completed in 12, 6, 4.5 and 3 days at 15, 20, 25 and 30°C on *Botrytis cinerea*, respectively, but development does not occur at 33°C. *Bursaphelenchus* (suprasp. *xylophilus*) *xylophilus* has higher developmental and reproductive rates at all temperatures tested on *Botrytis cinerea* on PDA than *Bursaphelenchus* (suprasp. *xylophilus*) *mucronatus* (Futai, 1980). *B.* (suprasp. *xylophilus*) *xylophilus* is heterosexual and is a facultative fungal/plant parasite capable of developing on many fungi (Kobayashi, 1987) or plant tissues such as alfalfa or pine callus (Mamiya, 1984). The mean fecundity is 79 eggs per female over 28 days at 25°C on *Botrytis cinerea* cultures (Mamiya and Furukawa, 1977). Oviposition is stimulated by n-alkanols such as ethanol, hexanol and octanol (Shuto and Watanabe, 1988). The second-stage juvenile (J2) ecloses from the egg, feeds and grows, moulting three times to the adult. Feeding occurs after each moult (Ishibashi *et al.*, 1978). The propagative phase will cycle as long as general conditions are suitable for growth and reproduction.

Bursaphelenchus (suprasp. *xylophilus*) *xylophilus* can be cultured axenically in defined and undefined media (Bolla and Jordan, 1982). However, routine culture is best accomplished on monoxenic cultures using *Botrytis cinerea* or *Monilinia fructicola* fungal hosts on unsupplemented or supplemented PDA (5–10% glycerol alone or in combination with 2 ml of 25% DL-lactic acid per litre of autoclaved PDA) (Giblin, 1987; Giblin and Kaya, 1984; Mamiya, 1984). The unsaturated fatty acids, oleic and linoleic acids, which constitute 97% of the fatty acids in the xylem of *Pinus densiflora*, can be used as supplements for increasing the reproductive rates and survivability of *B.* (suprasp. *xylophilus*) *xylophilus* in monoxenic culture on *Botrytis cinerea* (oleic at 3.5, 5.0, 10.0, 20.0 mg ml^{-1} hydrated PDA and linoleic acid at 10 mg ml^{-1} PDA) (Mamiya, 1990). Mass culture is easily done by introducing surface-sterilized nematodes onto autoclaved wheat and water in large flasks seeded with *Botrytis cinerea* or *M. fructicola* (Giblin, 1987).

Predauer stage

As conditions deteriorate, either in the Petri dish on a suitable host fungus (beginning about 90 days after inoculation) (Ishibashi and Kondo, 1977) or in the host conifer (after the death of the tree if the nematode was involved in primary infection or as conditions in the stressed tree or cut log become unsuitable), a developmental decision is made analogous to what has been observed in predauer (J_{2d}) formation in *Caenorhabditis elegans* (Riddle and Georgi, 1990). Developmental stages of *C. elegans* respond to titres of a 'food factor' in concert with a density-dependent pheromone. The environmental cues for predauer (J_{III}) formation in *Bursaphelenchus* (suprasp. *xylophilus*) have not been characterized. Concurrent with the overwintering stage of the insect most of the nematodes recovered from the tree are predauers (Mamiya, 1984).

The predauer (J_{III}) of *Bursaphelenchus* (suprasp. *xylophilus*) *xylophilus* has similar anterior morphology to the propagative J_3 but is larger (Ishibashi *et al.*, 1978), has larger depositions of densely packed materials in its hypodermal cells and intestine (lipid and glycogen), and has a thicker cuticle than the propagative J_3 (Kondo and Ishibashi, 1978).

Predauers of *Bursaphelenchus* (suprasp. *xylophilus*) *xylophilus* accumulate around the pupal chambers of the *Monochamus* host. Levels of saturated and unsaturated fatty acids are much higher around the pupal chamber than in free tisues of *Pinus densiflora* killed by *B.* (suprasp. *xylophilus*) *xylophilus*. The last instar larva of *M. alternatus* deposits its excretions on the walls of the chamber, and accumulations of fatty acids build up there. Analysis of the lipid composition of larvae of *M. alternatus* yields five fatty acids which have been tested in agar for their aggregative properties to mixed-stage populations of *B.* (suprasp. *xylophilus*) *xylophilus* (Miyazaki *et al.*, 1977, as cited in Kobayashi *et al.*, 1984). Nematodes do not aggregate under palmitic and stearic acids (saturated), but do aggregate under palmitoleic, oleic and linoleic acids (unsaturated).

The predauer can survive as long as a year and will moult to the pro-pagative J_4 if placed on a new fungal culture or if the insect pupa is removed from the pupal chamber (Ishibashi and Kondo, 1977). Some of the predauers will moult to the dauer (J_{IV}) if placed in water (Ishibashi and Kondo, 1977), but in nature predauers begin to moult to the dauer juvenile stage coincident with pupation of *Monochamus alternatus*. The chemical or environmental cues for initiation of this developmental stage are not known, but are suspected to be of insect origin (Ishibashi and Kondo, 1977).

Dauer juvenile stage

About the time the beetle ecloses to an adult a large proportion of the nematodes are in the dauer stage. This stage is attracted to carbon dioxide produced by the callow beetle adult in the chamber (Miyazaki *et al.*, 1978, as cited in Kobayashi *et al.*, 1984). The nematodes enter the tracheal system of the insect and the majority reside in the tracheae that connect to the metathoracic spiracles, which is where most of the carbon dioxide is emitted during respiration (Linit *et al.*, 1983). Some nematodes are found on the body of the beetle host (Kobayashi *et al.*, 1984). Under natural conditions the mean burden of dauers of *Bursaphelenchus* (suprasp. *xylophilus*) *xylophilus* ranges from 170 to 19 500 per *Monochamus alternatus* in Japan and *M. carolinensis* in the USA (Linit, 1988). The highest number of dauers of the pine wood nematode recorded from a single beetle is 289 000 (Linit, 1988).

The dauer juvenile stage has some resistance to desiccation and has been recovered alive from dry beetles that had been dead for as long as six months (Ishibashi and Kondo, 1977). Even though this stage may survive desicca-tion better than other stages of *Bursaphelenchus* (suprasp. *xylophilus*) *xylophilus*, nematodes require moisture to move. Togashi (1989) has reported that nematode density and water content of the xylem of *Pinus thunbergii* logs significantly affect the abundance of dauer juveniles of the pine wood nematode in emerging *Monochamus alternatus*. Nematode-free adult beetles are produced by placing nematode-infested logs under very dry conditions (Mamiya, 1984).

The dauer juvenile (J_{IV}) is morphologically distinct from the J_4 of the propagative phase. It has a non-offset head, apparent lack of labial sensillae, an indistinct stylet and oesophagus and intestine, a digitate tail, and the basal and cortical layers of the cuticle are thicker than the propagative J_4 (Ishibashi and Kondo, 1977). The dauer juvenile is slightly longer than the J_4, and the gonad of the predauer (J_{III}) and the dauer (J_{IV}) stages are much less developed than the corresponding stages in the propagative phase (Ishibashi *et al.*, 1978). It takes about five days at 25°C from when the dauer stage is placed onto a *Botrytis cinerea* fungal culture on PDA to reach the adult stage (Ishibashi *et al.*, 1978). This increase in developmental time may be due in part to the time required to catch up in gonad development.

Primary transmission

In primary transmission of *Bursaphelenchus* (suprasp. *xylophilus*) *xylophilus*, the dauer juvenile is carried in the respiratory system of its *Monochamus* host to a healthy susceptible conifer. The beetle feeds on new and one-year-old twigs of the tree to facilitate gonad maturation (Kobayashi *et al.*, 1984). Gas chromatographic/mass spectrophotometric (GC/MS) analyses of volatiles released from shoots of the pine wilt-susceptible pine, *Pinus densiflora*, reveal seven prevalent monoterpenes (α-pinene, camphene, β-pinene, β-myrcene, 3-carene, *l*-limonene and β-phellandrene) (Ishikawa *et al.*, 1986). Of these, the dauers or propagative stages of *B.* (suprasp. *xylophilus*) *xylophilus* are attracted to β-myrcene (Ishikawa *et al.*, 1986). Combinations of volatiles have not been evaluated, but this work suggests that one of the cues for transmigration of dauers from the beetle to the wounded conifer twig is β-myrcene. Furthermore, β-myrcene and the extracted resin of *P. densiflora* are equally capable of stimulating dauers of *B.* (suprasp. *xylophilus*) *xylophilus* to moult to the adult stage (Hinode *et al.*, 1987). For example, only about 10% of the dauers moult to adults within 48 hours at 23°C in the dark on water agar, compared with > 40% adults when dauers are exposed to vapours of β-myrcene (Hinode *et al.*, 1987). β-Myrcene also stimulates the multiplication of *B.* (suprasp. *xylophilus*) *xylophilus* on the fungus *Botrytis cinerea* as well as on fungi with poorer suitability as hosts for the nematode (Hinode *et al.*, 1987), but does not stimulate oviposition of the nematode (Shuto and Watanabe, 1988). A β-myrcene-conditioned atmosphere does not significantly affect the mycelial weight of *B. cinerea* produced after ten days at 23°C in the dark compared with a control (Hinode *et al.*, 1987). Therefore, a hormone-like activity has been ascribed to β-myrcene because it does not change the fungal host weight but does increase nematode densities. In addition, on control cultures with *B. cinerea* only, populations of the nematode begin to decline after two weeks. However, when vapour with β-myrcene is supplied to the cultures, the nematode populations continue to increase and do not crash as they do without β-myrcene.

Host volatiles like β-myrcene appear to be critical in primary transmission of *Bursaphelenchus* (suprasp. *xylophilus*) *xylophilus* because they elicit the active migration of the nematode from the insect transport host to the feeding wound in the pine. Once in the plant host, β-myrcene, and other chemicals that exude from damaged tree tissue, and perhaps the absence of inhibitory materials, stimulates moulting to the propagative phase and growth and reproduction within 48 hours of entry (Mamiya, 1984). Because of the importance of β-myrcene in primary transmission of *B.* (suprasp. *xylophilus*) *xylophilus*, it is not surprising that susceptibility of conifers to pine wilt disease appears to be correlated with the β-myrcene content of the wood from 1-year-old pine shoots (Ishikawa *et al.*, 1987).

Secondary transmission

As the tree wilts and dies it becomes a suitable host for *Monochamus* breeding and this allows for the ancestral mycophagous phase of the nematode association to take place. Primary transmission is probably a recent phenomenon based upon the ease with which cultures of *Bursaphelenchus* (suprasp. *xylophilus*) *xylophilus* grown on fungus revert to forms which are incapable of infecting the pine host (Bolla *et al.*, 1988). Non-virulent *B.* (suprasp. *xylophilus*) *xylophilus* apparently utilize different anaerobic metabolic pathways than virulent isolates (Bolla *et al.*, 1988). However, this trend has not been confirmed in all virulent and non-virulent isolates of the pine wood nematode.

In secondary transmigration, the cues for active movement from the tracheal system of the beetle into the brooding environment of the host are not clear. Matsumori *et al.* (1989) have isolated hormone-like volatile substances analogous to β-myrcene from the fungal host, *Botrytis cinerea*. Both materials (3-octanol and 1-octen-3-ol) are attractive to the propagative stages of *Bursaphelenchus* (suprasp. *xylophilus*) *xylophilus* but have not been tested as attractants for the dauer juvenile stage. Both volatile materials stimulate the moulting of dauer juveniles of *B.* (suprasp. *xylophilus*) *xylophilus* to adults relative to a deionized water control and cause increases in population density on cultures of *Botrytis cinerea* on PDA (Matsumori *et al.*, 1989).

Generalizations about the attractive and stimulative effects of these materials from *Botrytis cinerea* should be limited until further work is done on other fungi normally associated with the pine wilt diease. Although *Botrytis cinerea* is reportedly a very good host for *Bursaphelenchus* (suprasp. *xylophilus*) *xylophilus* (Kobayashi, 1987) and conifers have been reported as hosts (Farr *et al.*, 1989), it is not normally isolated from trees with pine wilt disease (Wingfield *et al.*, 1983; Kobayashi, 1987). *Ceratocystis minor* and *Macrophoma sugi* are the most likely fungi to be present and to serve as suitable hosts for the nematode in Japan (Kobayashi, 1987). *C. minor* and *M. sugi* are known to be associated through the life cycle of, and transmitted to shoots by *M. alternatus* during maturation feeding (Kobayashi, 1987).

Interestingly, *Ceratocystis minor* may even be pathogenic to *Monochamus alternatus*, infecting the beetle in the pupal chamber and causing deformities and discoloration of the wings, and reduced longevity of the adults (Kobayashi, 1987). *C. minor* and *M. sugi* are not usually isolated from healthy pines, but are the most prevalent colonizers in the succession of fungi isolated from pines that have been dead for up to three months (Kobayashi, 1987). *C. ips* is a good host for *Bursaphelenchus* (suprasp.

xylophilus) *xylophilus* Minnesota isolate from black spruce, *Picea mariana*, but is not a suitable host for the isolate from balsam fir. This is interesting because *C. ips* is not associated with balsam fir (Wingfield *et al.*, 1983).

There are a large number of species of *Bursaphelenchus* that are associated phoretically with scolytid bark beetles (Rühm, 1956) but these forms usually occur in surface galleries of these insects. Conversely, *B.* (suprasp. *xylophilus*) is a facultative anaerobe which is recovered from the anaerobic environment of the xylem and propagates well on fungi that colonize the xylem (Wingfield, 1987). This deep wood lifestyle may help *B.* (suprasp. *xylophilus*) avoid natural enemies and competition from nematodes, mites or other organisms in surface galleries (Wingfield, 1987).

15.1.4 Pathogenesis

Myers (1988) suggests that it is the invasion and migration of *Bursaphelenchus* (suprasp. *xylophilus*) *xylophilus* through host tissues which induce localized innate hypersensitive defence reactions and lead to the eventual death of the susceptible pine host. Invasion and dispersion of *B.* (suprasp. *xylophilus*) *xylophilus* occurs in two phases (Myers, 1988). First, nematodes are localized at the maturation site in the chlorophyllous cortical tissues while a few migrate throughout the host (Myers, 1988). Upon entry into the maturation feeding wound the nematodes straighten out after contact with oleoresin. Many recover and enter the cortical tissues. A few enter the phloem, cambium and xylem. Nematodes successful in phase one disperse randomly throughout the tree at a maximum speed of 40–50 cm per day (Myers, 1988). Oleoresin flow stops, parenchymal cells start to die, and the nematodes begin to reproduce.

The second phase of dispersion occurs as oleoresin pressure and flow and host transpiration decrease. Nematodes in the cortical tissues around the primary lesion disperse throughout the resin canals of the xylem. This movement concurs with abnormal leakage of oleoresin from adjacent radial and axial resin canals, which causes blocked tracheids (xylem). Tracheid blockage leads to whitewood formation, further restricts water conductivity and decreases transpiration. Pathogenesis proceeds more rapidly in young trees and seedlings than in mature conifer hosts (Myers, 1988).

Bursaphelenchus (suprasp. *xylophilus*) *xylophilus* and *B.* (suprasp. *xylophilus*) *mucronatus* each produce and exude different cellulase isozymes during movement and probably feeding in the propagative phase (Odani *et al.*, 1984). These cellulases are suspected to be one of the cell wall disturbing factors that cause oleoresin leaks from the resin canals to the tracheal zone, which eventually block sap flow in the susceptible pine host. Cellulase production and exudation are not the entire story. *B.* (suprasp. *xylophilus*) *mucronatus* produces cellulases but is not highly pathogenic to *Pinus densiflora* seedlings like *B.* (suprasp. *xylophilus*) *xylophilus*. Odani *et al.* (1984) suggest that the reason for this difference in pathogenicity is the restricted movement and reproduction of *B.* (suprasp. *xylophilus*) *mucronatus* relative to *B.* (suprasp. *xylophilus*) *xylophilus*.

The exudation of cellulases and the movement of the pine wood nematode from the point of transmission may also induce the release of host cell hydrolases which could cascade through the host, causing problems for cellulose biosynthesis (Bolla *et al.*, 1988). Phytotoxic resins, the result of monoterpene synthesis by the host in response to *Bursaphelenchus* (suprasp. *xylophilus*) *xylophilus* inoculation, have been recovered from highly suscep-tible *Pinus* seedlings (3 years old) and older trees (15–25 years old) (Oku, 1988). There is considerable variability to the resin composition of pines, and some of the differences observed in host responses to pine wood inocula-tion may be due to differences in the precursers that are available for monoterpene synthesis; i.e. mostly bicyclic monoterpenes (α- and β-pinene) occur in those trees producing phytotoxins in response to nematode inocula-tion versus 90% n-heptane for the *B.* (suprasp. *xylophilus*) *xylophilus*-resistant *P. jeffreyi* (Bolla *et al.*, 1988).

Research in Japan (Oku, 1988) and the USA (Bolla *et al.*, 1988) has shown that extracts of *Bursaphelenchus* (suprasp. *xylophilus*) *xylophilus*-infected *Pinus densiflora*, *P. thunbergii* and *P. sylvestris* yield at least five phyto-toxins which are not present in these pines when healthy or stressed by other abiotic or biotic agents. Oku (1988) isolated and identified benzoic acid and catechol in needles of nematode-infested pines, and benzoic acid, dihydroconiferyl alcohol and 8-hydroxycarvotanacetone (carvone hydrate) from branches of 3-year-old *P. densiflora*. Bolla *et al.* (1988) isolated and identified 10-hydroxyverbenone and 8-hydroxycarvotanacetone from resins extracted from pine wood nematode-infected *P. sylvestris* (10–20 years old).

Shaheen *et al.* (1984) have demonstrated that the phytotoxins present in a $CHCl_3$-base extract of pine wood nematode-inoculated *Pinus sylvestris* (2-year-old trees) (major components of the extract were 10-hydroxyverbenone and 8-hydroxycarvotanacetone) are recovered as early as three days post-inoculation and increase in concentration over time until 30 days post-inoculation. These phytotoxins affect 45-day-old *Pinus* spp. seedlings in a time and dose-dependent manner for each species examined, except *P. jeffreyi*, which does not wilt at any of the con-centrations tested. The wilt symptoms observed when the D-isomer of 8-hydroxycarvotanacetone is injected into susceptible pines are similar to the symptoms observed after pine wood nematode inoculation or induce-ment by the $CHCl_3$-base extract from a pine wood nematode-inoculated susceptible pine (Oku, 1988). No wilt symptoms have been observed when pine wilt disease-susceptible pines are challenged with extracts from healthy pines (Oku, 1988). Synergistic effects have been documented on *P. densi-flora* and *P. sylvestris* when the individually isolated phytotoxins are combined with each other or with several non-toxic abnormal metabolites unique to nematode-infected pines (Oku, 1988).

$CHCl_3$-base extracts from nematode-infected *Pinus sylvestris* cause temporary paralysis of *Bursaphelenchus* (suprasp. *xylophilus*) *xylophilus* in vitro, reduce final population densities of the nematode on *Botrytis cinerea*, and inhibit the growth of *Ceratocystis ips* but not *B. cinerea* (Bolla *et al.*, 1988). These observations suggest that the phytotoxins might have phytoalexin-like properties at high concentrations. Oku (1988) showed that catechol, benzoic acid, dihydroconiferyl alcohol and 8-hydroxycarvotanacetone do not inhibit *B. cinerea* growth and that catechol and benzoic acid do not reduce nematode growth and reproduction at $100 \mu g \ ml^{-1}$. However, dihydroconiferyl alcohol inhibits nematode multiplication at $10 \mu g \ ml^{-1}$, and the L-isomer of 8-hydroxycarvotanacetone is inhibitory at $30 \mu g \ ml^{-1}$, supporting the idea that these materials have phytoalexin-like properties. Shoot cuttings of *P. densiflora* (1 year old) which are treated with 300 p.p.m benzoic acid ($3.1 \ mg \ g^{-1}$ dry wt of wood) survived inoculation with *B.* (suprasp. *xylophilus*) *xylophilus*, whereas cuttings treated with less than 300 p.p.m do not, and application of 500 p.p.m benzoic acid was phytotoxic (Mamiya *et al.*, 1989). Benzoic acid at 300 p.p.m in these experiments may have altered water uptake dynamics to favourably offset the normal wilting associated with pine wilt disease (Mamiya *et al.*, 1989).

Nematodes are probably not responsible for monoterpene synthesis because they cannot metabolize mevalonic acid residues into bicyclic monoterpenes (Bolla *et al.*, 1988). However, *Bursaphelenchus* (suprasp. *xylophilus*) *xylophilus* may induce a host reaction involving monoterpene synthesis which gets out of control, diverts intermediates of carbohydrate metabolism away from normal sinks, and exacerbates problems with energy homeostasis in the conifer host, which is disrupted during the initial attack (Bolla *et al.*, 1987). Rapidly dispersing nematodes defeat the innate hyper-sensitive defence reaction of the susceptible pine host by continually trigger-ing and keeping ahead of the reaction front. Thus, the movement, feeding and exudation of cellulases by the nematode induce the host's own innate defence reaction to weaken and eventually kill itself (Myers, 1988). The inducement of host reactions appears to be complicated and a growing number of geographical and host-specific pathotypes of *B.* (suprasp. *xylophilus*) *xylophilus* have been identified (Bolla *et al.*, 1986; Kiyohara and Bolla, 1990; Wingfield *et al.*, 1983).

15.1.5 Biology of *Monochamus* spp.

Although there are 21 species of beetles in the Cerambycidae, two species in the Curculionidae and one genus in the Buprestidae known to carry dauer juveniles of *Bursaphelenchus* (suprasp. *xylophilus*) *xylophilus* worldwide, only five species in the genus *Monochamus* (Cerambycidae) have been documented to transmit the nematode (Linit, 1988). *M. alternatus* has been shown to be a primary vector in Japan, whereas *M. carolinensis*, *M. mutator*, *M. scutellatus* and *M. titillator* have been shown to be involved in primary or secondary transmission in North America (Linit, 1988). Most of the other reported insects are probably not involved in primary transmission of the pine wood nematode because they carry so few nematodes (mean number < 300 per insect) (Linit, 1988; Linit *et al.*, 1983) or they are not associated with healthy pines during their biology.

The biology of the species of *Monochamus* which have been documented to transmit pine wood nematodes has been reviewed (Kobayashi *et al.*, 1984; Linit, 1988). These beetles, which are mostly univoltine, are often called pine sawyers because of the loud noise the larvae make during gallery construction (Linit, 1988). The newly emerged adults of *Monochamus* orient to a healthy conifer host, where they feed on the bark of current and one-year-old twigs to reach reproductive maturity (reached in 16–30 days for females and 5–15 days for males of *M. alternatus*) (Togashi, 1990). The mean time to sexual maturity for females of *M. carolinensis* was 7–12 days (Walsh and Linit, 1985). Extracts from the bark of pine twigs which have been shown to stimulate beetle feeding include fructose, sucrose, β-sitosterol and several glycosides (Kobayashi *et al.*, 1984).

It is the maturation feeding behaviour of the beetle vector that allows *Bursaphelenchus* (suprasp. *xylophilus*) *xylophilus* dauer juveniles access to a healthy tree for primary infection. In fact, the maturation feeding requirement of *Monochamus* is probably what allowed the pine wood nematode to expand its biology from the ancestral secondary transmission-mycophagous life cycle to that of the more derived facultative plant/fungal parasite.

In Japan, *Monochamus alternatus* dispersion increases with age from emergence (one to five days) to reproductive maturation (15–19 days) and then decreases, suggesting that there is significant plasticity in flight behaviour to allow for long-distance dispersal to sites for maturation feeding or breeding (Togashi, 1990). In the southeastern USA, maximum tethered flight duration of different age groups of *M. carolinensis* coincides with the dispersion times associated with feeding and ovipositional activity (Humphrey and Linit, 1989a). Tethered flight experiments (beetles flown within 24 hours of emergence) with *M. carolinensis* reared from logs of *Pinus banksiana* which had been inoculated with *Bursaphelenchus* (suprasp. *xylophilus*) *xylophilus* and kept under constant conditions (30°C, 70% RH) indicate that flight duration is independent of beetle sex, but longer for larger beetles (Humphrey and Linit, 1989b).

The presence of large numbers of dauers of the pine wood nematode in the metathoracic tracheae of *Monochamus carolinensis* is visually impressive (Linit *et al.*, 1983) and suggests an impaired aerobic delivery system. The number of nematodes per beetle has a significantly negative effect on flight duration of tethered, newly emerged *M. carolinensis* (Humphrey and Linit, 1989b). However, total blockage of the tracheal system is probably rare and the actively directed air-flow system of the cerambycid pterothorax can accommodate a one-half to two-thirds blockage and still maintain aerobic metabolism during flight (Humphrey and Linit, 1989b).

Several authors have presented data suggesting that beetle size is positively correlated with the number of pine wood nematodes carried per insect (Linit, 1988; Humphrey and Linit, 1989b). One proposed explanation for this phenomenon is that large beetles respire more carbon dioxide than smaller beetles. Because carbon dioxide is attractive to dauers of *Bursaphelenchus* (suprasp. *xylophilus*) *xylophilus*, larger beetles could accumulate larger populations of nematodes than smaller beetles (assuming homogeneous nematode distribution and equal hydration levels and temperatures in the log at the time of beetle eclosion from the pupa) (Humphrey and Linit, 1989b). High densities of the pine wood nematode negatively affect longevity and the mean estimated fecundity of heavily infested *Monochamus alternatus* (Togashi and Sekizura, 1982).

Togashi (1985) grouped *Monochamus alternatus* into four potential vector classes contingent upon the initial number of *Bursaphelenchus* (suprasp. *xylophilus*) *xylophilus* dauers carried per beetle (> 10 000, 1000–9999, 100–999 and < 100). Averaged transmission curves had single respective peaks at *c.* 1500, 370, 38 and (no peak) nematodes for a five-day period between 20 and 35 days post-emergence of the beetle. The proportion of nematodes remaining in the host cadaver at beetle death increased as the initial nematode density increased, which might be an effect of reduced longevity related to larger initial densities of nematodes (Togashi and Sekizura, 1982). Togashi postulated that because a threshold of > 300 dauers is needed to kill a single *Pinus thunbergii* and that the mean maturation feeding duration of *M. alternatus* is 2.3 days, a single beetle with the initial nematode density class of > 10 000 nematodes per insect could initiate pine wilt in a susceptible host by depositing *c.* 700 nematodes per visit sometime between 20 and 35 days post-emergence. Pine wilt would have to be initiated by numerous attacks by beetles in the 1000–9999 density class and very large aggregations in the lower density classes. He suggested that the initial density classes may have ramifications for the population biology of both the pine wood nematode and the vector. The interaction between the beetle vector and the pine wood nematode in Japanese epiphytotics may function loosely as 'population mutualism' (Giblin, 1985). Thus, some individuals of the beetle population (those with > 10 000 nematodes per insect) may suffer a loss of reproductive fitness while increasing the chances for successful nematode transfer and pine wilt disease which will increase new beetle brooding environments for the rest of the beetle population (Togashi, 1985).

After maturation feeding (about three weeks after emergence), the adult beetles are attracted to host volatiles (ethanol and 11 different monoterpene hydrocarbons) released by stressed trees and cut logs (Kobayashi *et al.*, 1984). Ethanol apparently synergizes the attractancy of the monoterpenes, the most important being α-pinene, followed by β-pinene and β-phellandrene (Kobayashi *et al.*, 1984). Interestingly, ethanol or acetone injected into the trunk of a healthy tree induced attractancy to beetles. This kind of treatment coupled with treatment with paraquat may be useful for creating trap trees (Kobayashi *et al.*, 1984). Attractancy in stressed trees or cut logs facilitates congregation of bettles for mating and oviposition (Linit, 1988) and allows for secondary transmission of *Bursaphelenchus* (suprasp. *xylophilus*) *xylophilus* for fungal parasitism and association with the developing pine sawyers. The female prepares an egg niche (pit or slit) in the bark of the tree with her mandibles and then oviposits through the base of the pit into the phloem (Linit, 1988). Eggs can be laid singly or multiply, depending upon the species of *Monochamus*. The resulting larvae feed on the inner bark, cambium and outer sapwood, creating surface galleries filled with frass. As the larvae mature, they bore into the woody tissue and form a U-shaped gallery which ends just before the cambium. There can be three to eight larval instars for *Monochamus*, depending upon the species, and the last instar pupates in a chamber formed by closing the gallery off with packed excelsior (Linit, 1988). A few days after eclosion, the adult exits through the wood and bark above the pupal chamber, leaving a round hole. Developmental time from oviposition to adult emergence for *M. carolinensis* ranged from eight to 12 weeks when eggs were laid in May, whereas eggs laid in August and September gave rise to overwintering larvae (Pershing and Linit, 1986).

15.1.6 Distribution

All isolates of *Bursaphelenchus* (suprasp. *xylophilus*) recovered to date have been from the temperate Northern Hemisphere. *B.* (suprasp. *xylophilus*) *xylophilus* is probably indigenous to North America and *B.* (suprasp. *xylophilus*) *mucronatus* is probably native to Europe and Asia. *B.* (suprasp. *xylophilus*) *xylophilus* has been reported north of Mexico from at least 33 states of the USA and various provinces in southern Canada (Bergdahl, 1988). Reportedly, at least 27 species of *Pinus* in the continental USA and 38 species worldwide are hosts to *B.* (suprasp. *xylophilus*) *xylophilus* (Bergdahl, 1988). *Pinus* is a large genus (*c.* 90 species) endemic to the Northern Hemisphere (Bailey Hortorium, 1976). Several other non-pine conifers from the Northern Hemisphere have been reported as occasional hosts for *B.* (suprasp. *xylophilus*) *xylophilus* and include the fir, *Abies balsamea*, the spruces, *Picea glauca* and *P. pungens*, the cedars, *Cedrus atlantica* and *C. deodara*, the larches, *Larix decidua* and *L. larcina*, and *Pseudotsuga menziesii* (Bergdahl, 1988).

There is no evidence to suggest that *Bursaphelenchus* (suprasp. *xylophilus*) *xylophilus* has caused widespread pine wilt disease epiphytotics in conifer forests in North America to the scale observed during the last twenty years in Japan or southern China (Bergdahl, 1988; Rutherford *et al.*, 1990). In North America, pine wilt is an occasional disease of susceptible ornamental conifers, Christmas tree orchards, stressed or diseased pines in forest plantations, natural forest stands, windbreaks or conifer seed orchards (Bergdahl, 1988; Wingfield *et al.*, 1982). There have been no confirmed reports of pine wilt disease from Europe (Rutherford *et al.*, 1990). In fact, classical pine wilt disease symptoms are relatively rare outside of Japan and southern China (Rutherford *et al.*, 1990). Rutherford *et al.* (1990) suggest that a mean daily summer temperature below 20°C is the key limiting factor for the northern distribution of pine wilt disease. Mean daily summer temperatures above the 20°C isotherm are necessary for the pine wilt disease to become problematic in the presence of the disease. Host-plant resistance may limit its effects in warmer areas such as the southeastern USA, where the disease rarely occurs, except when susceptible conifers are grown or where pollution levels are high. Thus, pine wilt disease may be a major biotic factor influencing the natural distribution of some *Pinus* spp. in North America in areas where the mean daily summer temperatures are above 20°C (Rutherford *et al.*, 1990).

Recent pathotype development or introductions of *Bursaphelenchus* (suprasp. *xylophilus*) *xylophilus* into areas with moist summers which are warmer than the 20°C isotherm, such as southern Japan and southern China, with large forests of susceptible pines, probably allowed for the development of the current epiphytotics. Myers (1988) suggests that warm and wet summers stimulate pine-host growth rather than differentiation and that according to the growth–differentiation balance hypothesis this increases disease susceptibility. Susceptible *Pinus* (i.e. Monterey pine, *P. radiata*) which have been introduced into warmer climates where the pine wood nematode and its vector(s) do not occur could be in jeopardy if they were introduced into, e.g., Australia. Countries with climates similar to central to northern Europe and northern North America should have little to fear concerning the introduction or occurrence of pine wilt disease (Rutherford *et al.*, 1990), unless cold-tolerant pathotypes of the nematode develop or major changes in climate occur.

15.1.7 Management

Insecticidal treatment of felled trees is the most commonly used method for reducing potential *Monochamus alternatus* populations and for breaking the link between the nematode and the vector (Kobayashi *et al.*, 1984). There are a large number of insecticidal formulations registered for *M. alternatus* control which have as an active ingredient carbaryl, fenthion, fenitrothion, chlorpyrifosmethyl or diazinon. Autumn treatment of cut logs causes 95–100% mortality in *M. alternatus* larvae because the beetles are still just under the bark. However, oil emulsion of the insecticide is recommended for logs cut in winter or spring when the beetle larvae have bored deeper into the wood (Kobayashi *et al.*, 1984). Covering cut logs treated with insecticide with vinyl sheeting is also effective at controlling beetles (Kobayashi *et al.*, 1984).

Carbonizing wood to a 5 mm depth takes about 20 minutes and causes 100% *Monochamus alternatus* mortality. The felled pine wilt-infested trees can be harvested and used for wood or charcoal production. A portable kiln has been developed allowing for the production of charcoal from pine wilt-infested trees. Pine wilt and beetle-infested trees can also be felled and chipped for acceptable pulp wood (Kobayashi *et al.*, 1984).

Destruction of the pine wood nematode in infested trees is not a viable option in Japan. Attempts at chemical treatment, radio heating and use of natural enemies are not practical (Kobayashi *et al.*, 1984). However, prophylactic injection of or soil treatment with nematicides into valued susceptible hosts is used for disrupting the association between the nematode and its pine host (Kobayashi *et al.*, 1984). Mesulfenfos and marantel tartarate are registered in Japan for pine wilt disease control. Trunk injections at 150–200 p.p.m g^{-1} dry weight of tree are done at breast height about three months before challenge with nematodes (Kobayashi *et al.*, 1984).

Induced resistance is another factor that may have some application in protecting valuable susceptible trees. Kiyohara (1984) reported that a non-virulent isolate of *Bursaphelenchus* (suprasp. *xylophilus*) *xylophilus* can be used at concentrations of 3 × 10^4 nematodes to induce systemic resistance which lasts for at least 120 days in 2–13-year-old *Pinus thunbergii*. A period of 10–30 days post-induction is required before challenge with a virulent isolate of *B.* (suprasp. *xylophilus*) *xylophilus* and resistance can be induced in several species of *Pinus*. The basis for this phenomenon is not known, but population growth of virulent *B.* (suprasp. *xylophilus*) *xylophilus* is inhibited in pines which were induced with the non-virulent isolate. The practicality of such a treatment still needs to be demonstrated (Kobayashi *et al.*, 1984).

Some cultural practices to reduce stress on the pine hosts, such as thinning and watering, are used in Japan. However, genetically based host resistance appears to be the most promising method for establishing and maintaining host protection. Hybridization work and selection of resistant native pine hosts is practical and being applied in Japan. Also, the use of some pine wilt-resistant species of exotic pines for replanting is being tried (Kobayashi *et al.*, 1984).

Aerial treatments with insecticides may help protect stands of conifer hosts from *Monochamus alternatus* maturation feeding and reduce the general beetle population. Pine mortality caused by pine wilt disease declines with aerial treatments of insecticides relative to untreated control areas in most studies to date (Kobayashi *et al.*, 1984). Helicopters are the most effective means for delivery of either emulsified fenitrothion or hydrated carbaryl to the forest crown. Environmental contamination by such treatments has not been documented. However, there are some concerns for this method of treatment, such as phytotoxicity to several clones of Japanese cypress,

Chaemaecyparis abtusa, with fenitrothion, and hazards to non-target beneficial insects such as honey bees and silkworms. Arthropod populations are negatively affected in the zone from the forest crown area to the under layer, but appear to recover quickly (Kobayashi *et al.*, 1984) and pest resurgences have not been confirmed or correlated with carbaryl applications (Togashi, 1990). Comparison of life table data and mortality agents for *M. alternatus* suggests that carbaryl applications lower the mortality caused by predation to third instars through adults of *M. alternatus* by natural enemies and may help survival of *M. alternatus* populations depending upon pesticide application timing (Togashi, 1990).

The European and Mediterranean Plant Protection Organization (EPPO) has designated the pine wood nematode as an A1 quarantine pest because of its perceived potential for causing forest epiphytotics in Europe (Rutherford *et al.*, 1990). Also, the European Community (EC) is considering regulations for treatment of exotic forest tree products to prevent importation of pests such as those involved in the pine wilt disease (Dwinell, 1990). Finland in 1985 and the other Nordic countries in 1986 embargoed all raw softwood shipments from countries where the pine wilt disease is known to occur, i.e. North America and Japan, even though pine wood nematode introduction and establishment via green lumber or wood chips to live trees in northern Europe are highly unlikely (Dwinell, 1990; Rutherford *et al.*, 1990). As mentioned in the distribution section of this chapter, it is countries like Australia, where summer temperatures exceed 20°C and large areas of susceptible pine species are cultivated, that need to be alert to the introduction of the vector and pathogenic isolates of the pine wood nematode.

Although it would be expensive and require calibration of wood kilns, the pine wood nematode and *Monochamus* spp. can be eradicated from dimension lumber for exportation by heat-treating wood to 60°C. Heat-treating wood chips to 60°C for 1 hour also kills all of the pine wood nematodes present. Hot-water treatments are also effective and take considerably less time for nematode and insect mortality (Kinn, 1986). Prevention of *Monochamus* infestation in green lumber is another less costly but reliable method for producing and shipping raw softwood products without the risk of pine wilt disease introduction. This involves debarking trees soon after felling and water storage of logs to prevent beetle oviposition (Dwinell, 1990). Dwinell (1990) suggests that a mill certification programme ensuring that green wood is bark free and without borer holes would prevent exportation of endemic *Monochamus* spp. to foreign countries. Because pine sawyers cannot complete their life cycle in debarked logs or lumber, these products could not be used by foreign pine sawyers to become associated with pine wood nematodes which might be present in the wood.

15.2 RED RING DISEASE

15.2.1 Overview and symptomatology

Red ring disease was first reported from Trinidad in 1905. It is limited in distribution to the Neotropics, where it is one of the most important diseases of coconut, *Cocos nucifera*, and African oil palm, *Elaies guineensis* (Griffith and Koshy, 1990). Red ring disease losses in affected coconut and oil palm plantations are commonly in the range of 10–15% (Chinchilla, 1988). The disease is caused by the red ring or coconut palm nematode, *Bursaphelenchus cocophilus*. The American palm weevil, *Rhynchophorus palmarum* (Curculionidae: Coleoptera), is the main vector of this nematode. The dauer juvenile stage (J_{III}) of *B. cocophilus* survives poorly without an insect or plant host and appears to be parasitically associated with the palm weevil during metamorphosis (Griffith, 1968). The nematodes are probably deposited during weevil oviposition into the leaf bases or wounds in a susceptible palm tree, where they invade, feed on parenchymal cells, and can cause a lethal wilt in as little as two to four months (Fig. 15.2) (Griffith, 1987).

Symptoms vary widely with palm host species and age, cultivar and environmental factors. Classical red ring symptoms in *Cocos nucifera* often include premature nut fall (except for mature nuts), withering and necrosis of inflorescences, and yellowing, bronzing and death of progressively younger leaves. Yellowing of leaves usually starts at the tips of the pinnae and moves inward to the rachis and then to the base of the petiole. Several of the dying or dead leaves often break close to the petiole and remain hanging. A stem transverse section reveals a discrete brick to brownish-red ring which is 2–6 cm wide and occurs 2–6 cm within the stem periphery. The cortex of roots and leaf petioles can be discoloured and appear yellow to brownish-red. In longitudinal section, discoloration is usually continuous throughout the length of the stem, appearing as two bands which unite at the base and form discontinuous lesions near the crown. Coconut palms 3–10 years old usually die within several months of infection. Severe damage to the crown of red ring-diseased coconut palms is done by larval feeding of the weevil vector, often causing the palm crown to fall over under its own weight. Also, other insects, such as the bearded palm weevil, *Rhinostomus barbirostris*, are often attracted to and cause significant damage to palms dying from red ring disease.

In areas such as El Salvador, older palms (> 20 years old) have been reported with red ring disease displaying less definitive symptoms with a more prolonged death (Dean, 1979). Dauer juveniles can be harvested from the discoloured tissue of the ring (up to *c.* 11 000 nematodes per gram of tissue), from leaf petioles, or roots to confirm disease diagnosis in coconut palm (Blair, 1969). Little leaf symptoms in coconut palm can be caused by the red ring nematode (Hoof and Seinhorst, 1962) and may be more common

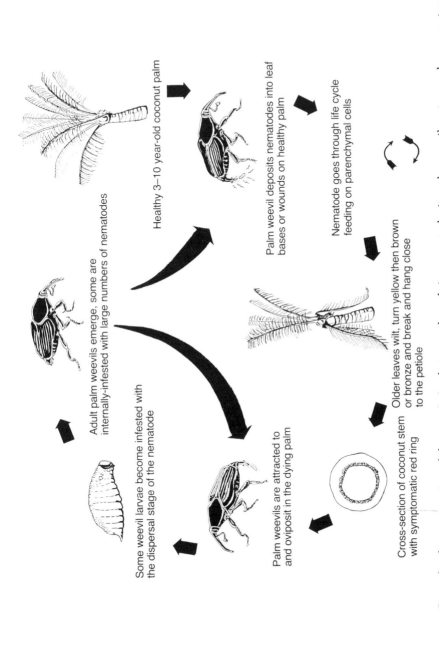

Some weevil larvae become infested with the dispersal stage of the nematode

Adult palm weevils emerge, some are internally-infested with large numbers of nematodes

Healthy 3–10 year-old coconut palm

Palm weevil deposits nematodes into leaf bases or wounds on healthy palm

Nematode goes through life cycle feeding on parenchymal cells

Palm weevils are attracted to and oviposit in the dying palm

Older leaves wilt, turn yellow then brown or bronze and break and hang close to the petiole

Cross-section of coconut stem with symptomatic red ring

Figure 15.2 Generalized representation of the association between the red ring nematode, its palm weevil vector, and a coconut palm host.

in older coconut palms (Kraaijenga and Ouden,1966). Coconut and African oil palms younger than 2 years old cannot be experimentally infected with the red ring nematode and red ring disease has not been observed in palms of this age in the field (Giblin-Davis, 1991).

Chinchilla (1988) has described red ring nematode-induced symptoms in African oil palm. Classical red ring symptoms include progressive premature yellowing and death of older leaves. As with coconut palm, these leaves often break at the petiole and hang for a long time while the symptoms progress to younger leaves. New leaves in diseased trees are usually more of a pale yellowish-green than those observed in healthy palms. Examination of a stem cross-section will reveal a brown, cream or rose-coloured ring which is a few centimetres in thickness and can be concentric to the periphery of the stem. Irregular-shaped rings and rings that are not continuous throughout the entire stem length are common. Feeding by the larval stage of *Rhynchophorus palmarum* often destroys the apical bud of the oil palm. The life cycle of *R. palmarum* can be completed in oil palm, but it may be a less suitable host than coconut palm (Schuiling and van Dinther, 1981). Generally, oil palms older than 5 years can be infected with the red ring nematode and death can occur within several months of red ring nematode introduction. Dauer juveniles of the red ring nematode can be harvested from tissues adjacent to and within the ring from the stem of African oil palm and range in number from 0 to 5000 nematodes per gram of tissue.

Little leaf symptoms and a combination of red ring and little leaf symptoms have been documented in red ring nematode-infested African oil palms. Little leaf appears in oil palm plantations in Central America regardless of the age (> 4 years old) and vigour of the host (Chinchilla, 1988). Little leaf symptoms can also be caused by a rot of the anterior stem, recovery from common and lethal spear rot (unknown cause), *Fusarium* wilt and boron deficiency. Also, different kinds of leaf malformations can be caused by feeding damage in the whorl by certain insects and rodents (Chinchilla, 1988). In African oil palms with typical little leaf symptoms, leaf colour commonly remains normal. However, palms begin abnormal production of very short leaves, which give crowns the unusual appearance of a feather duster. As the disease progresses, there is a decrease in leaf size and surface area to the point where an expressed leaf is reduced to a leafless rachis with suberized lesions over most of its surface. New leaf emission rate is reduced and inflorescences are aborted, making the infested palms unproductive. In contrast, healthy oil palms will present an average of two leaves per month. Red ring nematodes are most often recovered from necrotic lesions in the middle and distal parts of unpresented leaves in the region of rapid elongation (-11 to -2; counting backwards from the spearleaf ($=1$)) but can be recovered from leaves as young as -28 (Chinchilla, 1988). Internal symptoms in the stem base can include dark brown necrotic spots or a dark ring which only occurs for a short distance in a longitudinal section (Chinchilla, 1988). Little leaf is usually a chronic disease which can last for several years, producing trees of shorter stature than healthy palms of the same age (Chinchilla, 1988). Sometimes the little leaf symptomatic palm will recover and resume production of normal-sized leaves. These recoveries are often short lived and little leaf recurs (Chinchilla, 1988).

Symptoms such as little leaves with pale colour, necrotic inflorescences, irregular ripening of fruit, leaf necrosis or leaf arching can be associated with red ring nematode infestations in oil palm. Transverse cuts in the stem can reveal one to many discontinuous rings, random necrotic lesions, core necrosis or staining surrounded by a ring. Some of these stem core symptoms can, in chronic infestations, progress into a rot which leads to the formation of an empty cavity (Chinchilla, 1988).

15.2.2 Taxonomy of the red ring nematode

The red ring nematode was described by Cobb (1919) as *Aphelenchus cocophilus* from coconut roots from a red ring-diseased tree from Grenada, West Indies. Subsequently, it was placed in the genera *Chitinoaphelenchus*, *Aphelenchoides* and *Rhadinaphelenchus* (Goodey, 1960). Recently, the monotypic subfamily Rhadinaphelenchinae was eliminated and the genus *Rhadinaphelenchus* was synonomized with *Bursaphelenchus* (Baujard, 1989). The morphology of the male spicules and caudal alae, the number and position of the male caudal papillae, and the presence of a vulval flap in females of *B. cocophilus* appear to support its inclusion in the genus *Bursaphelenchus* (Giblin-Davis *et al.*, 1989c).

Bursaphelenchus cocophilus is a very long, thin nematode; the females and males are 60–139 and 65–179 times longer than wide, respectively, with the greatest body width being less than 15.5 μm and total length ranging from 775 to 965 μm from little leaf symptomatic African oil palm and 812–1369 μm from coconut or African oil palms with typical red ring symptoms (Dean, 1979; Gerber *et al.*, 1989; Giblin-Davis *et al.*, 1989c). It has been suggested that L and 'a' ratio differences between populations of *B. cocophilus* from coconut and populations from African oil palm in Costa Rica are divergent enough to be considered separate species (Salazar and Chinchilla, 1989). Gerber *et al.* (1989) suggest that divergences in morphometrics between different host and geographical isolates of *B. cocophilus* probably represent normal intraspecific variability, but further research is needed.

The metacorpus and stylet in the second-stage juveniles and adults are well developed. Stylet length is between 11 and 15 μm in adults. Females have a vulval flap which appears bowed posteriorly when viewed ventrally, a long post-uterine sac (extending about 75% of the vulva–anal distance), and an elongate tail (62–117 μm) with a rounded terminus. Males have seven caudal papillae: one ventral preanal papilla, one pair of subventral preanal or adanal papillae, and two pairs of subventral postanal papillae. The distal ends of the spicules in the males are heavily sclerotized and the caudal alae form a spade-shaped flap (= bursal flap) (Dean, 1979; Gerber *et al.*, 1989; Giblin-Davis *et al.*, 1989c).

15.2.3 Biology of the red ring nematode

Bursaphelenchus cocophilus has a generation time of nine to ten days when inoculated into the husk of an immature coconut seed (Blair, 1969). Reproduction may be by parthenogenesis (Griffith, 1968) although further research is needed to confirm this. Little is known of the effects of abiotic factors on generation time, fecundity or life history of *B. cocophilus* because it is difficult to culture in vitro (Giblin-Davis *et al.*, 1989a). Experiments with the culture of the red ring nematode suggest that it is an obligate phytoparasite incapable of growth and reproduction on monoxenic cultures of fungi such as *Monilinia fructicola* or *Botrytis cinerea* on GPDA (Giblin-Davis *et al.*, 1989a). In contrast, all other members of the genus *Bursaphelenchus* are mycophagous or facultative fungal and plant parasites. Aside from culturing the red ring nematode in coconut palms > 2.5 years old, husks of immature coconuts and coconut leaf stalks can be used for maintaining small cultures for study (Blair, 1969; Giblin-Davis *et al.*, 1989a). Preliminary evidence suggests that an autoclaved liquid oligidic medium using supplemented red ring infusion water could be optimized for axenic culture of *B. cocophilus* (Giblin-Davis *et al.*, 1989a).

In early stages of red ring disease of coconut palm, nematodes in the propagative phase disperse to new areas in the stem, petioles, and eventually to the cortex of roots where they feed on thin-walled parenchymal cells and exist as intercellular parasites. Adults, propagative juveniles and eggs of *Bursaphelenchus cocophilus* are usually recovered from areas of the stem where red or pink lesions have not yet coalesced into the red ring. Conversely, red ring tissue of the stem and petioles usually has a very high abundance of live dauer juveniles with virtually no adults or eggs. Blair (1969) has hypothesized that the accumulation of carbon dioxide concurrent with the nematode population explosion and the red ring formation in coconut palms prevents maturation of juveniles, thus causing an accumulation of dauers. Further work is needed in this area. It is just as likely that dauer juvenile formation is under partial or total control by nematode-produced or induced metabolites or a density-dependent pheromone and the disappearance of a food factor as in *Caenorhabditis elegans* (Riddle and Georgi, 1990).

Dauer juveniles of *Bursaphelenchus cocophilus* from coconut palm usually range from 700 to 920 μm and have a pointed tail with or without a mucron. The metacorpus is usually not well developed in dauers from the palm or the weevil vector and the stylet is not visible (Gerber *et al.*, 1989). Dauer juveniles and propagative forms of *B. cocophilus* persist very poorly in soil or in non-sterile water at room temperature (two to seven days) and they do not survive desiccation (Blair, 1969). Survival of dauer juveniles was prolonged when they were surface sterilized and incubated in autoclaved water (> 40 days) or in autoclaved red ring stem tissue infusion water supplemented with D-glucose, lactose or bacto-lactose broth (> 70 days) at 28°C (Giblin-Davis *et al.*, 1989a). The dauers of the red ring nematode persist in the degenerating coconut palm tissue for about three months after the death of the tree (Griffith, 1968).

In general, the highest concentrations of dauer juveniles occur in the petioles of coconut and African oil palm where *Rhynchophorus palmarum* often pupates, building its cocoon out of the palm fibre. The dauer juvenile stage of *Bursaphelenchus cocophilus* is apparently acquired either per os by larval feeding of red ring tissue or invades the tracheal system and/or other natural openings. The dauers do not moult within the insect host and are carried through metamorphosis into the adult weevil (Griffith, 1968; Gerber *et al.*, 1990; Gerber and Giblin-Davis, 1990a). Griffith (1968) reports that the dauer juveniles of the red ring nematode are associated in the tracheae of *R. palmarum* and are shed with the cuticle at each moult and reinvade the tracheal system of the new insect stage. A large number of dauers can aggregate in the region of the genital chamber of adult male and females of *R. palmarum* (Gerber and Giblin-Davis, 1990a; Griffith, 1968). The dauer stage is apparently parasitic and requires the weevil host to survive as conditions in the palm deteriorate.

Morphologically, the dauer stage isolated from red ring tissue of coconut palm and the hemocoel of *Rhynchophorus palmarum* are very similar. However, there is a significant shortening in body length and an increase in body width in dauers from the hemocoel of *R. palmarum* compared with those from red ring tissue (Gerber *et al.*, 1989). So far, no moult has been observed within the weevil and any stage of the nematode can be used to initiate red ring disease in a susceptible palm host. Gerber and Giblin-Davis (1990a) observed that close to 100% of newly emerged adult *R. palmarum* from field-collected cocoons from red ring-diseased coconut palms in Trinidad were internally infested with dauers of the red ring nematode and > 47% contained more than 1000 nematodes. In that study, < 5% of adult *R. palmarum* were externally contaminated or had red ring nematodes in their cocoons, supporting the observation that nematodes require a host to survive as conditions in the palm or cocoon become limiting. Further work with surface-disinfested larvae and pupae of *R. palmarum* supports the observation that dauer juveniles are acquired prior to metamorphosis and carried through to the adult internally (Gerber *et al.*, 1990; Griffith, 1968).

A minimum external inoculation of 5000 red ring dauer juveniles onto natural cracks on coconut petioles is necessary to initiate red ring disease (Griffith, 1968). However, only 10–50 nematodes are required to initiate red ring disease when inoculated into a small wound (Griffith, 1968). Considering the low densities of nematodes that persist in the soil (Blair, 1969), or contaminate the faeces or bodies of *Rhynchophorus palmarum* (Hagley, 1963), the more directed transmission by female *R. palmarum* into oviposition wounds seems more likely (Griffith, 1968). Infested root to healthy root transmission either directly or through soil, and contamination of wounds in a healthy coconut palm by red ring nematodes carried on the body or in faeces of *R. palmarum*, are still modes for the occasional spread of red ring disease. However, nematode inoculation by *R. palmarum* oviposition into healthy or pruned or wounded palms is probably the normal route for transmission (Griffith, 1968). Ninety-five per cent of red ring-diseased coconut palms in Trinidad had evidence of *R. palmarum* damage, and taking into account a six to eight-week incubation period there was a delayed correlation between the seasonal abundance of *R. palmarum* and red ring disease prevalence (Blair, 1969; Hagley, 1963). In African oil palm plantations in Central and South America, a delayed correlation (four to five-month incubation time for red ring disease) is apparent between increases in disease incidence and a rise in the proportion of *R. palmarum* infested with the red ring nematode (Chinchilla *et al.*, 1990; Morales and Chinchilla, 1990; Schuiling and Dinther, 1981). Relative abundance of *R. palmarum* and red ring disease was low in plantations of young (5-year-old) African oil palms and higher in middle-aged plantations (12–18 years old) (Chinchilla *et al.*, 1990; Morales and Chinchilla, 1990).

Recently, all newly emerged *Rhynchophorus palmarum* females examined from red ring-diseased coconut tissue in Trinidad were observed to be internally infested with dauer juveniles of *Bursaphelenchus cocophilus* and in 47% of these the genital capsule contained up to 5880 nematodes (Gerber and Giblin-Davis, 1990a). In that same study, > 70% of the newly emerged adult female weevils had more than 100 red ring dauers internally and > 47% had more than 1000. No transmission efficiency research has been done to date. However, assuming a 10% transfer rate of dauers by red ring nematode-infested *R. palmarum*, close to half of the newly emerged females would have been capable of initiating the disease (Gerber and Giblin-Davis, 1990a). Griffith (1968) collected six small, newly emerged females of *R. palmarum* from red ring-diseased coconuts and individually caged them on 4–5-year-old coconut palms for 48 hours after oviposition began. The number of nematodes per weevil was not known, but assumed to be high (> 1000). All six palms succumbed to red ring disease within ten weeks, providing good circumstantial evidence to support the hypothesis that oviposition by nematode-infested *R. palmarum* is the main mode of natural red ring disease transmission.

Blair (1969) demonstrated that 3-year-old royal palms, *Roystonea oleracea*, which have tightly sheathed leaf bases are susceptible to inoculation with red ring nematodes via damaged or undamaged roots or stem. However, in nature, this palm species usually does not get red ring disease where both the palm and the nematode co-occur or when red ring nematodes are inoculated into the leaf axils. Apparently, the tight adherence of the leaf bases prevents red ring nematodes from reaching the leaf base or the internode for penetration. In most palm species the leaf bases and internodes are accessible, suggesting that this is the most likely route of infection in the wild (Blair, 1969).

15.2.4 Pathogenesis

The fecundity and mode of reproduction of *Bursaphelenchus cocophilus* are not known, but inoculation studies to the stems of 5–6-year-old coconut palms have demonstrated that the nematode is highly motile and capable of exponential increases in numbers (Goberdhan, 1964). Coconut palms inoculated at different vertical levels on the stem have fully developed and indistinguishable internal symptoms at the time of the first external symptoms at about 30 days post-inoculation (Blair, 1964, 1969). Goberdhan (1964) inoculated ten coconut palms at four points (in cross-section) at the base of the stem with red ring-diseased tissue and harvested trees weekly to correlate internal symptoms and signs with external symptoms. At seven and 14 days post-inoculation, infection is limited to less than 5 cm from the inoculation points and there is no red ring formation or root infestation. At 21 days there are no external symptoms, but a red ring is observed and the infection has progressed to 60 cm above the inoculation point and 13 cm below, but does not involve the roots. Definite external symptoms are not present until 28 days post-inoculation and involve leaf yellowing at the tips of two leaves and the death of the oldest leaf. Internally, the red ring nematode infestation extends upwards for 109 cm and down to the bole. Several of the petioles have small infestations of nematodes but the roots are still uninfested. By 42 days, the entire stem is infested with red ring nematodes at close to peak population levels (Goberdhan, 1964). Some of the petioles and roots are also infested with nematodes. Blair and Darling (1968) reported that root inoculation in 3–7-year-old *Cocos nucifera* and the royal palm, *Roystonea oleracea*, without root-wounding causes typical red ring disease. However, when 12-year-old coconut trees are inoculated via unwounded roots the roots become infected and die without causing red ring disease.

Experimental injections of crystal violet into the stems of healthy and red ring-diseased coconut palms demonstrate that water movement in the vascular areas of the discoloured tissue is restricted compared with free movement in healthy palms (Blair, 1964). In healthy trees, the dye can be observed throughout the stem and in the xylem to a height of 3 m or more and into the leaves. The dye is observed in a large number of vascular bundles and is evenly distributed throughout the stem cross-section at 0.6 m. Conversely, in red ring-diseased palms the dye is severely restricted in its movement, suggesting xylem element occlusion.

The red ring nematode cannot penetrate or parasitize the anatomically similar tissues of the fruit stalk and the outer cortex of the stem of coconut (Blair, 1969). Thus, inoculations of red ring nematodes into nuts on healthy palms or unwounded trunks do not lead to red ring disease (Blair, 1969).

These tissues are characterized by lignified parenchyma, fibrous strands and small vascular bundles. It is hypothesized that the external limiting factor for the nematode distribution is the hardness of the cortical stem tissue (Blair, 1969) or small size of intercellular spaces (Griffith and Koshy, 1990). In the stem of *Cocos nucifera*, the narrow cortex surrounds a wide central cylinder of tissue composed of longitudinally running vascular bundles in ground parenchymatous tissue. The ground parenchymatous tissue serves as a storage reservoir for carbohydrates such as starch (Tomlinson, 1990). The bundles are composed of xylem and phloem, and are partially enclosed in sheathing fibrous tissue (Tomlinson, 1990). They are more numerous at the periphery of the stem than at the centre (Tomlinson, 1990).

The red ring nematodes occur intercellularly in the ground parenchyma in the red discoloured tissue of the stem, petioles and cortex of *Cocos nucifera* roots. They also occur for about 1.5 cm on the outside and 4.0 cm on the inside of the red ring of the stem (Blair and Darling, 1968). Intracellular pockets can be observed as parenchyma cells break down and fill up with dauer juveniles of *Bursaphelenchus cocophilus*. Nematodes do not occur in the xylem or phloem tissues but occur intercellularly in the bundle sheath (Blair and Darling, 1968). Xylem vessels become occluded by tyloses where they pass through red ring tissue. Because the sheathing fibres usually form a strand that partially encloses the phloem (Tomlinson, 1990), nematode damage to parenchyma cells near the xylem may initiate the formation of tyloses (growth of parenchymal cells into the xylem through pits). Vascular destruction in monocots such as palms is irreversible because there is no cambium tissue for vascular repair. Thus, nematode feeding and movement physically damage the ground parenchyma cells of the palm host. This damage compromises the storage capacity and energy dynamics of the palm and causes vascular occlusion which prevents normal water relations and kills the host.

The actual cause of the red ring is not known. The pigment is hypothesized to be polymerized anthocyanins resulting from the conversion of leuco-anthocyanins that are abundant in healthy coconut palm tissue (Blair, 1969). Leucoanthocyanins may be localized in the vacuolar sap and released due to damage caused directly or indirectly by feeding of the red ring nematode. The pigment is roughly confined to areas where nematode feeding damage occurs in *Cocos nucifera*. However, in diseased *Roystonea oleracea*, a red ring occurs and nematodes can be recovered in the discoloured zone, in the area just internal to the red ring, and in the central core (Blair, 1969).

Why are red ring nematodes not recovered from the central core of the coconut stem? In experiments where the cortex of the stem of a healthy coconut palm is trimmed away and the core inoculated with red ring nematodes, the tissue becomes infested with propagating nematodes and the tissue becomes discoloured (Blair, 1969). Also, when a red ring-diseased coconut is cut longitudinally and monitored the nematodes begin to pro-pagate rapidly in the central core and cause discoloured tissue (Blair, 1969).

Therefore, the central core of *Cocos nucifera* is suitable for red ring nematode reproduction. Blair (1969) suggests that carbon dioxide and oxygen levels in the stem control where the nematodes propagate. He found that the carbon dioxide and oxygen levels are inversely correlated across a transverse section of healthy coconut palm stem and that carbon dioxide is highest near the cortex and lowest at the centre. The gradient hypothetically exists because of higher levels of metabolism in the peripheral tissues. Red ring nematodes appear to be attracted to carbon dioxide and may be facultative anaerobes which migrate to the ground parenchymal tissue near the cortex for propagation (Blair, 1969). As red ring disease progresses, vascular occlusion prevents normal carbon dioxide exhaustion which leads to a rise in carbon dioxide levels and a breakdown in the carbon dioxide and oxygen gradient. Blair (1969) suggests that high carbon dioxide levels in the stem cause a decline and eventual cessation in propagation before the nematodes enter the central core of the stem. Other factors must be operating because during the progression of the disease the carbon dioxide/oxygen ratio in the central core should become acceptable for nematode reproduction. Further work is needed concerning the cause of red ring formation and limiting factors for nematode distribution in the stem.

The exact nature of the association between the red ring nematode and little leaf symptomatic coconut or African oil palms is unclear. Apparently, red ring nematodes propagate in the inner side of the stalk of unpresented leaves and inflorescences as ecto-or semi-endoparasites which cause dis-coloured swellings of tissue with abnormally small cells (Hoof and Seinhorst, 1962). Hoof and Seinhorst (1962) initiated little leaf symptoms and no red ring within 15 months of inoculation by pouring a suspension of *Bursaphelenchus cocophilus* from a little leaf symptomatic African oil palm onto the young leaves of one of five 10-year-old African oil palms. Red ring nematodes in little leaf symptomatic African oil palm tissue caused a red ring containing *B. cocophilus* in two of eight inoculated coconut palms (4–8 years old) (Kraaijenga and Ouden, 1966). Maas (1970) inoculated five healthy 4–7-year-old coconut, African oil palms, *Mauritia flexuosa*, and *Maximiliana maripa* each in Surinam with a stem piece from a red ring-diseased coconut palm. In that study, typical red ring disease was reported from two of the inoculated African oil palms and two of the coconut palms five months post-inoculation. Typical little leaf symptoms with *B. cocophilus* and a red ring remnant, 15 cm from the trunk periphery without nematodes, were observed from two of the African oil palms and one native palm, *Mauritia flexuosa*, one year after inoculation. The tissue within the ring of these little leaf symptomatic palms had disintegrated up to 30 cm below the bud. One of the *Maximiliana maripa* palms showed no external symptoms at harvest, but possessed similar internal symptoms to the little leaf symptomatic palms. Also, *B. cocophilus* was recovered from discoloured tissue in the stem and roots and from slightly discoloured areas in the bud of this palm. In addition, a survey of 13 African oil palms with typical little leaf symptoms in Surinam showed that 12 of these had hollow trunks (Maas, 1970). Interestingly, when red ring symptomatic coconut palms were injected with nemaphos (a systemic nematicide), the red ring nematodes died and the palm survived for an extra five months, but all tissue internal to the red ring decayed (Blair, 1969).

All of the foregoing suggests that little leaf symptomatology represents unsuccessful cases of lethal red ring disease which may be due to physiological or genetic differences in palm hosts. Palm host age (Kraaijenga and Ouden, 1966), soil nutritional balance and soil hydration levels could all interact with genetic factors in the host to modify disease expression. The above inoculation studies and other cross-inoculation studies (see Dean, 1979) suggest that *Bursaphelenchus cocophilus* from different hosts and symptoms are conspecific. The epidemiology of little leaf may be as follows: *B. cocophilus* initiates an infection through weevil wounds in a leaf base or the internode of the stem; nematode propagation occurs in the leaf and stem, causing some red ring tissue; nematode propagation is terminated before the vascular system is fully compromised; the nematodes migrate to the bud, and for some unexplained reason continue to propagate in the unpresented leaves without reinvading the stem. An especially quick response to nematode propagation by a host might prevent formation of a red ring.

15.2.5 Insect vector biology

Many different insects associated with palms in the Neotropics have been suggested as possible vectors for the red ring nematode (Dean, 1979). However, almost all of the insects listed carry less than 20 nematodes externally and are not associated with healthy palms for potential transmission of the disease. For example, weevils such as *Rhinostomus barbirotrus* and *Metamasius hemipterus* yield very few or no red ring nematodes upon dissection (Hagley, 1964; Morales and Chinchilla, 1990) and are attracted to palms late in their decline. Thus, if these weevils carried nematodes their biology would not put them in contact with healthy trees for a disease cycle. The most suitable candidates for vector status are those which will go through metamorphosis with the nematode and will carry large numbers to a healthy host during their life cycle, i.e. *Rhynchophorus* spp. or *Dynamis* spp. (Gerber *et al.*, 1990). *R. palmarum* and *D. borassi* carry as many as 13 613 and 1995 dauer juveniles of the red ring nematode through metamorphosis per adult, respectively (Gerber *et al.*, 1990).

Rhyncophorus spp. and *Dynamis* spp. are attracted to semiochemicals released by wounded palms or freshly cut palms (Wattanapongsiri, 1966; Weissling *et al.*, 1992) and because of moisture requirements may harbour in the moist leaf bases of healthy palms. Starting at about 4–7 years of age, coconut and African oil palms can receive regular pruning to remove leaves and harvest fruit inflorescences. Wounds made to healthy palms during these cultural practices may attract red ring-infested *R. palmarum*. In addition, males of *R. palmarum* produce a sex-specific pheromone which has been identified as (2*E*)-4-hydroxy-6-methyl-2-heptene (rhynchophorol) and is attractive to adults of both sexes (Rochat *et al.*, 1991). Thus, *R. palmarum* and probably most members of the subfamily Rhynchophorinae have the ability to locate a food source and call in conspecifics. The pheromone probably acts as a long-range attractant whereas host plant volatiles may serve as short-distance attractants or arrestants. *R. palmarum* is also attracted to stressed or dying palms (i.e. red ring-diseased palms). In fact, both *R. palmarum* and the red ring nematode can propagate in recently felled trunks of healthy coconut, African oil palm, and the wild palms, *Mauritia flexuosa* and *Maximiliana maripa* in Surinam (Maas, 1970).

Rhynchophorus palmarum is a pest of coconut, African oil palm and at least 20 other palm species. In addition, this weevil attacks banana, sugarcane, cacao and papaya (Wattanapongsiri, 1966). It is multivoltine, with trends for increased seasonal abundance at the end of the rainy season and throughout most of the dry season in coconut plantations in Trinidad (Hagley, 1963) and mostly in the dry season in oil palm plantations in Brazil, Costa Rica and Honduras (Chinchilla *et al.*, 1990; Morales and Chinchilla, 1990; Schuiling and Dinther, 1981). *R. palmarum* is a very large weevil that varies considerably in size (total length 38–60 mm) (Gerber and Giblin-Davis, 1990b). Weevil length has been reported as a reliable indicator of red ring disease vector status of *R. palmarum* (Griffith, 1987). Unfortunately, *R. palmarum* size has not been found to correlate with red ring nematode densities in most field studies (Chinchilla *et al.*, 1990; Gerber and Giblin-Davis, 1990a; Morales and Chinchilla, 1990; Schuiling and Dinther, 1981).

Rhynchophorus palmarum can be cultured in coconut stem tissue, sugarcane, or on pineapple–sugar-cane diets (Giblin-Davis *et al.*, 1989b). Females can lay between 90 and 400 eggs, with eclosion occurring after two to four days incubation. There are six to ten instars, which end in a prepupa about 42–62 days later. The last instar larvae are huge, often weighing between 3 and 14 g, and are considered a worthy food item by some people. The last instar larva makes a cocoon out of the fibre in the palm stem or petioles and enters the prepupal stage for 4–14 days before pupation, which can last 8–23 days. The life cycle takes 70–88 days in the laboratory and 59–85 days in the field. Longevity is 28–62 days for adult males and 25–56 days for females (Wattanapongsiri, 1966).

15.2.6 Distribution

The red ring nematode is co-distributed with its insect vector, *Rhynchophorus palmarum*, in the southern Antilles, Mexico southward through Central America into South America, where it has been reported from Colombia, Surinam, Guyana, Ecuador, Venezuela and Brazil. Reports from the Dominican Republic (Blair, 1969; Griffith, 1987) have not been confirmed, and are doubtful because of the distance to other confirmed disease sites and the lack of corroboration by other observers. The distribution of *R. palmarum* appears to be more extensive than the confirmed distribution of the red ring nematode. *R. palmarum* has been reported from the USA (California), Cuba, Jamaica, Bolivia, Argentina, Uruguay and Paraguay (Wattanapongsiri, 1966). The red ring nematode has not been confirmed to occur in these countries (Dean, 1979).

The association between the red ring nematode and the palm weevil may be a relatively recent phenomenon, with its seat of origin in northeastern South America. In the last century, the red ring nematode appears to have increased its range within the confirmed distribution of *Rhynchophorus palmarum*. For example, *R. palmarum* has been known from Ecuador for at least 70 years (Wattanapongsiri, 1966), whereas the first report of red ring disease was in 1967 (Gerber *et al.*, 1990). Similarly, *R. palmarum* has been known from Costa Rica since 1904 (Wattanapongsiri, 1966) whereas the first reports of red ring disease occurred in the 1970s (Chinchilla, 1988). In addition, *R. palmarum* and its congener the palmetto weevil, *R. cruentatus* from Florida and the southeastern United States, are alipatric but share at least two species of phoretic nematodes (Gerber and Giblin-Davis, 1990a; Gerber and Giblin-Davis, 1990b). However, *Bursaphelenchus cocophilus* has not been recovered from *R. cruentatus* (Giblin-Davis, unpublished). Also, red ring disease is not known from the Indo-Pacific region, which is the apparent area of origin for the genera *Rhynchophorus* and *Cocos* (Griffith, 1987).

15.2.7 Management

Phytosanitation is currently the best method of red ring disease management. This strategy is directed at reducing the vector population as well as the number of sources for nematode inoculum. As soon as palms with red ring disease or *Bursaphelenchus cocophilus*-induced little leaf symptoms have been detected they should be destroyed. In coconut, the disease can be confirmed by examining stem tissue extracted with a coring device for evidence of discoloured tissue and red ring nematodes. In African oil palm, stem coring for evidence of necrosis and red ring nematodes is not recommended because of the high probability of false-negative sampling (Chinchilla, 1988). Trees should be sprayed with an insecticide (e.g. methomyl) and killed with 100–150 ml (48.3% a.i.) of the herbicide monosodium acid methanearsonate (MSMA) or other herbicide, which is injected or placed into the trunk (Chinchilla, 1988; Griffith and Koshy, 1990). Occasionally trees injected with MSMA will harbour *Rhynchophorus palmarum* larvae. Therefore, once the tree is dry it should be cut and sectioned to make sure that weevils are not present. Palms that are heavily infested with *R. palmarum* should be cut, sectioned and treated with an insecticide such as methomyl, trichlorfon, monocrotophos, carbofuran, cabaryl or lindane (Chinchilla, 1988) or burned with kerosene.

Research is currently being conducted on the feasibility of using pesticide-laced traps with tissue-derived semiochemicals and a synthetically produced aggregation pheromone (rhyncophorol) specific to *Rhynchophorus palmarum*. Many different traps have been evaluated using cut palm tissue as bait (Weissling *et al.*, 1992). When tissue is used as a bait, a non-repellent insecticide can be used to kill the lured weevils (e.g. methomyl) (Chinchilla, 1988). Rochat (1987) has empirically tested a large number of compounds for positive electroantennagrams (EAG) for *R. palmarum*. Some of these materials, or EAG-positive compounds from freshly cut host plants (0–72 hours old) may work well with the synthetic aggregation pheromone and be economically feasible for a mass trapping strategy to reduce adult *R. palmarum* and potential vectors within coconut or African oil palm plantations. Because red ring disease is often clumped and patchily distributed, a trap-out strategy might be optimized by setting traps around disease hot spots and around the periphery of the plantation. Unfortunately, the trapping methodology and our understanding of the chemical ecology and management of *R. palmarum* is still in its infancy and it is not clear if mass trapping of the weevil vector reduces red ring disease prevalence.

Injections of systemic nematicides, such as fenamiphos, oxamyl or carbofuran, into little leaf symptomatic palms cause apparent palm recovery in some of the trees tested (Chinchilla, 1988). The recovery can take between six and eight months because of the damage to the very young leaves in little leaf palms. Interestingly, aldicarb was not effective for little leaf control in African oil palm in Honduras when applied on the soil, trunk or axilla, but was partially effective when applied directly to the whorl (Chinchilla, 1988).

The use of systemic nematicides/insecticides is not feasible for lethal red ring diseased coconut or African oil palms because of the late onset of external symptoms. By the time red ring symptoms are expressed, too much vascular damage has been done for host recovery or for the pesticide to be efficiently moved to the nematode feeding sites (Blair, 1969). At best, injection with a systemic pesticide would just postpone the death of the tree. Prophylactic injections or ground applications of systemic pesticides might be feasible in extremely valuable ornamental palms, but not for food crops with residue tolerances, such as coconuts which are harvested monthly. Soil applications of granular fenamiphos or aldicarb are active within the plant for as long as 14 weeks, whereas a trunk injection of fenamiphos lasts about four weeks (Hoyle, 1971, as cited in Chinchilla, 1988).

Multiple applications of endrin (at 50 and 70-day intervals during the rainy and dry seasons, respectively) to the leaf axils of healthy coconut palms were effective at reducing red ring disease prevalence in Trinidad (Hagley, 1963). This strategy is expensive in terms of labour and material and poses potential environmental risks. Also, regular prophylactic insecticidal applications to the crowns of palms do not always correlate with a reduction in red ring disease incidence (Fenwick, 1967, as cited in Chinchilla, 1988).

Biological control of *Rhynchophorus palmarum* has not been thoroughly investigated. Several natural enemies may hold some promise, including nematodes in the families Steinernematidae and Heterorhabditidae, the bacterium *Micrococcus roseus* (Griffith, 1987), a tachinid parasite, *Parabillaea rhyncophorae* from Bolivia (Candia and Simmonds, 1965), a cytoplasmic polyhedrosis virus (CPV) from *R. ferrugineus* from India (Gopinadhan *et al.*, 1990), and humans (Defoliart, 1990). Defoliart (1990) argues that development of a cottage industry for the production of the large *R. palmarum* last instar larvae for human consumption could lead to increased cash flow for farmers and incentives for weevil vector control. Timing of the cutting of dying trap trees for the production of larvae would be critical since harvest would need to be done at the right time (45–50 days after oviposition) to avoid the escape of potential vectors. An advertising blitz would be required to convince the world that *R. palmarum* larvae are a delicacy worthy of investment.

15.3 FIGS, FIG WASPS AND *SCHISTONCHUS* SPP.

Schistonchus caprifici, the first described member of the family Aphelenchoididae (described in 1864 by Gasparrini), was reported to be a phoretic associate of the fig wasp, *Blastophaga psenes* (Agaonidae) (Poinar, 1975). New evidence suggests that it is an obligate parasite of the fig wasp and that it also parasitizes florets in the fig syconia of *Ficus carica* (Vovlas *et al.*, 1992).

Adult females of the *Blastophaga psenes* carry 200–400 juveniles and adults of the nematode *Schistonchus caprifici* in their hemocoels to fig syconia (caprifigs or edible figs). The nematodes are apparently deposited during oviposition by the fig wasp into pistillate florets, where they develop and reproduce, causing necrosis and cavities in the cortical parenchyma (Vovlas *et al.*, 1992). Both male and female florets can be attacked and female florets in caprifigs are better for nematode reproduction than female florets in edible figs. Nematodes from a floret infect the fig wasp larva in its flower gall and apparently grow and reproduce within the insect as it goes through metamorphosis. Evidently, the wingless male wasps are not used as nematode hosts (Vovlas *et al.*, 1992). The male fig wasps emerge first and mate with female wasps prior to emergence from their gall flowers. Female fig wasps emerge and gather pollen from mature staminate (male) florets and exit through an emergence hole cut by the male fig wasps. The female disperses to a new caprifig or an edible fig to repeat the cycle.

Considering that there are more than 900 species of *Ficus* (Moraceae), each with a highly co-evolved association with its respective species-specific agaonid wasp pollinator (Frank, 1984), it is possible that there could be many highly co-evolved *Schistonchus* spp. For example, the fig wasps *Pegoscapus assuetes* and *P. jimenizi* (Agaonidae) from native *Ficus* spp. in southern Florida (*F. citrifolia* and *F. aurea*) are associated with new species of *Schistonchus* (Giblin-Davis, unpublished observations). The biology of the associations with these new species of *Schistonchus* are different from previously reported *Schistonchus–Ficus*-agaonid associations (Vovlas *et al.*, 1992), but the nematodes do appear to parasitize the fig florets during part of the interaction. An extensive survey of the nematode associates of fig wasps and fig syconia will probably greatly increase the number of known nematode species which interact with both insects and plants.

15.4 RAPE PLANT, BARINE WEEVIL AND *BEDDINGIA* SP.

There are two tylenchid nematode species which interact with insects and plants using dicyclic life cycles. In France, *Beddingia barisii* (Phaenopsitylen-chidae) parasitizes the univoltine barine weevil, *Baris caerulescens* (Curculionidae) (Remillet and Laumond, 1991). At the time weevils are mating and ovipositing on the collar of their rape plant host, *Brassica napus* var. *oleifera*, the third-stage juveniles of the nematode migrate out of the anus or genital tract of the insect and probably penetrate into tissues of the plant. It appears that two to three generations of *B. barisii* occur in the plant. The last generation results in mated pre-adult females of the insect-parasitic phase which penetrate the newly moulted larva of the insect. Once inside the insect host, the pre-adult female nematodes moult to adults and mature rapidly. Nematode eggs are laid and second and third-stage juveniles invade the hemocoel of the young adult weevil. As with other members of the genus *Beddingia*, adults of both sexes from the insect-parasitic cycle are morphologically different from the adults observed in the plant-parasitic cycle.

15.5 *EUCALYPTUS*, GALL FLIES AND *FERGUSOBIA* SPP.

Another dicyclic tylenchid genus which parasitizes both plants and insects is *Fergusobia* (Fergusobiidae) (Remillet and Laumond, 1991). Members of this genus have an amphimictic insect-parasitic cycle that is associated with very small eucalyptus gall flies in the genus *Fergusonia* (Fergusonidae) (Currie, 1937; Fisher and Nickle, 1968; Siddiqi, 1986). There is also a parthenogenetic plant-parasitic cycle with several generations of nematodes which may be associated with flower, leaf and stem gall production in *Eucalyptus* spp. Sometimes the galling can be so severe in *Eucalyptus* plantings in Australia that seed production for forestry applications and honey production are adversely affected (Currie, 1937).

  Fergusonia flies mate and oviposit into young leaves of *Eucalyptus stuartiana* or flower-buds of *E. macrorrhyncha* (Myrtaceae) in summer in Australia (Currie, 1937). Larval nematodes (*Fergusobia tumifaciens*) (1–50) are oviposited by the fly into the bud. The nematodes rapidly develop into parthenogenetic females (Fisher and Nickle, 1968). The fly egg undergoes embryonic development in the following six months while the nematodes feed on the primordia of the anthers, which form a circle around the inner wall of the flower-bud cavity. The nematodes hypothetically cause the anther primordia to proliferate rapidly, producing irregular masses of thin-walled, parenchymatous cells that can be full of sticky sap.

When the fly egg hatches, the larva cuts out a small crypt between two apposed masses of proliferating plant cells and begins feeding. The nematodes aggregate in the areas around the fly larva. Plant tissues in the crypt fuse together around each fly larva, producing a stalked 'gall-let'. Each flower-bud gall of *Eucalyptus macrorrhyncha* may have more than 20 'gall-lets' inside (Currie, 1937). The fly larva continues to feed in the 'gall-let' and develops through the second to the third instar (last instar) during autumn through summer in Australia. It is not clear how many parthenogenetic generations of nematodes may occur in the plant during this time. However, coincident with the appearance of the late third instar fly larva, fertilized female nematodes of the insect-parasitic cycle are available to infect the insect. The inner walls of the 'gall-let' are consumed by the third instar fly larva and all remaining nematodes die (Currie, 1937).

The fertilized females of the nematode *Fergusobia tumifaciens* infect the fly larvae in the late third instar, just prior to pupation. During the pupal stage of the fly, the nematode enlarges into a parasitic female, loses its stylet and gut, and becomes filled with ovary (Currie, 1937). Sexual maturation is apparently very rapid and the parasitic female nematode is associated intimately with the fat body of the pupal host (Currie, 1937; Fisher and Nickle, 1968). All dissected third instar larvae of the fly *Fergusonia tillyardi* had nematodes in their hemocoels but only adult female flies were observed with nematodes (Fisher and Nickle, 1968). Currie (1937) also observed that only adult female and no male flies of *F. carteri* were infested with nematodes. Nematode eggs are laid in the hemocoel of the adult female fly and the resulting larval nematodes migrate into the ovaries of the fly.

The cause of gall formation is not clear, but *Fergusonia–Fergusobia* associations were discovered to be associated with flower-bud, leaf, leaf-stem, stem-tip and axil-bud galls of many species of *Eucalyptus* (Currie, 1937). Currie (1937) hypothesized that the nematodes are solely responsible in the early phases of plant gall formation because galling started where nematodes occurred and not around the fly eggs. After the fly larva hatches the nematodes and insect somehow interact for continued stimulation of galled tissue. Using these observations together with the fact that all female flies have nematodes, Currie suggested that the association between the flies and the nematodes is mutualistic. This would further exemplify the uniqueness of the interaction because no other mutualistic relationship is known between nematodes and insects (Poinar, 1975; Giblin, 1987). Demonstrating the causal agent(s) of galling is difficult because no adult female flies or galls have been observed without nematodes. Thus, the obvious question of whether or not galls are initiated at oviposition by secretions from the female fly cannot be addressed. Currie's (1937) attempts to inoculate nematodes into flower buds on *Eucalyptus macrorrhyncha* failed.

These *Eucalyptus–Fergusonia–Fergusobia* interactions are probably highly co-evolved. Thus, like the *Ficus*–fig wasp–nematode interactions discussed above, there is a good chance that with further research many more species will be characterized. Siddiqi (1986) has recently described two new species from India and Australia and there are probably many more to be discovered.

REFERENCES

Bailey Hortorium, Cornell University (1976) *Hortus Third: A Concise Dictionary of Plants Cultivated in the United States and Canada*, Macmillan, New York.

Baujard, P. (1980) Trois espèces nouvelles de *Bursaphelenchus* et remarques sur le genre. *Revue de Nématologie*, 3, 167–77.

Baujard, P. (1989) Remarques sur les genres des sous-familles Bursaphelenchina Paramonov, 1964 et Rhadinaphelenchinae Paramonov, 1964 (Nematoda: Aphelenchoididae). *Revue de Nématologie*, 12, 323–4.

Bergdahl, D.R. (1988) Impact of pinewood nematode in North America: Present and future. *Journal of Nematology*, 20, 260–5.

Blair, G.P. (1964) Red ring disease of the coconut palm. *Journal of Agricultural Society of Trinidad and Tobago*, 64, 31–49.

Blair, G.P. (1969) Studies of red ring disease of coconut palm. *Proceedings of the Symposium on Tropical Nematology*, 1967, University of Puerto Rico, pp. 89–106.

Blair, G.P. and Darling, H.M. (1968) Red ring disease of the coconut palm, inoculation studies and histopathology. *Nematologica*, 14, 395–403.

Bolla, R.I. and Jordan, W. (1982) Cultivation of the pine wilt nematode, *Bursaphelenchus xylophilus*, in axenic culture media. *Journal of Nematology*, 14, 377–81.

Bolla, R.I., Winter, R.E.K., Fitzsimmons, K. and Linit, M.J. (1986) Pathotypes of the pinewood nematode *Bursaphelenchus xylophilus*. *Journal of Nematology*, 18, 230–8.

Bolla, R.I., Fitzsimmons, K. and Winter, R.E.K. (1987) Carbohydrate concentration in pine as affected by inoculation with *Bursaphelenchus xylophilus*. *Journal of Nematology*, 19, 51–7.

Bolla, R.I., Kozlowski, P. and Fitzsimmons, K. (1988) Carbohydrate catabolism in populations of *Bursaphelenchus xylophilus* and in *B. mucronatus*. *Journal of Nematology*, 20, 252–9.

Candia, J.D. and Simmonds, F.J. (1965) A tachinid parasite of the palm weevil, *Rhynchophorus palmarum*. *Commonwealth Institute of Biological Control, Technical Bulletin No. 5*, pp. 127–8.

Chinchilla, C. (1988) El Sindrome del anillo rojo-hoja pequeña en palma aceitera y cocotero. *Boletin Technico*, 2, 113–36.

Chinchilla, C., Menjívar, R. and Arias, E. (1990) Picudo de la palma y enfermedad del anillo rojo/hoja pequeña en una plantación comercial en Honduras. *Turrialba*, 40, 471–7.

Cobb, N.A. (1919) A newly discovered nematode (*Aphelenchus cocophilus*, n. sp.) connected with a serious disease of the coco-nut palm. *West Indies Bulletin*, 17, 203–10.

Currie, G.A. (1937) Galls on *Eucalyptus* trees: A new type of association between flies and nematodes. *Linnean Society of New South Wales*, 62, 147–74.

Dean, C.G. (1979) Red ring disease of *Cocos nucifera* L. caused by *Rhadinaphelenchus cocophilus* (Cobb, 1919) Goodey, 1960. An annotated bibliography and review. *Technical Communication No. 47 of Commonwealth Inst. Helminthol.*, 70 pp.

Defoliart, G. (1990) Hypothesizing about palm weevil and palm rhinoceros beetle larvae as traditional cuisine, tropical waste recycling, and pest and disease control on coconut and other palms: Can they be integrated? *Food Insects Newsletter*, **3**, 1-6.

Dwinell, D.L. (1990) Heat-treating and drying southern pine lumber infested with pinewood nematodes. *Forest Products Journal*, **40**, 53-6.

Farr, D.F., Bills, G.F., Chamuris, G.P. and Rossman, A.Y. (1989) *Fungi on Plants and Plant Products in the United States*, APS Press, St Paul, Minnesota, 1252 pp.

Fisher, J.M. and Nickle, W.R. (1968) On the classification and life history of *Fergusobia curriei* (Sphaerulariidae: Nematoda). *Proceedings of the Helminthological Society of Washington*, **35**, 40-6.

Frank, S.A. (1984) The behavior and morphology of the fig wasps *Pegoscapus assuetes* and *P. jimenezi*: Descriptions and suggested behavioral characters for phylogenetic studies. *Psyche*, **91**, 289-308.

Futai, K. (1980) Developmental rate and population growth of *Bursaphelenchus lignicolus* (Nematoda: Aphelenchoididae) and *B. mucronatus*. *Applied Entomology and Zoology*, **15**, 115-22.

Gerber, K. and Giblin-Davis, R.M. (1990a) Association of the red ring nematode, *Rhadinaphelenchus cocophilus*, and other nematode species with *Rhynchophorus palmarum* (Coleoptera: Curculionidae). *Journal of Nematology*, **22**, 143-9.

Gerber, K. and Giblin-Davis, R.M. (1990b) *Teratorhabditis palmarum* n. sp. (Nemata: Rhabditidae): An associate of *Rhynchophorus palmarum* and *R. cruentatus*. *Journal of Nematology*, **22**, 337-47.

Gerber, K., Giblin-Davis, R.M., Griffith, R., Escobar-Goyes, J. and D'Ascoli Cartaya, A. (1989) Morphometric comparisons of geographic and host isolates of the red ring nematode, *Rhadinaphelenchus cocophilus*. *Nematropica*, **19**, 151-9.

Gerber, K., Giblin-Davis, R.M. and Escobar-Goyes, J. (1990) Association of the red ring nematode, *Rhadinaphelenchus cocophilus*, with weevils from Ecuador and Trinidad. *Nematropica*, **20**, 39-49.

Giblin, R.M. (1985) Association of *Bursaphelenchus* sp. (Nematoda: Aphelenchoididae) with nitidulid beetles (Coleoptera: Nitidulidae). *Revue de Nématologie*, **8**, 369-75.

Giblin, R.M. (1987) Culture of nematode associates and parasites of insects, in *Vistas on Nematology* (eds J. Veech and D.W. Dickson), Society of Nematologists, Hyattsville, pp. 408-13.

Giblin-Davis, R.M. (1991) The potential for introduction and establishment of the red ring nematode in Florida. *Principes*, **35**, 147-53.

Giblin, R.M. and Kaya, H.K. (1984) Host, temperature, and media additive effects on the growth of *Bursaphelenchus seani*. *Revue de Nématologie*, **7**, 13-17.

Giblin-Davis, R. M., Gerber, K. and Griffith, R. (1989a) *In vivo* and *in vitro* culture of the red ring nematode, *Rhadinaphelenchus cocophilus*. *Nematropica*, **19**, 135-42.

Giblin-Davis, R.M., Gerber, K. and Griffith, R. (1989b) Laboratory rearing of *Rhynchophorus cruentatus* and *R. palmarum* (Coleoptera: Curculionidae). *Florida Entomologist*, **72**, 480-8.

Giblin-Davis, R.M., Mundo-Ocampo, M., Baldwin, J.G., Gerber, K. and Griffith, R. (1989c) Observations on the morphology of the red ring nematode, *Rhadinaphelenchus cocophilus*. *Revue de Nématologie*, **12**, 285-92.

Goberdhan, L.C. (1964) Observations on coconut palms artificially infested by the nematode *Rhadinaphelenchus cocophilus* (Cobb, 1919) Goodey, 1960. *Journal of Helminthology*, **38**, 25-30.

Goodey, J.B. (1960) *Rhadinaphelenchus cocophilus* (Cobb, 1919) n. comb., the nematode associated with 'red-ring' disease of coconut. *Nematologica*, **5**, 98-102.

Gopinadhan, P.B., Mohandas, N. and Vasudevan, K.P. (1990) Cytoplasmic polyhedrosis virus infecting redpalm weevil of coconut. *Current Science*, 59, 577–80.

Griffith, R. (1968) The mechanism of transmission of the red ring nematode. *Journal of the Agricultural Society of Trinidad and Tobago*, 68, 437–57.

Griffith, R. (1987) Red ring disease of coconut palm. *Plant Disease*, 71, 193–6.

Griffith, R. and Koshy, P.K. (1990) Nematode parasites of coconut and other palms, in *Plant Parasitic Nematodes in Subtropical and Tropical Agriculture* (eds M. Luc, R.A. Sikora and J. Bridge), CAB International, Wallingford, pp. 363–86.

Guiran, G. de and Bruguier, N. (1989) Hybridization and phylogeny of the pine wood nematode (*Bursaphelenchus* spp.). *Nematologica*, 35, 321–30.

Guiran, G. de, Lee, M.J., Dalmasso, A. and Bongiovonni, M. (1985) Preliminary attempt to differentiate pinewood nematodes (*Bursaphelenchus* spp.) by enzyme electrophoresis. *Revue de Nématologie*, 8, 88–9.

Hagley, E.A.C. (1963) The role of the palm weevil, *Rhynchophorus palmarum*, as a vector of red ring disease of coconuts. I. Results of preliminary investigations. *Journal of Economic Entomology*, 56, 375–80.

Hagley, E.A.C. (1964) Role of insects as vectors of red ring disease. *Nature*, 204, 905–6.

Hinode, Y., Shuto, Y. and Watanabe, H. (1987) Stimulating effects of β-myrcene on molting and multiplication of the pine wood nematode, *Bursaphelenchus xylophilus*. *Agricultural and Biological Chemistry*, 51, 1393–6.

Hoof, H.A. van and Seinhorst, J.W. (1962) *Rhadinaphelenchus cocophilus* associated with little leaf of coconut and oil palm. *Tijdschrift over Plantenziekten*, 68, 251–6.

Humphrey, S.J. and Linit, M.J. (1989a) Tethered flight of *Monochamus alternatus* (Coleoptera: Cerambycidae) with respect to beetle age and sex. *Environmental Entomology*, 18, 124–6.

Humphrey, S.J. and Linit, M.J. (1989b) Effect of nematode density on tethered flight of *Monochamus alternatus* (Coleoptera: Cerambycidae). *Environmental Entomology*, 18, 670–3.

Ishibashi, N. and Kondo, E. (1977) Occurrence and survival of the dispersal forms of pine wood nematode, *Bursaphelenchus lignicolus* Mamiya and Kiyohara. *Applied Entomology and Zoology*, 12, 293–302.

Ishibashi, N., Aoyagi, M. and Kondo, E. (1978) Comparison of the gonad development between the propagative and dispersal forms of pine wood nematode, *Bursaphelenchus lignicolus* (Aphelenchoididae). *Japanese Journal of Nematology*, 8, 28–31.

Ishikawa, M., Shuto, Y. and Watanabe, H. (1986) β-myrcene, a potent attractant component of pine wood for the pine wood nematode, *Bursaphelenchus xylophilus*. *Agricultural and Biological Chemistry*, 50, 1863–6.

Ishikawa, M., Kaneko, A., Kashiwa, T. and Watanabe, H. (1987) Participation of β-myrcene in the susceptibility and/or resistance of pine trees to the pine wood nematode, *Bursaphelenchus xylophilus*. *Agricultural and Biological Chemistry*, 51, 3187–91.

Kinn, D.N. (1986) Heat-treating wood chips: A possible solution to pinewood nematode contamination. *Tappi*, 69, 97–8.

Kiyohara, T. (1984) Pine wilt resistance induced by prior inoculation with avirulent isolate of *Bursaphelenchus xylophilus*, in *Proceedings of the United States–Japan Seminar: The Resistance Mechanisms of Pines Against Pine Wilt Disease* (ed. V. Dropkin), National Science Foundation, Washington DC, pp. 178–84.

Kiyohara, T. and Bolla, R.I. (1990) Pathogenic variability among populations of the pinewood nematode, *Bursaphelenchus xylophilus*. *Forest Science*, 36, 1061–76.

Kobayashi, T. (1987) Microorganisms associated with the pine wood nematode in

Japan, in *Pathogenicity of the Pine Wood Nematode* (ed. M.J. Wingfield), APS Press, St Paul, pp. 91–101.

Kobayashi, F., Yamane, A. and Ikeda, T. (1984) The Japanese pine sawyer beetle as the vector of pine wilt disease. *Annual Review of Entomology*, **29**, 115–35.

Kondo, E. and Ishibashi, N. (1978) Ultrastructural differences between the propagative and dispersal forms in pine wood nematode, *Bursaphelenchus lignicolus*, with reference to the survival. *Applied Entomology and Zoology*, **13**, 1–11.

Kraaijenga, D.A. and Ouden, H. den. (1966) 'Red ring' disease in Surinam. *Tijdschrift over Plantenziekten*, **72**, 20–7.

Linit, M.J. (1988) Nematode–vector relationships in the pine wilt disease system. *Journal of Nematology*, **20**, 227–35.

Linit, M.J., Kondo, E. and Smith, M.T. (1983) Insects associated with the pine-wood nematode, *Bursaphelenchus xylophilus* (Nematoda: Aphelenchoididae), in Missouri. *Environmental Entomology*, **12**, 467–70.

Maas, P.W. Th. (1970) Contamination of the palm weevil (*Rhynchophorus palmarum*) with the red ring nematode (*Rhadinaphelenchus cocophilus*) in Surinam. *Nematologica*, **16**, 429–33.

Mamiya, Y. (1975) The life history of the pine wood nematode, *Bursaphelenchus lignicolus*. *Japanese Journal of Nematology*, **5**, 16–25.

Mamiya, Y. (1984) The pine wood nematode, in *Plant and Insect Nematodes* (ed. W.R. Nickle), Marcel Dekker, New York, pp. 589–626.

Mamiya, Y. (1990) Effects of fatty acids added to media on the population growth of *Bursaphelenchus xylophilus* (Nematoda: Aphelenchoididae). *Applied Entomology and Zoology*, **25**, 299–309.

Mamiya, Y. and Enda, N. (1979) *Bursaphelenchus mucronatus* n. sp. (Nematoda: Aphelenchoididae) from pine wood and its biology and pathogenicity to pine trees. *Nematologica*, **25**, 353–61.

Mamiya, Y. and Furukawa, M. (1977) Fecundity and reproductive rate of *Bursaphelenchus lignicolus*. *Japanese Journal of Nematology*, **7**, 6–9.

Mamiya, Y. and Kiyohara, T. (1972) Description of *Bursaphelenchus lignicolus* n. sp. (Nematoda: Aphelenchoididae) from pine wood and histopathology of nematode-infested trees. *Nematologica*, **18**, 120–4.

Mamiya, Y., Ikeda, T. and Shoji, T. (1989) Inoculation of the pine wood nematode, *Bursaphelenchus xylophilus*, to *Pinus densiflora* shoot cuttings treated with benzoic acid. *Annals of the Phytopathological Society of Japan*, **55**, 303–8.

Matsumori, M., Izumi, S. and Watanabe, H. (1989) Hormone-like action of 3-octanol and 1-octen-3-ol from *Botrytis cinerea* on the pine wood nematode, *Bursaphelenchus xylophilus*. *Agricultural and Biological Chemistry*, **53**, 1777–81.

Morales, J.L. and Chinchilla, C. (1990) Picudo de la palma y enfermedad del anillo rojo/hoja pequeña en una plantación comercial en Costa Rica. *Turrialba*, **40**, 478–85.

Myers, R.F. (1988) Pathogenesis in pine wilt caused by pinewood nematode, *Bursaphelenchus xylophilus*. *Journal of Nematology*, **20**, 236–44.

Nickle, W.R. (1970) A taxonomic review of the genera of the Aphelenchoidea (Fuchs, 1937) Thorne, 1949 (Nematoda: Tylenchida). *Journal of Nematology*, **2**, 375–92.

Nickle, W.R., Golden, A.M., Mamiya, Y. and Wergin, W.P. (1981) On the taxonomy and morphology of the pine wood nematode, *Bursaphelenchus xylophilus* (Steiner & Buhrer 1934) Nickle 1970. *Journal of Nematology*, **13**, 385–92.

Odani, K., Yamamoto, N., Nishiyama, Y. and Sasaki, S. (1984) Action of nematodes in the development of pine wilt disease, in *Proceedings of the United States–Japan Seminar: The Resistance Mechanisms of Pines Against Pine Wilt Disease* (ed. V. Dropkin), National Science Foundation, Washington DC, pp. 128–40.

Oku, H. (1988) Role of phytotoxins in pine wilt disease. *Journal of Nematology*, **20**, 245–51.

Pershing, J.C. and Linit, M.J. (1986) Biology of *Monochamus carolinensis*

(Coleoptera: Cerambycidae) on scotch pine in Missouri. *Journal of the Kansas Entomological Society*, **59**, 706–11.

Poinar, G.O., Jr (1975) *Entomogenous Nematodes*, E.J. Brill, Leiden, 317 pp.

Remillet, M. and Laumond, C. (1991) Sphaerularioid nematodes of importance in agriculture, in *Plant and Insect Nematodes* (ed. W.R. Nickle), Marcel Dekker, New York, pp. 967–1024.

Riddle, D.L. and Georgi, L.L. (1990) Advances in research on *Caenorhabditis elegans*: Application to plant parasitic nematodes. *Annual Review of Phytopathology*, **28**, 247–69.

Rochat, D. (1987) Etude de la communication chimique chez un coleoptere curculionidae: *Rhynchophorus palmarum* L., M.Sc. thesis, Universite Paris VI, Institut National Agronomique, 30 pp.

Rochat, D., Malosse, C., Lettere, M., Ducrot, P.-H., Zagatti, P., Renou, M. and Descoins, C. (1991) Male-produced aggregation pheromone of the American palm weevil, *Rhynchophorus palmarum* L. (Coleoptera, Curculionidae): Collection, identification, electrophysiological activity, and laboratory bioassay. *Journal of Chemical Ecology*, **17**, 2127–40.

Rühm, W. (1956) Die Nematoden der Ipiden. *Parasitologische Schriftenreihe*, **6**, 1–437.

Rutherford, T.A., Mamiya, Y. and Webster, J.M. (1990) Nematode-induced pine wilt disease: Factors influencing its occurrence and distribution. *Forest Science*, **36**, 145–55.

Salazar, F.L. and Chinchilla, C. (1989) Caracterizacion morpfometrica de cuatro poblaciones de *Rhadinaphelenchus* spp. obtenidas de *Cocos nucifera* y *Elaeis guineensis*. *Nematropica*, **19**, 18.

Schauer-Blume, M. (1990) Preliminary investigations on pathogenicity of European *Bursaphelenchus* species in comparison to *Bursaphelenchus xylophilus* from Japan. *Revue de Nématologie*, **13**, 191–5.

Schuiling, M. and Dinther, J.B.M. van (1981) 'Red ring disease' in the Paricatuba oilpalm estate, Para, Brazil. *Zeitschrift für angewandte Entomologie*, **91**, 154–61.

Shaheen, F., Winter, R.E.K. and Bolla, R.I. (1984) Phytotoxin production in *Bursaphelenchus xylophilus*-infested *Pinus sylvestris*. *Journal of Nematology*, **16**, 57–61.

Shuto, Y. and Watanabe, H. (1988) Stimulating effect of ethanol on oviposition of the pine wood nematode, *Bursaphelenchus xylophilus*. *Agricultural and Biological Chemistry*, **52**, 2927–8.

Siddiqi, M.R. (1986) A review of the nematode genus *Fergusobia* Currie (Hexatylina) with descriptions of *F. jambophila* n. sp. and *F. magna* n. sp., in *Plant Parasitic Nematodes of India: Problems and Progress* (eds G. Swarup and D.R. Dasgupta), Indian Agricultural Research Institute, India, pp. 264–78.

Steiner, G. and Buhrer, E.M. (1934) *Aphelenchoides xylophilus*, n. sp., a nematode associated with blue-stain and other fungi in timber. *Journal of Agricultural Research*, **48**, 949–51.

Togashi, K. (1985) Transmission curves of *Bursaphelenchus xylophilus* (Nematoda: Aphelenchoididae) from its vector, *Monochamus alternatus* (Coleoptera: Cerambycidae), to pine trees with reference to population performance. *Applied Entomology and Zoology*, **20**, 246–51.

Togashi, K. (1989) Factors affecting the number of *Bursaphelenchus xylophilus* (Nematoda: Aphelenchoididae) carried by newly emerged adults of *Monochamus alternatus* (Coleoptera: Cerambycidae). *Applied Entomology and Zoology*, **24**, 379–86.

Togashi, K. (1990) Change in activity of adult *Monochamus alternatus* Hope (Coleoptera: Cerambycidae) in relation to age. *Applied Entomology and Zoology*, **25**, 153–9.

Togashi, K. and Sekizura, H. (1982) Influence of the pine wood nematode,

Bursaphelenchus xylophilus (Nematoda: Aphelenchoididae), on longevity of its vector, *Monochamus alternatus* (Coleoptera: Cerambycidae). *Applied Entomology and Zoology*, 17, 160–5.

Tomlinson, P.B. (1990) *Structural Biology of Palms*, Oxford University Press, Oxford.

Vovlas, N., Inserra, R.N. and Greco, N. (1992) *Schistonchus caprifici* parasitizing caprifig (*Ficus carica sylvestris*) florets and the relationship with its vector fig wasp (*Blastophaga psenes*). *Nematologica*, 38, 215–26.

Walsh,. K.D. and Linit, M.J. (1985) Oviposition biology of the pine sawyer, *Monochamus carolinensis* (Coleoptera: Cerambycidae). *Annals of the Entomological Society of America*, 78, 81–5.

Wattanapongsiri, A. (1966) A revision of the genera *Rhynchophorus* and *Dynamis* (Coleoptera: Curculionidae). *Department of Agriculture Science Bulletin 1*, Department of Agriculture, Bangkok, Thailand, 1–328.

Webster, J.M., Anderson, R.V., Baillie, D.L., Beckenbach, K., Curran, J. and Rutherford, T.A. (1990) DNA probes for differentiating isolates of the pinewood nematode species complex. *Revue de Nématologie*, 13, 255–63.

Weissling, T.J., Giblin-Davis, R.M., Scheffrahn, R.H. and Marban, M.N. (1992) A trap for quantifying adult *Rhynchophorus cruentatus* (Coleoptera: Curculionidae) response to semiochemicals. *Florida Entomologist*, 75, 212–21.

Wingfield, M.J. (1987) A comparison of the mycophagous and the phytophagous phases of the pine wood nematode, in *Pathogenicity of the Pine Wood Nematode* (ed. M.J. Wingfield), APS Press, St Paul, pp. 81–90.

Wingfield, M.J., Blanchette, R.A. and Robbins, K. (1982) Association of pine wood nematode with stressed trees in Minnesota, Iowa, and Wisconsin. *Plant Disease*, 66, 934–7.

Wingfield, M.J., Blanchette, A. and Kondo, E. (1983) Comparison of the pine wood nematode, *Bursaphelenchus xylophilus* from pine and balsam fir. *European Journal of Forest Pathology*, 13, 360–72.

16

Management of disease complexes

Rashid M. Khan and P. Parvatha Reddy

Awareness and need for new directions and priority of research are probably influenced by new scientific perceptions based on changing scenarios and archaic scientific knowledge on one hand and economic needs, ecological repercussions and technical compulsions on the other. Protection of crop plants from disease-causing agents has been the focal point of the scientist's concern in dealing with these organisms. However, the concept of protection itself has successively undergone a sea change. In-depth analysis and realization about the startling features of microbial ecology have impelled workers to replace the term 'control' with 'management'. Prohibitive costs of chemicals and their adverse ecological impacts are major compulsions for diversion of research priorities from chemical methods to other alternatives. The successive shifting of priorities from chemical to cultural and currently to integrated management demonstrates the elasticity of scientific ideas. Efforts directed towards formulation of management systems by and large have been aimed against monopathogenic situations, although the occurrence of disease complexes involving plant-parasitic nematodes is not uncommon in nature (Powell, 1979). Social and economic needs and scientific curiosity warrant that management strategies for complex diseases are devised with scientific brilliance. Except for Brodie's (1970) brief review, there is hardly any comprehensive effort to highlight the significance of the multipathogenic scenario vis-à-vis their management. In this chapter, we discuss various methods adopted to deal with management of different plant pathogens involved in complex diseases, the difficulties encountered and the ecological outcome of the measures.

16.1 MANAGEMENT BY RESISTANCE BREEDING

For many reasons this method of management has an edge over other measures. The pathogenesis of fungi, bacteria and other organisms and the role of nematodes as a modifier of the whole process have been dealt with in preceding chapters. Plant-parasitic nematodes facilitate other pathogens to establish on the varieties bred to keep them at bay (Powell, 1979). Breeding programmes for all crops should be fashioned in such a way as to include the potent pathogenic flora and fauna of the region. Undoubtedly breeding of crop cultivars for multipathogenic adversity is no less challenging and laced with enormous impediments; however, it relieves farmers from costly chemical protection schedules. The major problem with such a scheme is pooling information about all the eco-pathogenic attributes of two or more pathogens. Furthermore, field populations of fungi and nematodes quite often consist of a mixture of different races/pathotypes with varying pathogenic abilities. Assembling data on the genes providing resistance to a pathogen or a group of pathogens and to their races and to incorporate them in cultivars equipped with genes for higher yield and acceptable agronomical attributes demands strenuous efforts. The possibility of losing one or the other resistant gene for one or other pathogen is always present during the course of incorporation (Fassuliotis, 1987). It is possible that an agronomically acceptable cultivar may have an undesirable character in terms of its susceptibility to some pathogen. In cotton, *Verticillium* wilt susceptibility was always found to be linked with small seeds having a slow germination rate but high establishing stand and higher yield (Bird, 1982).

Sidhu and Webster (1974, 1979, 1983), while working on the *Meloidogyne–Fusarium* complex or *Fusarium–Verticillium* complex, pleaded that in the disease complex process the epistasis phenomenon operates wherein the genes responsible for imparting resistance to one organism were found to be modified when the other pathogenic forms were present. Abawi and Barker (1984), however, found that the monogenic resistance in tomato against *Fusarium* wilt remained unaffected in the presence of *M. incognita*, which multiplied well on tomato cultivars Florida, MH-1, Manapal etc., resistant to *F. oxysporum* f. sp. *lycopersici* race 1. Gwyne *et al.* (1986) found that it was possible to transfer the resistance in tobacco against tobacco mosaic virus, cyst nematode, root-knot nematode and bacteria from *Nicotiana repanda* to *N. tabacum*. Similarly a cucumber variety exhibiting resistance against a variety of fungal, bacterial and viral diseases has been developed (Peterson *et al.*, 1982).

16.2 MANAGEMENT BY CHEMICAL METHODS

16.2.1 Nematode–fungal complexes

Chemicals have been considered as by far the most effective and efficient means of managing the pathogenic population of fungi, bacteria, nematodes, etc. Chemicals such as nematicides, fungicides and herbicides have been formulated and are being successfully used against their respective targets (Van Gundy and McKenry, 1977; Dekker, 1977; Wright, 1981). These chemobiocides, especially systemics, are formulated with the specific target organism in view. The discovery and occurrence of complex diseases and the use of fumigants against some nematodes and their efficacy against associated wilt fungi resulted in accrual of data which led to a foundation for multiple control strategy. A brief review of Brodie (1970) on the imperatives of tackling the fungus–nematode complexes and subsequent realization of the multiferous activities of systemic chemicals, coupled with a better understanding of complex diseases, impressed upon the workers the need to evolve a multiple treatment plan with a single or variety of chemicals in varying ratios to achieve optimal minimization of the potentially damaging population. The reviews of Katan and Eshel (1973), Altman and Campbell (1977), Rodriguez-Kabana and Curl (1980) and Trappe *et al.* (1984) can be referred to in exploring the modalities for evolving chemical formulations exhibiting true features of 'broad-spectrum biocides', which Brodie (1970) has termed 'super granule'.

In this section we do not intend to review the work done on multiple control strategy, but to collate the information so as to stimulate an interest in working out a multiple management system for which scientists from different disciplines have to work in unison.

Soil fumigants

Use of soil fumigants, their negative impact on nematode populations and direct correlation with plant growth led nematologists to realize the roles nematodes play either solely as a pathogen or in modifying the host plant to facilitate the establishment of other pathogens leading to complex diseases. Soil fumigants have been used singly or in combination against pathogens involved in complex diseases. Usually combination of fumigants have been found to have chemical compatibility coupled with toxicity against soil biota like insect pests, fungi and weeds capable of causing substantial damage.

Good and Rankin (1964) explored the possibility of working out a common treatment schedule for nematodes and fungi posing a serious complex production problem for cotton and legume fields. Of 11 soil fumigants representing specific nematode toxicity and a broad-spectrum nature, only vapam and mylone individually and a mixture of DD and vorlex were found to be effective against the nematode *Pratylenchus brachyurus*, the fungus *Sclerotium rolfsii* and weeds like pursley, carpet weed and crab grass. Application methods and timing significantly influence the performance of the chemicals. In the nematode–fungal complex of cotton variety Coker 100w having *Meloidogyne* sp., *Pratylenchus* sp., *Tylenchorhynchus* sp., *Trichodorus* sp. and *Fusarium oxysoprum* f. sp. *vasinfectum* as constituents of the complex, only wilting at initial stages of EDB application (85% at the rate of 2 gallons per acre) was reduced. However, at the end of the growing season there was ample reduction in root-knot and root-rot indices (Newsom and Martin, 1953). In selection of chemicals, compatibility, method and time of application, the pathogens and crop involved are important determinants. Good (1964) found that amongst all fumigants, i.e. MBr, SMDC, DMTT, DD and DBCP, tested at 10 and 20 inches depth against root-knot–*Fusarium* wilt of okra and soft-seeded weed, only MBr significantly reduced the disease complex. DD, on the other hand, could slightly check the nematode–fungal complex but only when applied at a 10-inch depth, whereas DBCP was much more effective than DD at the same depth.

Meagher and Jenkins (1970) found that MCP (methyl bromide and chloropicrin), Ditrapex and EDB were effective against *Verticillium* wilt of strawberry and MCP and EDB against *M. hapla*. Only EDB exhibited dual efficacy against nematode–fungus complex. The authors concluded that in the case of nematode–fungus complex and short-duration crops only low-cost nematicides like EDB should be selected. For durable control MCP was found to be effective against *Pratylenchus–Verticillium* wilt of tobacco. Chloropicrin was effective only against *Verticillium*, and Telone only against *Pratylenchus* (Taylor *et al.*, 1970). Methyl bromide, chloropicrin, DD mixture, dazomet and mercury salts tested against *Heterodera avenae* and *Gaeumannomyces* (*Ophiobolus*) *graminis* were effective in the first year against both organisms. Dazomet was effective against the fungus in the second year also but in the third year the fungus population increased in all the treated plots except those that received two successive applications of DD mixture. Chloropicrin double application was effective against the nematode. On the other hand, two formalin applications caused a population build-up of *H. avenae* which at the end of third year almost trebled (Williams and Salt, 1970).

Farley and Riedel (1971) working on the efficacy of DD and D140 against *Pratylenchus–Verticillium* complex of tomato showed that DD at pre-planting or D140 at transplanting time reduced the nematode population but was ineffective against the wilt fungus. DD was found to be effective against *Heterodera schachtii* and *Fusarium* wilt of sugar-beet (Altman and Fitzgerald, 1960).

In *Verticillium dahliae–Globodera rostochiensis* complex of potato, a mixture of methyl bromide and chloropicrin was found to have comparatively more fungicidal and nematicidal efficacy than either benomyl (fungicide) or aldicarb (nematicide) tested alone. However, its residual effect could not last up to the third crop (Hide and Corbett, 1974). DBCP in addition to nematicidal action possesses fungicidal properties also. Its twin potential in combination with fungicides like sodium azide is synergized. This combination was found to be potentially effective against nematode-fungus complex in soya bean (Kinloch and Schenck, 1978). In *Pratylenchus–Verticillium* complex of strawberry also, a combination of ditrapex and benlate was very effective (Szczygiel and Reabandal, 1982).

Systemics

Jones and Overman (1976) noticed an improvement in the yield of tomato affected by nematodes like *Belonolaimus longicaudatus*, *Trichodorus christiei, Meloidogyne acrita* and wilt fungi *Fusarium oxysporum* f. sp. *lycopersici* race 2 and *Verticillium albo-atrum*, when corbofuran was applied at a higher soil pH (6.5–7.5). They suggested that although carbofuran is not fungicidal, its negative influence should be seen in the light of nematode–fungus interaction. Aldicarb and phenamiphos also have been reported to check the *Fusarium* wilt of maize effectively although indirectly through their toxic action against nematodes (Minton *et al.*, 1985). Kimpinski *et al.* (1987) recorded a positive correlation between aldicarb application and yield of barley and wheat; however, they did not observe any significant relation between the treatment and decline in nematode population and fungal disease.

The non-fumigant nematicides have been tried in combination with fungicide and herbicides so as to attain the goal of multiple pest control. Brodie and Hauser (1970) used a compatible mixture of aldicarb (a nematicide), pentachloronitrobenzene (a fungicide) and trifluraline (a herbicide) in a ratio of 8 : 2 : 1. The mixture was reported to be effective against pest complex of cotton that involved *Belonolaimus longicaudatus, Fusarium* wilt and weeds like *Richardia scabra* and *Digitaria sanguinalis.* Replacement of aldicarb with fensulfothion was not effective against the nematode. Similarly, the herbicidal efficacy of trifluraline too seemed to be synergized in combination. Effective wilt control was also related to decline in nematode population rather than merely due to the efficacy of fungicide against wilt fungus. The study clearly indicates the possibility of amalgamation of various chemicals with proven biocidal efficacy against the pest–disease complex of a given crop.

Meloidogyne incognita–Phytophthora infestans f. sp. *nicotianae* complex was managed by a combination of nemacur and dasanit (Fortnum and Curin, 1984). Abu-El-Amayem *et al.* (1985) also explored a suitable nematicide–fungicide combination against *M. incognita–Rhizoctonia solani* complex on soya bean. They found a potent action of carbofuran–benomyl combination against the nematode, and of phenamiphos/carbofuran-benomyl against the fungus. Hasan (1989) demonstrated that aldicarb, carbofuran and phorate significantly reduced the severity of the disease complex involving *Heterodera cajani* and *Fusarium udum* on pigeonpea.

16.2.2 Nematode–virus complexes

Management of nematodes which vector plant viruses can effectively reduce the viral diseases. Introduction of nematode vectors along with plant material poses a threat of introducing viruses also. These nematodes have been found to survive over a wider range of eco-stresses, viz. temperature, moisture, fallowing, etc., for a varying period of time. Furthermore, the wide host range of these nematodes, the common weed host of both nematode vectors and viruses and the occurrence of nematodes in comparatively deeper layers ranging from 0.3 and 0.9 m to as deep as 3.9 m excerbates the barriers encountered in managing the population of vectors (Norton, 1978; Taylor, 1978; Lamberti, 1981). Placed under such limitations it would be perilous to ignore the significance of the virus–vector relationship and the damage it causes.

DD, dazomet and quintozone effectively checked virus transmission by *Longidorus elongatus* and *Xiphinema diversicaudatum* to ryegrass at temperature, ranging from 7 to 22°C. DD and quintozone were most effective (Taylor and Gordon, 1970). Spraing disease of potato caused by tobacco rattle virus, which is transmitted by *Trichodorus* sp., was found to be checked by methomyl used at 8 lb per acre and dazomet at 150 lb per acre only for one year. However, DD at 200 lb per acre checked the spread of the virus for two years. Its efficacy was slightly increased when applied in autumn and was effective only against the nematode. Methomyl, on the other hand, was reported to be effective against the virus rather than the nematode (Cooper and Thomas, 1971).

Dose, type of soil and soil preparation seem to have a decisive influence on chemical efficiency. 1,3-D at 250 gallons per acre was a suitable recommendation for heavy soil against *Xiphinema index*, a vector of grapevine fan leaf virus, while only 200 gallons per acre was found to be efficient in lighter soil (Rankin *et al.*, 1971).

Some studies show that chemicals were effective in reducing the population of vector nematode, but this failed to have any impact on virus infection (Pitcher and McNamara, 1973). DD, ditrapex and dazomet reduced *Xiphinema diversicaudatum* population on hops; however, transmission of arabis mosaic virus remained unaffected. Quintozone (PCNB), aldicarb and methomyl neither influenced population of the vector nor the virus transmission (Pitcher and McNamara, 1973).

Maas (1974) emphasized that soil fumigation for controlling *Trichodorus*, a vector of spraing of potato, is the most effective. The presence of the nematode, its survival below the tillage surface down to 120 cm, and its ability to retain the virus for up to three years hamper its management. Even small populations that may survive the fumigation can transmit the virus.

16.2.3 Nematode–bacteria complexes

Information on nematode–bacteria complex management is meagre, possibly because of the small number of bacterial plant pathogens. Apt *et al.* (1960) used malic hydrazide to suppress the onset of inflorescence to control annual ryegrass toxicity jointly caused by *Anguina funesta* and *Clavibacter* sp. Since the nematode matures and reproduces in flowers, suppression of flower development controls the nematode and concurrently the bacterium is also managed in an effective way.

16.2.4 Management of insect vectors

The role of insects like palm weevil, *Rhynchophorus palmarum*, in transmitting *Bursaphelenchus cocophilus*, the causal organism of the red ring disease of coconut, and *Monochamus alternatus* in transmitting the nematode *Bursaphelenchus* spp., causal organisms of pine wilt (Blair, 1969; Mamiya, 1983), is well known. If the vector population is effectively managed, the chances of nematode infestation will be minimized to a great extent. The palm weevil population was successfully managed when tree crowns were sprayed with 1–10% palmoral (Blair, 1969). A 0.5% emulsion of insecticides like fentirothion, fenthion or carbaryle if sprayed at the rate of 2.3 litres per tree during May to June, can effectively reduce the vector population (Mamiya, 1984).

16.2.5 Lateral activities of plant protection chemicals

The influence of chemicals (insecticides, fungicides, nematicides or herbicides), besides affecting the primary organism against which they are synthesized, is also radiated towards other constituents of the biotic complex. Information on diverse toxicity of fungicides, nematicides, insecticides and herbicides has been documented and discussed at length (Katan and Eshel, 1973; Altman and Campbell, 1977; Rodriguez-Kabana and Curl, 1980; Menge, 1982; Trappe *et al.*, 1984). This diverse form of toxicity has been termed variously non-target, unintended activities, side effects, etc.

The lateral toxic influence of the goal-specific pesticides, however, can be effectively utilized in dealing with complex disease situations. Keeping the diversity of biological forms in view and their interdependence in juxtaposition with edaphic factors, the most vital variable, the crop itself, especially requires an arduous study of the complex system and close and meaningful coordinated efforts in developing a desirable 'super granule', at least for widely grown and globally important crops of high economic return. In this section, we discuss briefly the lateral toxic properties, both beneficial and harmful, in terms of population/growth reduction or increase and their resultant indirect influence on crop growth.

Nematicides

Nematicides, although they target the nematodes, may display toxic properties against soil fungi indiscriminately. In such an event, antagonists as a check for keeping up a delicate balance would also be affected, eventually leading to dislocation of the population configuration. Among fumigant nematicides, probably DBCP is the most broad spectrum, and it is toxic against a number of fungal plant pathogens. *Pythium* spp. and *Rhizoctonia solani* are important pathogens of a variety of crops especially at seedling stage. DBCP at doses ranging between 9 and 14 litres per hectare has been found to be effective against *R. solani* (Ashworth *et al.*, 1964) and *P. ultimum* (Brodie, 1961). In Aligarh, India, *R. solani* causes damping-off of cauliflower seedlings. The disease assumed serious proportions in the presence of an ectoparasitic nematode, *Tylenchorhynchus brassicae* (Khan *et al.*, 1971). DBCP, in in vitro studies, inhibited the cauliflower isolate of *R. solani*. Saprophytes like *Aspergillus* sp. and *Penicillium* sp., however, remained largely unaffected (Khan *et al.*, 1980). 1,3-Dichloropropane at higher doses was found to be effective against *Sclerotium rolfsii* (Clayton *et al.*, 1949), *Fusarium* sp.(Stark and Lear, 1947) and *Phytophthora* sp. (Zentmyer and Kendrich, 1949). Growth inhibition of pathogenic fungi like *R. solani* and *S. rolfsii* was obtained in the presence of ethoprop and at some concentrations it was comparable to the efficacy of PCNB, a fungicide. *Aspergillus* sp. and *Rhizopus stolonifer* remained unaffected (Rodriguez-Kabana *et al.*, 1976b). Fensulfothion was toxic to *S. rolfsii* and *R. solani* but seemed to be non-toxic to *Trichoderma harzianum* (Rodriguez-Kabana *et al.*, 1976a).

Nematicides also cause negative lateral influences. They may influence the growth of even fungal pathogens negatively. This influence, however, may predispose plants to other pathogens, thereby nullifying the effect of nematicides. Although cotton seeds treated with phorate were protected against *Rhizoctonia solani*, the treatment made it susceptible to *Pythium* sp. (Erwin *et al.*, 1961). The positive aspect of secondary biotoxicity of DBCP must be fully appreciated. Its chemical properties at the other extreme seem to favour the new sclerotia of *Sclerotium rolfsii* (Rodriguez-Kabana *et al.*, 1979). Application of EDB and DD reduced the population of *Hirschmanniella* spp. in rice; however, an increase in leaf-hopper population and damage caused by stem borer was noticed in treated plots (Iyatomi and Nishizawa, 1970). Application of oxamyl increased the incidence of downy mildew (Minton *et al.*, 1977).

Soil is also the home of beneficial microbes like mycorrhizae, mycorrhizal fungi and root-nodule bacteria, etc. Observations on the impact of nematicides on mycorrhizae and mycorrhizal fungi are many and incidentally contradictory. DBCP increased the root infection by mycorrhizal fungi on cotton roots (Bird *et al.*, 1974), whereas on peanut no effect was observed (Backman and Clark, 1977). DBCP when applied 30–60 days after inoculation of mycorrhizal fungi at comparatively lower concentrations enhanced root colonization of sudangrass. On the other hand, simultaneous treatment and higher concentrations failed to improve the growth of mycorrhiza. Corn also failed to show any response to endogenous mycorrhizal infection when treated with phorate (Ocampo and Hayman, 1980) and grape when treated with oxamyl (Atilano and Van Gundy, 1979). Similarly, root nodulation on peanut was found to be adversely affected by oxamyl and phenamiphos at low doses, whereas at higher doses it was increased (Clarkson *et al.*, 1982). In contrast, Kaul *et al.* (1986) observed that at higher doses of nematicides root nodulation on pea was decreased, whereas at lower doses it was increased.

Use of sublethal doses of nematicides like carbofuran, phenamiphos and oxamyl have been found to induce resistance in *Xiphinema index* and *Meloidogyne incognita* (Yamashita and Viglierchio, 1987).

Fungicides

Fungicides, when used for management of the fungal pathogen in question, also encounter the polytaxon situation. Like nematicides, this group of chemicals too, based on practical experience, seems to be endowed with multiple biocidal properties. The role of fungi is of paramount significance in disease complexes.

Pentachloronitrobenzene (PCNB) has been found to effectively check the population increase of nematodes like *Longidorus elongatus* and *Xiphinema diversicaudatum*. This efficacy would eventually influence the spread of viruses transmitted by them (Murant and Taylor, 1965; Taylor and Gordon, 1970). *Meloidogyne incognita* and *Rotylenchulus reniformis* are the other important nematodes which are affected by this chemical (Adams *et al.*, 1979). Ethazole and benomyl too, in addition to their established fungitoxicity, have been reported to possess nematicidal properties. Benomyl application inhibited feeding of *X. americanum* on cucumber, which was reflected in a marked reduction of its transmission ability of tobacco ring spot virus (McGuire and Good, 1970; Rodriguez-Kabana and King, 1977). The fungicides that seem to be effective against nematodes present a utility in management strategies against complex diseases. This aspect needs to be explored with due care since they may lead to an increase in the population of some nematodes. PCNB has been found to favour the population of some plant-parasitic nematodes. Application of terraclor (Rich and Miller, 1964) and PCNB (Miller and Waggoner, 1963; Boswell, 1968; Adams *et al.*, 1979) increased the population of *Pratylenchus* spp. Besides, there are a number of reports available wherein use of fungicides has been implicated to pave the way for other diseases (Horsfall, 1979).

Fungicides have been found to alter the mycorrhizal association with plants. Fungicides like captan were found to reduce the mycorrhizal infection of corn (Nesheim and Linn, 1969), while on wheat and citrus, etc. (Jalali and Domsch, 1975; Timmer and Leyden, 1978), it failed to show any effect.

Herbicides

Although designed to act specifically against weeds, the application of herbicides affects other organisms directly or indirectly. From the point of view of the pathogen population, attention is drawn to how the herbicides should be chosen to overcome an unwarranted eventuality.

Malic hydrazide, naptalam, propham and dalapon were toxic to

Alternaria solani (Richardson, 1959); diphenamid was toxic to *Rhizoctonia solani* and *Pythium aphanidermatum* (Cole and Batson, 1975). The presence of root-knot nematodes with damping-off fungi (*R. solani* and *P. aphanidermatum*) takes a heavy toll of tomato seedlings in nurseries (R.M. Khan, personal observation). If the efficacy of these herbicides is proved against this group of nematodes also, it may open other avenues for testing chemicals with wider toxic potentialities. Application of CDEC on spinach was found to reduce the population of *Meloidogyne incognita* (Good and Taylorson, 1964). Gall formation on tomato caused by *Meloidogyne* spp. was significantly reduced when tomato seedlings were dipped in oryzaline (Osman and Viglierchio, 1981). Application of flurodifen in a soya bean field was found to eliminate the population of *Tylenchus* sp. and *Aphelenchoides* sp. (Mohammad, 1987).

Herbicides seem to play a negative role from the plant-protection strategy viewpoint, i.e. their use results in an increase in plant diseases. Growth/ population of fungi like *Fusarium oxysporum* f. sp. *vasinfectum, Rhizoctonia solani* and *Sclerotium rolfsii* was found to be increased after application of the herbicides atrazine, prometryne and EPTC respectively (Rodriguez-Kabana and Curl, 1970; Beam and Curl, 1971; Rodriguez-Kabana *et al.*, 1970). Nematode population has been reported to be increased by herbicide application. Application of EPTC at lower dosages was found to increase gall formation by *Meloidogyne arenaria* (King *et al.*, 1977). Numbers of *Ditylenchus dipsaci* in oats were highly enhanced when 1,4-D was applied as a post-inoculation spray (Webster, 1967). Similarly Webster and Lowe (1966) noticed an increase in the population of *Aphelenchoides ritzemabosi* after 2,4-D application. Schmidt and Nelson (1987) have discussed at length the combination of nematicides and herbicides and their influence on nematode population. In the majority of the cases nematicides like fensulfothion or phenamiphos when used simultaneously with herbicides like alchlor resulted in an increase in *Heterodera glycines* population (Schmitt and Corbin, 1981).

16.3 MANAGEMENT BY PHYSICAL METHODS

Soil solarization is a soil disinfestation method for disease control. It involves the use of heat as a lethal agent for pest control through the use of traps for capturing solar energy by means of polyethylene soil mulches (Katan, 1981). In Israel, Italy and the USA, the use of 0.03 mm thick polyethylene (plastic) transparent sheets in moist and well-cultivated soil continuously for 30–50 days has been found to be effective against pathogenic nematodes, fungi and weeds. *Ditylenchus dipsaci, Pratylenchus thornei* among nematodes, *Fusarium* spp., *Verticillium dahliae* among fungi and *Anagallis coeruella, Avena fatua, Chenopodium album* and *Digitaria sanguinalis*, etc., among weeds were suppressed (Cartia, 1985). In South Africa, soil solarization reduced populations of *Pratylenchus pratensis*,

Rotylenchus incultus and *Paratrichodorus lobatus* and a fungus, *Phytophthora cinnamomi*. However, the reduction in *Meloidogyne incognita* population was not consistent. In such a soil the growth of tomato and grape was significantly enhanced (Barbercheck and Broembsen, 1986). Significant reduction in root galling on tomato by *Meloidogyne* sp. and population of a weed *Cyperus* spp. occurred in soil exposed to summer heat for about nine weeks. The population reduction in *Rotylenchulus reniformis*, however, was not consistent (McSorley and Parrado, 1986). The major deficiency in this is its limited application. This measure can be effective in tropical climates and the recommendation can be useful only during summer. Besides, beneficial flora also have to be taken into consideration. Rahim *et al.* (1988) noticed an adverse effect of soil solarization on root nodulation on broad bean.

16.4 MANAGEMENT BY CULTURAL METHODS

Cultural methods like fallowing, planting date manipulation, weed management and crop rotation are currently practised for management of plant diseases. These too carry some demerits. The major drawbacks in most cultural practices are their haphazard approaches and ignorance of the contemporaneous pathogenic flora/fauna. Stupendous efforts are required for engineering a novel cultural practice which may be able to keep most of the major pests/diseases below economic threshold level. Comprehensive efforts to work out common cultural strategies are very few. Formulations of either cultural method are based on the exploitation of any bio- and/or eco-event(s) of the microorganism. Since all the organisms involved in the complex disease situation are made up of various genetic pools, it is quite possible that one strategy of cultural practices may not be effective for all.

16.4.1 Fallowing

Fallowing, which attempts to check the upward population trend, is directed mostly only against individual organisms. The vulnerability period of all or a few organisms in the system may not necessarily be of equal length and a very long period of fallow would not be advisable, mainly from economic considerations. Willis and Thompson (1979) found in *Pratylenchus penetrans* and *Fusarium* sp. complex of alfalfa that fallowing checked the population increase of nematodes only in one growing season, whereas *Fusarium* sp. remained unaffected. Such a practice may provide some degree of relief especially when the fungus involved is a secondary invader or a weak pathogen. Klingler and Zwicky (1981) noted that fallowing was not effective for *Xiphinema* sp. in Germany because the nematode survives at deeper layers. Consequently the spread of grapevine fan virus also remained unaffected.

16.4.2 Weed management

Keeping the field free from weeds is a normal agricultural practice. Since weeds harbour a number of pests/pathogens their removal may greatly reduce the population of respective pests/pathogens. In the *Heterodera avenae–Gaeumannomyces graminis* complex of wheat, drilling in conjunction with weed control reduced the damage caused by these organisms (Boer and Kollmorgenj, 1988).

16.4.3 Fertilizer application

The role of various components of the complete fertilizer regime in predisposing plants to diseases is well known (Yarwood, 1959; Colhoun, 1979). Fertilizer application has been found to increase or decrease the population of plant pathogens (fungi, nematodes), leading to an increase or decrease in disease intensity (Wallace, 1963, 1973; Norton, 1978; Colhoun, 1979). Development of fertilizer schedules that are capable of striking a suitable balance among various biotic components of an agro-ecosystem and minimizing the chances of an upsurge in the population of plant pathogenic organisms is needed. While adapting the fertilizer schedule for a particular crop, besides encompassing the faunistic and floristic ecology in its totality with emphasis on the organism of pathogenic potential, the impact on crop either in terms of toxicity or crop yield should never be ignored. Application of anhydrous ammonia $(NH_3$ and NH_3 formulated with potassium azide) at the rate of 500 mg N/Can in the glasshouse resulted in 99% reduction in the population of *Rotylenchulus reniformis* and significantly suppressed fungi such as *Fusarium* sp. Urea could not influence either of the organisms (Birchfield and Parr, 1969).

16.4.4 Crop rotation

Notwithstanding certain inherent disadvantages, crop rotation has been successfully tested against fungal and nematode diseases. However, to work out a scientifically and economically viable system of cropping sequence against complex disease involving nematodes with either fungi, bacteria or any other organism requires enormous data, particularly on host range and survival ability under a variety of ecological conditions. For the single pathogen host range itself may be quite large. Overlapping of host range is one of the main hurdles. In France, to save grapevines from attack by *Xiphinema index* and subsequently infection by grapevine fan leaf virus, a seven-year rotation with cereal and alfalfa crop has been recommended (Lamberti, 1981).

To work out a crop rotation scheme for a particular disease complex, information on comparative population build-up of other microorganisms on the crops also needs to be evaluated. Economic acceptability of the crop ping sequence and allelopathic influence on subsequent crops are the other important criteria which need proper attention.

16.4.5 Organic amendment and biocontrol

Use of plant residues and organic amendment has been recognized as an effective way of achieving substantial population reduction of plant-pathogenic life forms like fungi, bacteria, nematodes, etc. (Patrick and Toussoun, 1965; Sayre *et al.*, 1964). Like other management measures, most of the organisms have been addressed using the compartmental approach. Plant residues or organic amendments such as oil cakes have been reported to check the population of either of the organisms through a variety of mechanisms (Patrick and Toussoun, 1965; Sayre *et al.*, 1964; Cook, 1977; Sitaramaiah, 1990). There are indications that quite often during the decomposition process of one or other kind of organic material an identical mechanism may operate, leading to population reduction of the pathogenic organisms. Schippers and Palm (1973) identified ammonia and its fungistatic role in chitin-amended soil. Khan, Alam and Ahmad (1974) too noticed the release of ammonia and its nematicidal potency during the process of oil cake decomposition. Similarly there are reports in the literature about the enhanced activity of antagonistic fungi due to soil amendments by various organic material which hampers the infection process of pathogenic fungi, nematodes, etc. The literature on this aspect has been reviewed (Cook, 1977; Jatala, 1985; Sitaramaiah, 1990). Khan, Khan and Saxena (1973, 1974) showed that incorporation of oil cakes, viz. margosa, castor, groundnut, in natural field soil resulted in population decline of parasitic fungi and nematodes in the rhizosphere of tomato and eggplant. Rahman and Khan (1988) identified the deleterious effect of oil cakes on seed germination of a weed, *Chenopodium album*, along with nematodes and fungi. A common strategy for management of disease complexes by the use of organic materials can be evolved.

Recent work by Shahzad and Ghaffar (1989) indicates that *Paecilomyces lilacinus*, a biocontrol agent of endoparasitic nematodes, can be effectively used against *Rhizoctonia solani–Meloidogyne incognita* complex of okra and mungbean. Similarly, *Trichoderma harzianum* has been found to be biocidal against *M. arenaria* (Windham *et al.*, 1989) and *M. incognita* (R.M. Khan, unpublished). These observations indicate the feasibility of utilizing biocontrol agents for managing nematode–fungus complexes.

The pathogenicity of fungi like *Beaveria bassiana* on *Monochamus alternatus*, an insect vector of *Bursaphelenchus xylophilus*, has been proved (Katagiri and Shimazu, 1980). On the other hand, DD-136 has been found effective against *B. xylophilus* (Mamiya, 1984). An attempt to combine both *Baeveria bassiana* and DD-136 (*Steinernema feltiae*, an insect parasitic nematode) may open some exciting avenues for managing both organisms simultaneously. However, it needs sufficient information on the basic aspects of all of the four organisms involved.

16.5 INTEGRATED MANAGEMENT OF COMPLEX DISEASES

The integrated management concept is of comparatively recent origin. In multipathogenic situations it would still take time for workers to appreciate its importance, validity and above all the pooling of background information gathered from other traditional management methods and relevant fields. However, stray reports are available in the literature wherein some workers have successfully attempted to blend a few management practices with a some degree of success. Integration of two or more methods for arresting the damage wrought by multipathogenic situations demands a comprehensive look not only at the individual pathogens involved in the complex but also a close coordination among scientists of various disciplines. Devising an Integrated Pest Management (IPM) prescription for the complex disease situation, however, is far more difficult and challenging.

Yellow disease of *Piper nigrum* caused by *Radopholus similis*, *Meloidogyne incognita* and *Fusarium oxysporum* was managed by an integration of fertilizer and nematicide(s). Application of NPK (15 : 15 : 15) at the rate of 250 g per plant per year in combination with aldicarb at a rate of 50 g per plant and/or mancozeb at a rate of 12 g per plant was able to reduce the severity of the complex. Additionally berry yield was also increased (Mustika *et al.*, 1984).

A combination of soil solarization with certain chemicals was attempted against the fungus–nematode complex of cotton. Application of nematicides like 1, 3-D, ethoprop and metham sodium along with solarization reduced the populations of *Pythium ultimum* and *Meloidogyne incognita*; it was, however, no better than solarization alone (Stapleton *et al.*, 1987).

Application of metham sodium through drip irrigation was found to have an edge over 1,3-D for root-knot, *Phyhium ultimum* and *Fusarium* sp. complex of carrot and tomato (Roberts *et al.*, 1988). Similarly compatible synchronization of chemicals and plant nutrition with irrigation can also be conceived against a nematode–fungal complex. In early potato dying disease, *Pratylenchus penetrans* predisposes plants to *Verticillium dahliae* infection (Rowe *et al.*, 1985). In such a complex, severity of *V. dahliae* is compounded if fertilizer is applied through gravity furrow irrigation, whereas fertigation through a sprinkler system lessens the severity (Davis and Everson, 1986). If the nematode population is high disease severity will multiply. A chemo-fertigation scheme would be an ideal proposition in such a situation.

16.6 CONCLUSION

Involvement of plant-parasitic nematodes in synergistic interactions with other plant pathogens and its agricultural impacts are now established. Despite this, matching management efforts encompassing the full ecological arena of the complex soil ecosystem are lacking. Breeding for resistant varieties against a complex pathogenic situation is the most acceptable form of management system. However, keeping in view the enormity coupled with complexity of relationship among microbes, the task is discouraging. Quick and palliative action of a variety of chemotherapeutic agents, however, offers a spontaneous solution to the problems; recognition of their lateral actions add another dimension to the problem. Indiscriminate use of chemicals leads to upsetting the delicate eco-balance of the soil, which is witnessed in a variety of forms like iatrogenic diseases, resistant biotypes, mortality of beneficial life forms, etc. But they also show toxicity against several pathogens and at some doses seem to favour beneficial and antagonistic microbes also. This aspect also needs due attention. Work on the physical, cultural and integrated management aspects dealing with the complex disease situation requires proper attention and needs to be tackled under a broad ecoperspective. To evolve a viable management practice the foundation ought to be laid on a sound ecological basis encompassing all edaphic and biotic factors vis-à-vis the pathogen complex, the crop and therapeutic agents. The natural processes favouring any antagonist at the cost of pathogenic ones should be exploited. Plant residues/organic amendments serve a triple purpose. They favour antagonists, improve crop growth and act as a deterrent for pathogenic forms. However, their influence on other beneficial microbes like nodule-forming bacteria and mycorrhizae also needs to be worked out with precision. Recently the emergence of biocontrol agents like *Paecilomyces lilacinus*, due to their compatibility with chemicals, seems to hold promise. This aspect needs to be addressed from the ecological angle, taking the complex interdependent and interacting web of factors into account. This kind of study currently seems to be lacking.

REFERENCES

Abawi, G.S. and Barker, K.R. (1984) Effects of cultivar, soil temperature and population levels of *Meloidogyne incognita* on root necrosis and *Fusarium* wilt of tomatoes. *Phytopathology*, 74, 433–8.

Abu-El-Amayem, M.M., El-Shoura, M.Y., Radwan, M.A., Ahmad, A.H. and Abd-El-All, A. (1985) Joint action effects of some nematicides and benomyl against *Meloidogyne incognita* and *Rhizoctonia solani* on soybean. *Mededelingen van de Faculteit Landbouwwetenschappen Rijksuniversiteit, Gent*, 50, 839–50.

Adams, J.R., Rodriguez-Kabana, R. and King, P.S. (1979) The nontarget effects of pentachloronitrobenzene on plant parasitic and free living nematodes. *Nematropica*, 9, 110–28.

Altman, J. and Campbell, L.E. (1977) Effect of herbicides on plant diseases. *Annual Review of Phytopathology*, 15, 361–85.

Altman, J. and Fitzgerald, B.J. (1960) Late fall application of fumigants for control of sugarbeet nematodes, certain soil fungi and weeds. *Plant Disease Reporter*, 44, 868–71.

Apt, W.J., Austenson, H.M. and Courtney, W.D. (1960) Use of herbicides to break the life cycle of bentgrass nematode, *Anguina agrostis* (Stein, 1799) Filipjev 1936. *Plant Disease Reporter*, 44, 524–6.

Ashworth, L.J., Langley, B.C. and Thames, W.H. (1964) Long term inhibition of *Rhizoctonia solani* by a nematocide, 1,2-dibromo-3-chloropropane. *Phytopathology*, 54, 187–91.

Atilano, R.A. and Van Gundy, S.D. (1979) Effect of some systemic, nonfumigant and fumigant nematicides on grape mycorrhizal fungi and citrus nematode. *Plant Disease Reporter*, **63**, 729–33.

Backman, P.A. and Clark, E.M. (1977) Effect of carbofuran and other pesticides on vesicular–arbuscular mycorrhizae in peanuts. *Nematropica*, **7**, 13–17.

Barbercheck, M.E. and Broembsen, S.L. Von (1986) Effects of soil solarization on plant parasitic nematodes and *Phytophthora cinnamomi* in South Africa. *Plant Disease*, **70**, 945–50.

Beam, H.W. and Curl, E.A. (1971) Effects of fluometuron and prometryne on *Rhizoctonia solani* in soil. *Phytopathology*, **61**, 884.

Birchfield, W. and Parr, J.F. (1969) Nematicidal and fungicidal effects of some soil applied nitrogen compounds. *Phytopathology*, **59**, 1018.

Bird, G.W., Rich, J.R. and Glover, S.U. (1974) Increased endomycorrhizae of cotton roots in soil treated with nematicides. *Phytopathology*, **64**, 48–51.

Bird, L.S. (1982) The MAR (multiadversity resistance) system. *Plant Disease*, **66**, 172–6.

Blair, G.P. (1969) The problem of control of red ring disease, in *Nematodes of Tropical Crops* (ed. J.E. Peachy), Technical Communication No. 40, Commonwealth Bureau of Helminthology (Commonwealth Agricultural Bureau), pp. 99–108.

Boer, R.F. De and Kollmorgenj, J.F. (1988) Effects of cultivation and stubble retention on soil and stubble borne pathogens of wheat in Victoria: An overview. *Plant Protection Quarterly*, **3**, 3–4.

Boswell, T.E. (1968) Pathogenicity of *Pratylenchus brachyurus* to spanish peanuts. PhD thesis, Texas A & M University, College Station, Texas.

Brodie, B.B. (1961) Use of 1,2-dibromo-3-chloropropane as a fungicide against *Pythium ultimum*. *Phytopathology*, **51**, 798–9.

Brodie, B.B. (1970) Use of non-selective and mixtures of selective pesticides for multiple pest control. *Phytopathology*, **60**, 12–15.

Brodie, B.B. and Hauser, E.B. (1970) Multiple pest control in cotton with a mixture of selective pesticides. *Phytopathology*, **60**, 1609–12.

Cartia, G. (1985) [Solar heating of soil pests and perennial weeds] La 'Solarizzazione' del terreno nella lotta ai parassiti ipogei delle piante erbacee. *Colture Protte*, **14**, 37–42.

Clarkson, D., Bull, P.B. and Moles, D.J. (1982) Effect of two granular nematicides on growth and nodulation of *Arachis hypogea*. *Plant and Soil*, **66**, 413–16.

Clayton, E.E., Gains, J.W. and Todd, F.A. (1949) Soil treatments with chemicals for control of tobacco parasites. *Phytopathology*, **39**, 4–5.

Cole, A.W. and Batson, W.E. (1975) Effects of diphenamid on *Rhizoctonia solani*, *Pythium aphanidermatum* and damping-off of tomato. *Phytopathology*, **65**, 431–4.

Colhoun, J. (1979) Predisposition by the environment, in *Plant Disease*, Vol. IV (eds J.G. Horsfall and E.B. Cowling), Academic Press, New York, pp. 75–96.

Cook, R.J. (1977) Management of the associated microbiota, in *Plant Disease*, Vol. I (eds J.G. Horsfall and E.B. Cowling), Academic Press, New York, pp. 145–60.

Cooper, J.I. and Thomas, P.R. (1971) Chemical treatment of soil to prevent transmission of tobacco rattle virus to potatoes by *Trichodorus*. *Annals of Applied Biology*, **69**, 23–4.

Davis, J.R. and Everson, D.O. (1986) Relation of *Verticillium dahliae* in soil and potato tissue, irrigation method and N-fertility to Verticillium wilt of potato. *Phytopathology*, **76**, 430–6.

Dekker, J. (1977) Chemotherapy, in *Plant Disease*, Vol. I (eds J.G. Horsfall and E.B. Cowling), Academic Press, New York, pp. 307–25.

Erwin, D.C., Reynolds, H.T. and Garber, M.J. (1961) Predisposition to *Pythium* seedling disease and an activated charcoal fungicide interaction as factors influencing emergence of cotton seed treated with phorate. *Journal of Economic Entomology*, **54**, 855–8.

Farley, J.D. and Riedel, R.M. (1971) Effects of *Pratylenchus penetrans* on *Verticillium* wilt of tomato. *Plant Disease Reporter*, **55**, 673–5.

Fassuliotis, G. (1987) Genetic basis of plant resistance to nematodes, in *Vistas on Nematology* (eds J.A. Veech and D.W. Dickson), Society of Nematologists, Hyattsville, pp. 364–71.

Fortnum B.A. and Curin, R.E. (1984) Evaluation of nematicides and Ridomil for control of root-knot and black shank complex (Nematicide Reports, section editor R.A. Kinloch), *Fungicide and Nematicide Test APS*, **40**, 114.

Good, J.M. (1964) Effect of soil application and sealing methods on the efficacy of row applications of several soil nematicides for controlling root-knot nematodes, weeds and Fusarium wilt. *Plant Disease Reporter*, **48**, 199–203.

Good, J.M. and Rankin, H.W. (1964) Evaluation of soil fumigants for control of nematodes, weeds and soil fungi. *Plant Disease Reporter*, **48**, 194–9.

Good, J.M. and Taylorson, R.B. (1964) Interaction of EDB, CDEC and irrigation on control of *Meloidogyne incognita acrita*. *Phytopathology*, **54**, 622.

Gwyne, G.R., Barker, K.R., Reilly, J.J., Komn, D.A., Burk, L.G. and Reed, S.M. (1986) Genetic resistance to tobacco mosaic virus, cyst nematodes, root-knot nematodes, and wild fire from *Nicotiana repanda* into *N. tabacum*. *Plant Disease*, **70**, 958–62.

Hasan, A. (1989) Efficacy of certain nonfumigant nematicides on the control of pigeonpea wilt involving *Heterodera cajani* and *Fusarium udum*. *Journal of Phytopathology*, **126**, 335–42.

Hide, G.A. and Corbett, D.C.M. (1974) Field experiments in the control of *Verticillium dahliae* and *Heterodera rostochiensis* on potatoes. *Annals of Applied Biology*, **78**, 295–307.

Horsfall, J.G. (1979) Iatrogenic disease: Mechanism of action, in *Plant Disease: An Advanced Treatise*, Vol. IV (eds J.G. Horsfall and E.B. Cowling), Academic Press, New York, pp. 343–55.

Iyatomi, K. and Nishizawa, T. (1970) Growth response of rice to soil fumigation, in *Root Diseases and Soil-Borne Pathogens, Second International Symposium on Factors Determining the Behaviour of Plant Pathogens in Soil* (eds J.A. Toussoun, R.V. Bega and P.E. Nelson), University of California Press, Berkeley, pp. 226–8.

Jalali, B.L. and Domsch, K.H. (1975) Effect of systemic fungitoxicants on the development of endotrophic mycorrhiza, in *Endomycorrhizas* (eds F.E. Sanders, B. Moss and P.B.H. Tinker), Academic Press, New York, pp. 619–26.

Jatala, P. (1985) Biological control of nematodes, in *An Advanced Treatise on Meloidogyne, Vol. I: Biology and Control* (eds J.N. Sasser and C.C. Carter), North Carolina State University Graphics, Raleigh, pp. 303–8.

Jones, J.P. and Overman, A.J. (1976) Tomato wilts, nematodes and yields as affected by soil reaction and persistent contact nematicide. *Plant Disease Reporter*, **60**, 913–17.

Katagiri, K. and Shimazu, M. (1980) Search for pathogenic bacteria and fungi to *Monochamus alternatus*. *Forest Pests*, **29**, 28–33.

Katan, J. (1981) Solar heating (solarization) of soil for control of soilborne pests. *Annual Review of Phytopathology*, **19**, 211–36.

Katan, J. and Eshel, Y. (1973) Interaction between herbicides and plant pathogens. *Residue Review*, **45**, 145–77.

Kaul, V.K., Bhandari, S.C. and Khurana, A.S. (1986) Interaction of *Meloidogyne incognita*, nematicides and *Rhizobium leguminosarum* on *Pisum sativum*. *Annals of Biology*, **2**, 77–82.

Khan, A.M., Saxena, S.K. and Khan, M.W. (1971) Interaction of *Rhizoctonia solani* and *Tylenchorhynchus brassicae* in pre-emergence damping-off of cauliflower seedlings. *Indian Journal of Nematology*, 1, 95–8.

Khan, A.M., Alam, M.M. and Ahmad, R. (1974) Mechanism of the control of plant parasitic nematodes as a result of the application of oil-cakes to the soil. *Indian Journal of Nematology*, 41, 93–6.

Khan, M.W., Khan, A.M. and Saxena, S.K. (1973) Influence of certain oil cake amendments on nematodes and fungi in tomato field. *Acta Botanica Indica*, 1, 49–54.

Khan, M.W., Khan, A.M. and Saxena, S.K. (1974) Rhizosphere fungi and nematodes of eggplant as influenced by oil cake amendments. *Indian Phytopathology*, 26, 480–4.

Khan, R.M., Singh, B., Ansari, A.R., Khan, M.W. and Saxena, S.K. (1980) Effect of D.B.C.P. on certain rhizospheric fungi. *Indian Journal of Nematology*, 10, 245.

Kimpinski, J., Johnston, H.W. and Martin, R.A. (1987) Influence of aldicarb on root lesion nematode, leaf disease and root-rot in wheat and barley. *Plant Pathology*, 36, 333–8.

King, P.S., Rodriguez-Kabana, R. and Ingram, E.G. (1977) Effects of two thiocarbamate herbicides on severity of disease caused by *Meloidogyne arenaria*. *Journal of Nematology*, 9, 274.

Kinloch, R.A. and Schenck, N.C. (1978) Nematodes and fungi associated with soybeans grown under various pesticide regimes. *Proceedings of Soil and Crop Science Society, Florida*, 37, 224–7.

Klingler, J. and Zwicky, P. (1981) [Reappearance of nematodes of the genus *Xiphinema* and virus disease symptoms in vines after soil treatments and fallow] Mitteilungen Wiederauftreten von Nematoden der Gattung *Xiphinema* und von Nematodenubertragenen Virus Krankheiten an Reben nach Bodenbehandlung and Brache. *Klosterneuburg Rebe und Wein Obstbau und Fruchteverwertung*, 31, 89–97.

Lamberti, F. (1981) Combating nematode vectors of viruses. *Plant Disease*, 65, 113–17.

Maas, P.W.T. (1974) Soil disinfection to prevent transmission of tobacco rattle virus to potatoes and flower bulbs. *Agriculture and Environment*, 1, 329–38.

Mamiya, Y. (1983) Pathology of the pine wilt disease caused by *Bursaphelenchus xylophilus*. *Annual Review of Phytopathology*, 21, 201–20.

Mamiya, Y. (1984) The pine wood nematode, in *Plant and Insect Nematodes* (ed. W.R. Nickle), Marcel Dekker, New York, pp. 587–626.

McGuire, J.M. and Good, M.J. (1970) The effect of benomyl on *Xiphinema americanum* and tobacco ringspot virus infection. *Phytopathology*, 60, 1150–1.

McSorley, R. and Parrado, J.L. (1986) Application of soil solarization to rockdale soil in subtropical environment. *Nematropica*, 16, 125–40.

Meagher, J.W. and Jenkins, P.T. (1970) Interaction of *Meloidogyne hapla* and *Verticillium dahliae* and chemical control of wilt in strawberry. *Australian Journal of Experimental Agriculture and Animal Husbandry*, 10, 493–6.

Menge, J.A. (1982) Effect of soil fumigants and fungicides on vesicular–arbuscular fungi. *Phytopathology*, 72, 1125–32.

Miller, P.M. and Waggoner, P.E. (1963) Interaction of plastic mulch, pesticides and fungi in control of soil borne nematodes. *Plant and Soil*, 18, 45–52.

Minton, N.A., Parker, M.B. and Flowers, R.A. (1977) Infection of downy mildew on soybeans following applications of several nematicides to the soil and foliar sprays. *Plant Disease Reporter*, 61, 436–8.

Minton, N.A., Parker, M.B. and Summer, D.R. (1985) Nematode control related to Fusarium wilt in soybean and root rot and zinc deficiency in corn. *Journal of Nematology*, 17, 314–21.

Mohammad, G. (1987) Effect of herbicides upon dynamics of nematode population in soybean var. Bragg. *Pesticides*, **21**, 30-1.

Murant, A.F. and Taylor, C.E. (1965) Treatment of soil with chemicals to prevent transmission of tomato black ring and raspberry ring spot virus by *Longidorus elongatus*. *Annals of Applied Biology*, **55**, 227-37.

Mustika, I., Suardgat, D. and Wikanda, A. (1984) [Control of pepper yellow disease with fertilizer and pesticides] Penanggulangan penyakit kuning lada dengan pupuk dan pestisida. *Pemberitan Penelitan Tanaman Industri*, **9**, 37-43.

Nesheim, O.N. and Linn, M.B. (1969) Deleterious effect of certain fungitoxicants in the formation of mycorrhiza on corn by *Endogone fasciculata* and corn root development. *Phytopathology*, **59**, 297-300.

Newsom, L.D. and Martin, W.J. (1953) Effect of soil fumigation on populations of plant parasitic nematodes, incidence of Fusarium wilt and yield. *Phytopathology*, **43**, 292-3.

Norton, D.C. (1978) *Ecology of Plant Parasitic Nematodes*, John Wiley, New York.

Ocampo, J.A. and Hayman, D.S. (1980) Effects of pesticides on mycorrhiza in field grown barley, maize and potatoes. *Transactions of the British Mycological Society*, **74**, 413-16.

Osman, A.A. and Viglierchio, D.R. (1981) Herbicide effects in nematode diseases. *Journal of Nematology*, **13**, 544-6.

Patrick, Z.A. and Toussoun, T.A. (1965) Plant residue and organic amendments in relation to biological control, in *Ecology of Soil-Borne Plant Pathogens* (eds K.F. Baker and W.C. Snyder), John Murray, London, pp. 440-59.

Peterson, C.E., Williams, P.H., Palmer, M. and Louward, P. (1982) Wisonsin 2757 Cucumber. *Horticulture Science*, **17**, 268.

Pitcher, R.S. and McNamara, D.G. (1973) The control of *Xiphinema diversicaudatum*, the vector of arabis mosaic virus in hops. *Annals of Applied Biology*, **75**, 468-9.

Powell, N.T. (1979) Internal synergisms among organisms inducing disease, in *Plant Diseases*, Vol. IV (eds J.G. Horsfall and E.B. Cowling), Academic Press, New York, pp. 113-31.

Rahim, A.M.F., Satour, M.M., Mickail, K.Y., El-Eraki, S.A., Grinstein, A., Chen, Y. and Katan, J. (1988) Effectiveness of soil solarization in furrow-irrigated Egyptian soil. *Plant Disease*, **72**, 143-6.

Rahman, R. and Khan, M.W. (1988) Effect of oilcake amendments on seed germination of *Chenopodium album* and population of fungi and nematodes. *Acta Botanica Indica*, **16**, 106-10.

Rankin, D.J., Hewitt, W.B. and Schmitt, R.V. (1971) Controlling fan leaf virus-dagger nematode disease complex in vineyards by soil fumigation. *California Agriculture*, **25**, 11-14.

Rich, S. and Miller, P.M. (1964) Verticillium wilt of strawberries made worse by fungicides that stimulate meadow nematode populations. *Plant Disease Reporter*, **48**, 246-8.

Richardson, L.T. (1959) Effect of insecticides and herbicides applied to soil on the development of plant disease. II. Early blight and Fusarium wilt of tomato. *Canadian Journal of Plant Science*, **39**, 30.

Roberts, P.A., Magyarosy, A.C., Mathews, W. and May, D.W. (1988) Effects of metham sodium applied by drip irrigation on root-knot nematodes, *Pythium ultimum* and *Fusarium* sp. in soil and carrot and tomato roots. *Plant Diseases*, **72**, 213-17.

Rodriguez-Kabana, R. and Curl, E.A. (1970) Effect of atrazine on growth of *Fusarium oxysporum* f. sp. *vasinfectum*. *Phytopathology*, **60**, 65.

Rodriguez-Kabana, R. and Curl, E.A. (1980) Nontarget effects of pesticides on soilborne pathogens and disease. *Annual Review of Phytopathology*, **18**, 311-32.

Rodriguez-Kabana, R. and King, P.S. (1977) Nematicidal activity of the fungicide ethazole. *Journal of Nematology,* **9,** 203–6.

Rodriguez-Kabana, R., Curl, E.A. and Peeples, J.L. (1970) Growth response of *Sclerotium rolfsii* to the herbicide EPTC in liquid culture and soil. *Phytopathology,* **60,** 431–6.

Rodriguez-Kabana, R., Backman, P.A., Karr, G.W., Jr and King, P.S. (1976a) Effects of the nematicide fensulfothion on soil borne pathogens. *Plant Disease Reporter,* **60,** 521–4.

Rodriguez-Kabana, R., Backman, P.A. and King, P.S. (1976b) Antifungal activity of the nematicide ethoprop. *Plant Disease Reporter,* **60,** 255–9.

Rodriguez-Kabana, R., Beute, M.K. and Beckman, P.A. (1979) Effect of dibromochloropropane fumigation on the growth of *Sclerotium rolfsii* and on the incidence of southern blight in field grown peanuts. *Phytopathology,* **69,** 1219–22.

Rowe, R.C., Riedel, R.M. and Martin, M.J. (1985) Synergistic interactions between *Verticillium dahliae* and *Pratylenchus penetrans* in potato early dying disease. *Phytopathology,* **75,** 412–18.

Sayre, R.M., Patrick, Z.A. and Thorpe, J.H. (1964) Substances toxic to plant parasitic nematodes in decomposing plant residue. *Phytopathology,* **54,** 905.

Schippers, B. and Palm, L.C. (1973) Ammonia, a fungistatic volatile in chitin amended soil. *Netherlands Journal of Plant Pathology,* **79,** 279–81.

Schmitt, D.P. and Corbin, F.T. (1981) Interaction of fensulfothin and phorate with pre-emergence herbicides on soybean parasitic nematodes. *Journal of Nematology,* **13,** 37–41.

Schmitt, D.P. and Nelson, L.A. (1987) Interactions of nematicides with other pesticides, in *Vistas on Nematology* (eds J.A. Veech and D.W. Dickinson), Society of Nematologists, Hyattsville, 455–60.

Shahzad, S. and Ghaffar, A. (1989) Use of *Paecilomyces lilacinus* in the control of root rot and root-knot complex of okra and mungbean. *Pakistan Journal of Nematology,* **7,** 47–53.

Sidhu, G.S. and Webster, J.M. (1974) Genetic resistance in the tomato to root-knot nematode–wilt fungus complex. *Journal of Heredity,* **65,** 153–6.

Sidhu, G.S. and Webster, J.M. (1979) Genetics of tomato resistance to the *Fusarium–Verticillium* complex. *Physiological Plant Pathology,* **15,** 93–8.

Sidhu, G.S. and Webster, J.M. (1983) Horizontal resistance in tomato against the *Meloidogyne–Fusarium* complex: An artefact of parasitic epistasis. *Crop Protection,* **2,** 205–10.

Sitaramaiah, K. (1990) Mechanism of reduction of plant parasitic nematodes in soils amended with organic materials, in *Progress in Plant Nematology* (eds S.K. Saxena, M.W. Khan, A. Rashid and R.M. Khan), CBS Publishers and Distributors, New Delhi, pp. 263–95.

Stapleton, J.J., Lear, B. and De Vay, J.E. (1987) Effect of combining soil solarization with certain nematicides on target and non-target organisms and plant growth. *Annals of Applied Nematology,* **1,** 107–12.

Stark, F.L. and Lear, B. (1947) Miscellaneous greenhouse tests with various soil fumigants for the control of fungi and nematodes. *Phytopathology,* **37,** 698–711.

Szczygiel, A. and Reabandal, Z. (1982) [Control of root-knot nematode and *Verticillium dahliae* on strawberries using Di-Trapex and Benlate] Zwalczanie pasozytnichczych nicieni korzeniowych i werticky liozy truskawki za pomoca preparatow Di Trapex i Benlate. *Roczniki Nauk Rolnezych E,* **12,** 163–74.

Taylor, C.E. (1978) Plant-parasitic Dorylaimida: Biology and virus transmission, in *Plant Nematology* (ed. J.F. Southey), Her Majesty's Stationery Office, London, pp. 232–43.

Taylor, C.E. and Gordon, S.C. (1970) A comparison of four nematicides for the control of *Longidorus elongatus* and *Xiphinema diversicaudatum* and the virus they transmit. *Horticultural Research*, **10**, 133–4.

Taylor, J.B., Canter-Visscher, J.W., Newhook, F.W. and Grandioson, G.S. (1970) Prolonged control of *Verticillium* wilt and lesion nematodes by soil fumigations. *New Zealand Journal of Science*, **13**, 591–602.

Timmer, L.W. and Leyden, R.F. (1978) Relationship of seed bed fertilization and fumigation to infection of sour orange seedlings by mycorrhizal fungi and *Phytophthora parasitica*. *Journal of American Society of Horticultural Science*, **103**, 537–41.

Trappe, J.M., Molina, R. and Castellano, M. (1984) Reactions of mycorrhizal fungi and mycorrhiza formation to pesticides. *Annual Review of Phytopathology*, **22**, 331–59.

Van Gundy, S.D. and McKenry, M.V. (1977) Action of nematicides, in *Plant Disease: An Advanced Treatise*, Vol. I (eds J.G. Horsfall and E.B. Cowling), Academic Press, New York, pp. 263–83.

Wallace, H.R. (1963) *The Biology of Plant Parasitic Nematodes*, Edward Arnold, London.

Wallace, H.R. (1973) *Nematode Ecology and Plant Diseases*, Edward Arnold, London.

Webster, J.M. (1967) Some effects of 2,4-dichlorophenoxyacetic acid herbicides on nematode infested cereals. *Plant Pathology*, **16**, 23–5.

Webster, J.M. and Lowe, D. (1966) The effect of synthetic plant growth substances 2,4-dichlorophenoxy acetic acid, on host parasite relationships of some plant parasitic nematodes in monoxenic callus cultures. *Parasitology*, **56**, 313–22.

Williams, J.D. and Salt, G.A. (1970) The effects of soil sterilants on the cereal cyst nematode (*Heterodera avenae*), take all fungus (*Ophiobolus graminis*) and yields of spring wheat and barley. *Annals of Applied Biology*, **66**, 329–38.

Willis, C.B. and Thompson, L.S. (1979) Effects of phenamiphos, methyl bromide and fallowing on *Pratylenchus penetrans*, yield of *Medicago sativa* and *Fusarium* infection. *Journal of Nematology*, **11**, 265–9.

Windham, G.L., Windham, M.T. and Williams, W.P. (1989) Effects of *Trichoderma* spp. on maize growth and *Meloidogyne arenaria* reproduction. *Plant Disease*, **73**, 493–6.

Wright, D.J. (1981) Nematicides: Mode of action and new approaches to chemical control, in *Plant Parasitic Nematodes*, Vol. III (eds B.M. Zuckerman and R.A. Rohde), Academic Press, New York, pp. 421–49.

Yamashita, T.T. and Viglierchio, D.R. (1987) Field resistance to nonfumigant nematicides in *Xiphinema index* and *Meloidogyne incognita*. *Revue de Nematologie*, **10**, 327–32.

Yarwood, C.E. (1959) Predisposition, in *Plant Pathology*, Vol. I (eds J.G. Horsfall and A.E. Dimond), Academic Press, London, pp. 521–62.

Zentmyer, G.A. and Kendrich, J.B. (1949) Fungicidal action of volatile soil fumigants. *Phytopathology*, **39**, 864.

Index

Page numbers in *italics* refer to tables